ION-SELECTIVE ELECTRODES IN ANALYTICAL CHEMISTRY

VOLUME 2

MODERN ANALYTICAL CHEMISTRY

Series Editor: David Hercules
University of Pittsburgh

ANALYTICAL ATOMIC SPECTROSCOPY
By William G. Schrenk

PHOTOELECTRON AND AUGER SPECTROSCOPY
By Thomas A. Carlson

MODERN FLUORESCENCE SPECTROSCOPY, VOLUME 1
Edited by E. L. Wehry

MODERN FLUORESCENCE SPECTROSCOPY, VOLUME 2
Edited by E. L. Wehry

APPLIED ATOMIC SPECTROSCOPY, VOLUME 1
Edited by E. L. Grove

APPLIED ATOMIC SPECTROSCOPY, VOLUME 2
Edited by E. L. Grove

TRANSFORM TECHNIQUES IN CHEMISTRY
Edited by Peter R. Griffiths

ION-SELECTIVE ELECTRODES IN ANALYTICAL CHEMISTRY, VOLUME 1
Edited by Henry Freiser

ION-SELECTIVE ELECTRODES IN ANALYTICAL CHEMISTRY, VOLUME 2
Edited by Henry Freiser

ION-SELECTIVE ELECTRODES IN ANALYTICAL CHEMISTRY

VOLUME 2

Edited by

Henry Freiser

University of Arizona
Tucson, Arizona

PLENUM PRESS · NEW YORK AND LONDON

Library of Congress Cataloging in Publication Data

Main entry under title:

Ion-selective electrodes in analytical chemistry.

(Modern analytical chemistry)
Includes bibliographical references and index.
1. Electrodes, Ion selective. I. Freiser, Henry, 1920-
QD571.I59 543'.087 78-16722
ISBN 0-306-40500-8

A Division of Plenum Publishing Corporation
227 West 17th Street, New York, N.Y. 10011

Printed in the United States of America

Contributors

Richard P. Buck, Department of Chemistry, University of North Carolina, Chapel Hill, North Carolina

Henry Freiser, Department of Chemistry, University of Arizona, Tucson, Arizona

Robert J. Huber, Department of Electrical Engineering, University of Utah, Salt Lake City, Utah

Jiří Janata, Department of Bioengineering, University of Utah, Salt Lake City, Utah

Robert K. Kobos, Department of Chemistry, Virginia Commonwealth University, Richmond, Virginia

Owen R. Melroy, Graduate Student, Department of Chemistry, University of North Carolina, Chapel Hill, North Carolina

James C. Thompsen, Graduate Student, Department of Chemistry, University of North Carolina, Chapel Hill, North Carolina

Preface

We continue in this second volume the plan evident in the first; i.e., of presenting a number of well-rounded up-to-date reviews of important developments in the exciting field of ion-selective electrodes in analytical chemistry. In this volume, in addition to the exciting applications of ISE's to biochemistry systems represented by the description of enzyme electrodes, there is featured the most recent development in ISE's, namely, the joining of the electrochemical and solid state expertise, resulting in CHEMFETS. The scholarly survey of the current status of ISE's will undoubtedly be welcomed by all workers in the field.

Tucson, Arizona Henry Freiser

Contents

Chapter 1

Potentiometric Enzyme Methods

Robert K. Kobos

Chapter 2

Coated Wire Ion-Selective Electrodes

Henry Freiser

Chapter 3

Chemically Sensitive Field Effect Transistors

Jiří Janata and Robert J. Huber

Chapter 4

A Compilation of Ion-Selective Membrane Electrode Literature

Richard P. Buck, James C. Thompsen, and Owen R. Melroy

Chapter 1

Potentiometric Enzyme Methods

Robert K. Kobos

1. INTRODUCTION

Over the past decade ion-selective electrode-based potentiometry has become a well-established electroanalytical method. One of the most exciting and fastest growing areas of research has been the use of membrane electrodes in enzymatic analysis. Using this approach, the applicability of the potentiometric method has been greatly extended, enabling the simple and accurate determination of many biologically and clinically important substances.

Enzymes are proteins which function as highly efficient biological catalysts. They have found widespread use in analytical chemistry due to the specificity and sensitivity of their reactions.[1–3] The specificity will vary depending on the particular enzyme. Some enzymes are highly specific to a single optical isomer, e.g., glucose oxidase, while other enzymes catalyze reactions for several related substances, e.g., L-amino acid oxidase, which reacts with L-cysteine, L-leucine, L-tyrosine, L-tryptophan, L-phenylalanine, and L-methionine at varying rates. Since enzymes act as catalysts they alter the speed of chemical reactions without being consumed in the process. Enzyme reactions involving a single substrate can be formulated in a general way as

$$E + S \underset{k_2}{\overset{k_1}{\rightleftarrows}} ES \xrightarrow{k_3} E + P \tag{1}$$

where E is the enzyme, S is the substrate, ES is the enzyme substrate complex, and P is the product. The rate of the enzyme-catalyzed reaction is

Robert K. Kobos • Department of Chemistrv, Virginia Commonwealth University, Richmond, Virginia 23284.

given by the Michaelis–Menten equation[4]:

$$v = \frac{V_m[S]}{K_m + [S]} \tag{2}$$

where v is the rate of the reaction, V_m is the maximum velocity given by k_3E_t, E_t is equal to the total amount of enzyme, $[S]$ is the initial substrate concentration, and K_m is the Michaelis constant given by $(k_2 + k_3)/k_1$. K_m is equal to the value of $[S]$ when $v = V_m/2$. The Michaelis–Menten equation can be applied to enzyme reactions involving more than one substrate if the concentration of all other subtrates is held constant. This would be the case for many analytical applications. When K_m is small compared to $[S]$ the reaction is zero order with respect to substrate and first order with respect to enzyme concentration. If K_m is large compared to $[S]$, the reaction is first order with respect to enzyme and substrate concentration, and with the amount of enzyme constant, the rate is directly proportional to the substrate concentration. Equation (2) has been shown to be applicable to a wide variety of enzyme-catalyzed reactions. However, the actual mechanism is frequently more complicated than is shown in reaction (1), involving the formation of two or more complexes.[3,5]

Often in enzymatic methods of analysis, the enzyme is used to convert the substance to be determined, which is not directly measurable, into a substance which can be determined by some analytical technique. Manometry, spectrophotometry, fluorescence, and electrochemistry are commonly used methods. Alternatively, the concentration of a second reactant, which is directly measurable, can be monitored. For the determination of substrates, the reaction must be first order or pseudo-first-order with respect to substrate, i.e., $[S] < K_m$. Two basic techniques may be used in enzymatic analysis of substrates: kinetic and equilibrium. In the equilibrium method the reaction is allowed to proceed to completion, i.e., until equilibrium is reached, before measurements are made. The kinetic method involves the measurement of the initial rate of the reaction. Rate measurements can be made in several ways: the slope method, the fixed-time method, and the variable-time method.[3] With the slope method, the concentration of the measured species or the signal which is linearly related to the concentration is plotted as a function of time. The initial rate of the reaction is obtained from the slope of a tangent drawn to the curve at time zero, or if a lag phase is present to a point after the lag time. The fixed-time method involves the measurement of the change in signal over some fixed time interval near the initiation of the reaction. The variable-time method involves the measurement of the time interval for the signal to change from one specified value to another. Kinetic methods offer advantages in terms of speed and the possibility of circumventing competing side reactions that might occur if longer time periods were involved. However, kinetic measurements are inherently less precise than the equilibrium method[6–8] and require careful temperature control. Besides the measurement of substrate concentrations, enzymes themselves, as well as activators and inhibitors of the enzymatic reaction, can be assayed using the kinetic method. For enzyme activity measurements, conditions are adjusted so that the reaction is first order or pseudo-first-order with respect to the enzyme and zero order with respect to substrate, i.e., $[S] \gg K_m$.

The utilization of enzymes in conjunction with potentiometric membrane electrodes offers several advantages in terms of cost, speed, accuracy, and convenience over other methods. Furthermore, potentiometric methods are not affected by turbidity so that measurements can be made in suspensions with no loss of accuracy. Both equilibrium and kinetic methods can be used. Kinetic measurements are somewhat complicated by the fact that the measured potential is logarithmically related to the concentration of the measured species. However, it has been shown that the kinetic method can be simplified by the following considerations.[9,10] The relationship between the product formed and the measured potential is given by the Nernst equation:

$$E = \text{constant} + \frac{RT}{ZF}\ln[\text{product}] \tag{3}$$

Differentiation of this equation with respect to time yields

$$\frac{dE}{dt} = \frac{RT}{ZF} \times \frac{1}{[\text{product}]} \times \frac{d[\text{product}]}{dt} \tag{4}$$

It can be assumed that the $1/[\text{product}]$ term changes relatively more slowly than the $d[\text{product}]/dt$ term in the initial stage of the reaction so that it can be included in the constant term. The change of the electrode potential with time is then directly proportional to $d[\text{product}]/dt$ or the reaction rate.

There are various configurations in which enzymes can be used with ion-selective electrodes for analytical purposes. The simplest approach is to use soluble enzymes under stationary (nonflow) conditions. The enzyme solution is added to the substrate solution and the potential change is monitored with the appropriate electrode. Either the substrate or the enzyme by using excess substrate, can be determined. In addition, by keeping the substrate and enzyme levels constant inhibitors or activators can be determined using the kinetic method. The system is easily automated. The major disadvantage of using soluble enzymes is that large amounts of enzyme are needed since it cannot be easily recovered from the reaction mixture. In addition, the substrate is consumed in the analysis.

The use of immobilized enzymes solves the first problem since the enzyme is easily recovered and reused. Immobilized enzymes can be used in stationary solutions with the immobilized enzyme added to the reaction vessel or with the enzyme immobilized in a reactor column. In the latter case, measurements are made in the collected column effluent. Automated flow-through analysis is also possible with the enzyme immobilized in a reactor column or reaction coil. These immobilized methods can be used in much the same manner as soluble enzymes to determine substrates, the activity of the immobilized enzymes, and enzyme activators and inhibitors. An advantage of using a configuration in which the electrode and the enzyme are separated, e.g., automated flow-through systems, is that each can be operated under optimum conditions which are often not the same for both.

Another possible configuration for using immobilized enzymes is the enzyme electrode. In this case, the enzyme is held or immobilized at the surface of a suitable membrane electrode. Unlike the other arrangements described, enzyme electrodes consume only negligible amounts of substrate since the enzymatic reaction occurs only in the small volume of the enzyme layer. A steady state is built up in the enzyme layer instead of reaching equilibrium. The response of the electrode is governed by a complex mechanism.[10] First, the substrate diffuses into the enzyme layer. Then the enzymatic reaction occurs at a rate given by the Michaelis–Menten equation. The products formed diffuse in two directions toward the indicator electrode and out of the enzyme layer into the solution. The concentration of the measured species at the steady state depends on the diffusion coefficients of the substrate and product, the K_m of the enzyme, the enzyme concentration in the enzyme layer, the substrate concentration in the bulk solution, and the thickness of the enzyme layer. There are several advantages to using the enzyme electrode configuration. It offers the simplicity and convenience of a dipping probe, making measurements as simple as pH determinations. Also, as was mentioned earlier, the sample is not appreciably changed during the analysis. Furthermore, these probes can be scaled down to micro proportions for *in vivo* measurements. A disadvantage of this configuration is that since the electrode and enzyme are directly coupled, a compromise of operating conditions is frequently necessary. Another often cited disadvantage is the relatively long response times encountered. Response times from 1 to 10 min have been reported depending on the substrate concentration, the sensing electrode, and the thickness of the enzyme layer. However, due to the selectivity of many of these sensors, sample pretreatment can be eliminated making the time of analysis favorable compared to other techniques.

There have been several excellent reviews on potentiometric-based enzyme analysis and enzyme electrodes.[3,10–15] Since the time of these reviews there have been many advances in this rapidly growing field. In this review, although early developments are discussed, primary emphasis is placed on recent advances and state-of-the-art technology. The discussion is limited to potentiometric-based methods.

2. SOLUBLE ENZYME SYSTEMS

Ion-selective electrodes were first used in enzymatic analysis in conjunction with soluble enzymes. Substrates, enzymes, and enzyme inhibitors have been determined under stationary and flow-stream conditions.

2.1. Substrate Determinations

The pH glass electrode was the first ion-selective electrode used in enzymatic analysis. Since many enzymatic reactions involve the consumption or production of hydrogen ions, the pH electrode can be used to monitor the pH change during the course of the reaction.[16] The change in pH that occurs, however, can affect enzyme activity and therefore the rate of the reaction. A more frequently used approach is the "pH stat" method[17–19] in which the pH is kept constant during the reaction by the addition of acid or base. The rate at which the reagent is added gives the rate of the reaction. This method is well established and several automatic "pH stat" instruments are commercially available.

The first report of the analytical application of an enzyme in conjunction with an ion-selective electrode, other than the pH electrode, was described by Katz and Rechnitz.[20] A monovalent cation glass electrode was used for the determination of urea by measuring the ammonium ion produced by the urease-catalyzed reaction in solution, i.e.,

$$\text{urea} + 3H_2O \xrightarrow{\text{urease}} 2NH_4^+ + HCO_3^- + OH^- \tag{5}$$

The equilibrium method was used in which the potential measured after the completion of the reaction was proportional to the amount of ammonium ion produced and therefore to the initial concentration of urea. Katz also applied this method to the determination of urease[21] and to the study of the effect of various ions on the urease reaction.[22]

Guilbault *et al.*[9] further applied this approach to the assay of several deaminase enzyme systems: glutamine–glutaminase, asparagine–asparaginase, and D- and L-amino acids–D- and L-amino acid oxidases. The general reaction is given by

$$\text{substrate} \xrightarrow[\text{enzyme}]{\text{deaminase}} NH_4^+ \tag{6}$$

The kinetic method was used. Calibration plots of the initial rate of the reaction, $\Delta E/\Delta t$, were linear with both enzyme and substrate concentrations. All substrates and enzymes were assayed with a precision and accuracy of about 2.5%.

Since the monovalent cation electrode was used in these early deaminase-based systems, other monovalent cations, e.g., potassium and sodium, were serious interferences. The development of a liquid-membrane ammonium-selective electrode[23] based on the antibiotic nonactin helped to alleviate this problem somewhat. This electrode has improved selectivity characteristics for ammonium ion over other monovalent cations. Neubecker and Rechnitz[24] used this liquid-membrane electrode in conjunction with the enzymes arginase and urease to determine the amino acid L-arginine. Ammonium ion was produced from arginine by a two-step process. First, arginine was converted to urea and ornithine by the reaction

$$\text{L-arginine} + H_2O \xrightarrow{\text{arginase}} \text{ornithine} + \text{urea} \tag{7}$$

The urease-catalyzed reaction (5) was used to convert the urea produced to ammonium ions. An equilibrium method with an incubation time of 3 hr was used.

Hussein and Guilbault[25] used a silicone rubber-based nonactin ammonium electrode in developing a method for the determination of nitrate. The two enzymes, nitrate and nitrite reductases, isolated from *Escherichia coli* strains B and Bn, respectively, were used to reduce nitrate to ammonium ion by the following reaction sequence:

$$NO_3^- \xrightarrow[\text{reductase}]{\text{nitrate}} NO_2^- \tag{8}$$

$$NO_2^- + 6e^- + 8H^+ \xrightarrow[\text{reductase}]{\text{nitrite}} NH_4^+ + 2H_2O \tag{9}$$

A kinetic method was used in the analysis. A problem encountered was that the enzymes used had relatively low activity and could be isolated only in small quantities.

The introduction of gas-sensing membrane electrodes greatly improved potentiometric-based enzymatic analysis in systems involving the production of gaseous species, e.g., NH_3 and CO_2. The use of these electrodes eliminates many of the interference problems previously encountered in biological samples. However, interfering levels of NH_3 in urine, for systems based on the ammonia electrode, and CO_2 in blood, for systems based on the carbon dioxide electrode, require the removal of these substances before measurements can be made. The gas sensor consists of an internal ion-selective electrode, generally a pH electrode, in contact with an internal electrolyte solution. This internal solution is separated from the sample solution by a gas-permeable membrane[26-28] or an air gap.[29] The measured gaseous species diffuses through the gas-permeable membrane or across the air gap causing a shift in equilibrium in the internal electrolyte solution. The resulting activity change is measured by the internal electrode. Since ions cannot cross the membrane or the air gap, these electrodes are unaffected by ionic species.

Hansen and Ruzicka[30] used an ammonia air gap electrode in conjunction with soluble urease to measure urea levels in whole blood and plasma. The sample was incubated with urease for 2 min after which the reaction was quenched with sodium hydroxide to raise the pH to 11. At this pH essentially all of the ammonium ion produced is converted to ammonia, which is sensed by the air gap electrode. Since the sensing electrode never actually comes in contact with the sample solution, there are no problems caused by the presence of proteins, blood cells, or ionic species.

Thompson and Rechnitz[31] used a gas-sensing ammonia electrode for the enzymatic determination of creatinine. A partially purified creatinine deiminase enzyme was used to catalyze the formation of ammonia from creatinine according to the reaction

$$\text{creatinine} + H_2O \xrightarrow[\text{deiminase}]{\text{creatinine}} N\text{-methylhydantoin} + NH_3 \tag{10}$$

The solutions were incubated at 37 °C for 1 hr after which the reaction was quenched with sodium hydroxide and the ammonia level measured.

The carbon dioxide gas-sensing electrode has been used in enzymatic analysis in conjunction with soluble decarboxylase enzymes. The general reaction here is given by

$$\text{substrate} \xrightarrow[\text{enzyme}]{\text{decarboxylase}} CO_2 \tag{11}$$

Substances determined by this method include oxalic acid,[32] L-glutamic and pyruvic acids,[33] and urea in human serum.[34]

Ion-selective electrode-based enzymatic analysis can be easily automated. A diagram of a typical autoanalysis flow-through system is shown in Fig. 1. Basically, the system consists of an automatic sampler, a proportioning pump, a water bath, a flow-through electrode assembly, and a pH/mV meter with a strip-chart recorder. Automated sample handling components which are already available in many laboratories can readily be used, e.g., the Technicon autoanalyzer system. Various flow-through electrode assemblies are possible, several of which are shown in Fig. 2. In flow systems that use soluble enzymes, the enzyme solution is used as a consumable reagent, so that relatively large amounts of enzyme are needed.

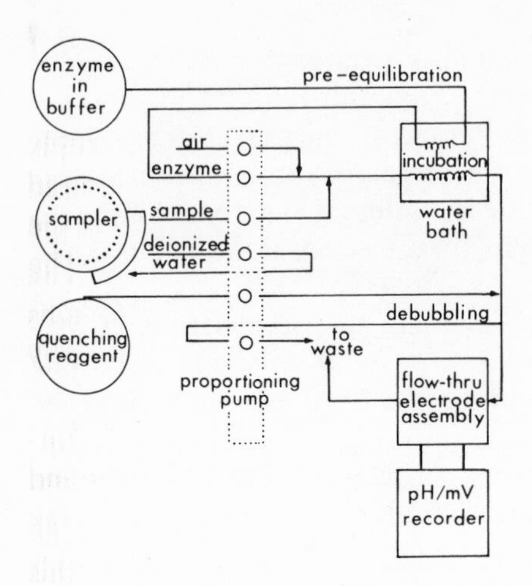

Fig. 1. Schematic diagram of an automated flow-through apparatus. Reprinted with permission from reference 65.

There have been several systems developed based on enzymatic reactions which produce hydrogen peroxide. The peroxide produced is measured indirectly by the following reaction:

$$H_2O_2 + 2I^- + 2H^+ \xrightarrow{\text{catalyst}} I_2 + 2H_2O \tag{12}$$

where the amount of iodide consumed is determined by a flow-through precipitate-based iodide electrode. The catalyst of this reaction can either be Mo(VI) or the enzyme peroxidase.

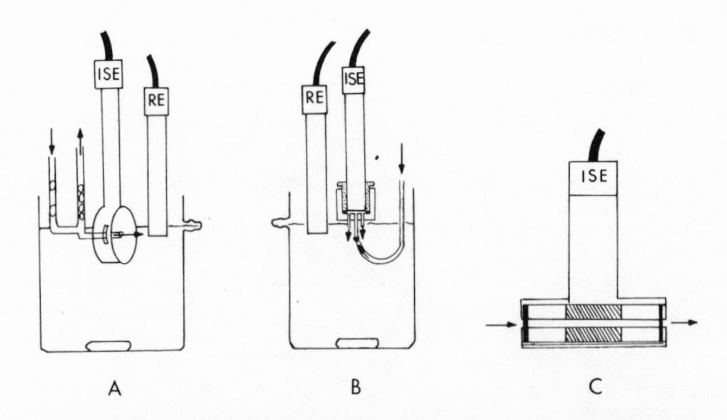

Fig. 2. Flow-through electrode assemblies: (A) Flow-through "lollipop" electrode. Reprinted with permission from reference 35. (B) Conventional flow-through cap. Reprinted with permission from reference 35. (C) Flow-through tubular electrode. Reprinted with permission from reference 39.

Llenado and Rechnitz[35] reported an automated method for the determination of glucose based on this approach. The enzyme glucose oxidase was used to catalyze the following reaction:

$$\beta\text{-D-glucose} + H_2O + O_2 \xrightarrow[\text{oxidase}]{\text{glucose}} \text{gluconic acid} + H_2O_2 \qquad (13)$$

and reaction (12) catalyzed by Mo(VI) was employed to measure the amount of hydrogen peroxide produced. The fixed-time kinetic method was used, enabling sampling rates of up to 70 determinations per hour. A novel flow-through "lollipop" electrode developed by Thompson and Rechnitz[36] (Fig. 2) was used to eliminate any problems caused by protein poisoning in protein-loaded solutions, including serum. The system functioned well in the physiological glucose range. Nagy *et al.*[37] developed a similar system for the determination of glucose. A second iodide electrode served as the reference electrode to cancel out any interferences. Catalase was used to catalyze reaction (12). Possible interferences encountered in blood were investigated. Reducing agents, such as ascorbic acid, tyrosine, and uric acid, were found to interfere. Therefore sample pretreatment was required to remove these substances.

Papastathopoulos and Rechnitz[38] used a similar system for the determination of total cholesterol. The enzyme cholesterol ester hydrolase was used to convert cholesterol esters into free cholesterol:

$$\text{cholesterol esters} + H_2O \xrightarrow[\text{ester hydrolase}]{\text{cholesterol}} \text{free cholesterol} + \text{fatty acids} \qquad (14)$$

The free cholesterol was then broken down by cholesterol oxidase:

$$\text{free cholesterol} + O_2 \xrightarrow[\text{oxidase}]{\text{cholesterol}} H_2O_2 + \text{cholest-4-en-3-one} \qquad (15)$$

The hydrogen peroxide was determined using reaction (12), catalyzed by Mo(VI), in conjunction with a flow-through iodide electrode. A sampling rate of 20 samples per hour was possible. Results obtained in standard reference serum samples and actual patient serum samples showed the method to be a satisfactory alternative to conventional spectrophotometric-based methods.

Mascini and Palleschi[39] reported a flow system for L-amino acids and alcohols that uses a tubular iodide electrode (Fig. 2). The reactions used are as follows:

$$\text{L-amino acid} + O_2 \xrightarrow[\text{oxidase}]{\text{L-amino acid}} H_2O_2 + \text{keto acid} + NH_3 \qquad (16)$$

or

$$\text{alcohol} + O_2 \xrightarrow[\text{oxidase}]{\text{alcohol}} H_2O_2 + \text{aldehyde} \tag{17}$$

Reaction (12) catalyzed by molybdate served as the indicator reaction in both cases. Sampling rates of up to 20 samples per hour were possible. The enzymes used are not highly specific so that response was obtained to a variety of substrates. In the case of the amino acids, various degrees of response were obtained for L-phenylalanine, L-leucine, L-arginine, and L-isoleucine between 10^{-4} and 10^{-2} *M*. The alcohols, methanol and ethanol, could be determined in the range 10^{-3}–10^{-2} *M*. The precision obtained in both cases was between 5 and 10%.

In general, these oxidase-enzyme-based systems are better handled with polarographic methods. The system is much simpler since only one reaction is involved, and offers greater flexibility since either the oxygen consumed or the hydrogen peroxide produced can be directly measured. In addition, problems of interferences can be easily eliminated. The most serious problem in the potentiometric-based method is that of interferences. It has been shown[37] that oxidizable compounds present in blood such as uric acid, tyrosine, ascorbic acid, and Fe(II), which compete in the oxidation of iodide to iodine in the indicator reaction, are serious interferences. These compounds have to be removed by sample pretreatment before measurements can be made. Polarographic-based systems involving oxidase enzymes have been reported for several substances including glucose,[40–43] L-amino acids,[44–46] alcohols,[46–49] and cholesterol.[50–52]

The gas sensing membrane electrodes have also been advantageously used under flow-stream conditions. Llenado and Rechnitz[53] developed an automated procedure for the determination of urea using soluble urease in conjunction with a flow-through ammonia gas-sensing electrode. After an incubation time of 10 min, the reaction was quenched with sodium hydroxide. Using this approach both the enzymatic reaction and the ammonia gas sensor can be operated under optimal pH conditions, i.e., pH = 7.2–7.4 for the enzyme reaction and pH = 11 for the ammonia measurement. Good precision was easily obtained in the physiological range 10^{-3}–10^{-1} *M* in both aqueous and serum samples. Recently, a similar system was developed for the continuous monitoring of urea levels during hemodialysis.[54] A flow-through system for urea determination using an ammonium ion electrode, based on the neutral carrier nonactin in a PVC matrix, has also been reported.[55]

A flow-injection method was reported by Ruzicka *et al.*[56] for the enzymatic determination of urea. A flow-through capillary pH electrode was used to measure the pH change resulting from the urease-catalyzed hydrolysis reaction. A linear relationship between the change in pH and the urea concentration was obtained in the range 4–15 m*M* urea by maintaining a constant buffer capacity of the carrier stream solution. Therefore greater sensitivity and reproducibility can be obtained than in the case of the normal logarithmic relationship. With sampling rates of 60 samples per hour, the reproducibility was ±0.52%. In the analysis of serum the presence of bicarbonate resulted in elevated urea values. Consequently, a blank had to be run for each sample and the change in pH taken with respect to this blank level. Since the soluble enzyme was used, moderate amounts of enzyme were consumed in the analysis. However, it was suggested that by using a merging-zone flow-injection technique, in which the sample and enzyme are simultaneously injected by means of a double injector, the consumption of enzyme could be greatly reduced.

An automated flow-through system for the determination of L-arginine and L-lysine was reported by Tong and Rechnitz.[57] The system is based on the respective decarboxylase enzymes which catalyze the production of CO_2 from the amino acid according to reaction (11). The CO_2 produced was measured with a newly developed liquid-membrane carbonate-selective electrode[58,59] after the pH had been adjusted to 8.4. Excellent results were obtained in the concentration range 10^{-3}–10^{-2} M in aqueous samples having physiological saline levels.

A list of some substrates determined by using soluble enzymes in conjunction with ion-selective electrodes can be found in Table 1.

TABLE 1. Substrates Determined Using Soluble Enzymes in Conjunction with Ion-Selective Electrodes

Substrate	Enzyme	Electrode	Mode of operation	Concentration range (M)	Reference
Urea	Urease (E.C. 3.5.1.5)	Monovalent cation	Stationary solution	—	20
		Monovalent cation	Stationary solution	1×10^{-5}–2×10^{-3}	9
		NH_3 air gap	Stationary solution	1×10^{-4}–1×10^{-2}	30
		CO_2 gas sensing	Stationary solution	3×10^{-4}–1×10^{-3}	34
		NH_3 gas sensing	Flow stream	1×10^{-3}–1×10^{-1}	53
		NH_3 gas sensing	Flow stream	3×10^{-3}–2×10^{-1}	54
		Ammonium ion	Flow stream	2×10^{-3}–2×10^{-2}	55
		pH	Flow stream	4×10^{-3}–2×10^{-2}	56
L-Glutamine	Glutaminase (E.C. 3.5.1.2)	Monovalent cation	Stationary solution	3×10^{-6}–7×10^{-5}	9

continued overleaf

TABLE 1 (*Continued*)

Substrate	Enzyme	Electrode	Mode of operation	Concentration range (M)	Reference
L-Asparagine	Asparaginase (E.C. 3.5.1.1)	Monovalent cation	Stationary solution	4×10^{-6}–8×10^{-5}	9
D-Amino acids	D-Amino acid oxidase (E.C. 1.4.3.3)	Monovalent cation	Stationary solution	6×10^{-5}–6×10^{-4}	9
L-Amino acids	L-Amino acid oxidase (E.C. 1.4.3.2)	Monovalent cation	Stationary solution	6×10^{-6}–2×10^{-4}	9
	L-Amino acid oxidase (E.C. 1.4.3.2) and peroxidase (E.C. 1.11.1.7)	Iodide	Flow stream	1×10^{-4}–1×10^{-2}	39
L-Arginine	Arginase (E.C. 3.5.3.1) and urease (E.C. 3.5.1.5)	Ammonium ion	Stationary solution	3×10^{-5}–3×10^{-3}	24
	Arginine decarboxylase (E.C. 4.1.1.19)	Carbonate	Flow stream	1×10^{-4}–1×10^{-2}	57
L-Lysine	Lysine decarboxylase (E.C. 4.1.1.18)	Carbonate	Flow stream	1×10^{-4}–1×10^{-2}	57
Nitrate	Nitrate reductase (E.C. 1.9.6.1) and nitrite reductase (E.C. 1.6.6.4)	Ammonium ion	Stationary solution	1×10^{-4}–1×10^{-3}	25
Creatinine	Creatinine deiminase (E.C. 3.5.4.21)	NH_3 gas sensing	Stationary solution	9×10^{-5}–9×10^{-3}	31
Oxalate	Oxalate decarboxylase (E.C. 4.1.1.2)	CO_2 gas sensing	Stationary solution	3×10^{-5}–1×10^{-4}	32
L-Glutamate	Glutamate decarboxylase (E.C. 4.1.1.15)	CO_2 gas sensing	Stationary solution	3×10^{-5}–3×10^{-4}	33
Pyruvate	Pyruvate decarboxylase (E.C. 4.1.1.1)	CO_2 gas sensing	Stationary solution	3×10^{-4}–2×10^{-3}	33
β-D-Glucose	Glucose oxidase (E.C. 1.1.3.4)	Iodide	Flow stream	1×10^{-3}–3×10^{-2}	35
	and peroxidase (E.C. 1.11.1.7)	Iodide	Flow stream	1×10^{-4}–1×10^{-3}	37

TABLE 1 (*Continued*)

Substrate	Enzyme	Electrode	Mode of operation	Concentration range (M)	Reference
Cholesterol	Cholesterol oxidase (E.C. 1.1.3.6) and peroxidase (E.C. 1.11.1.7)	Iodide	Flow stream	80–400 mg/dl	38
Alcohol	Alcohol oxidase (E.C. 1.1.3.13) and peroxidase (E.C. 1.11.1.7)	Iodide	Flow stream	1×10^{-3}–1×10^{-2}	39

2.2. Enzyme Determinations

Following the pioneering work of Katz and Rechnitz[20,21] and Guilbault *et al.*[9] in using ion-selective electrodes to measure enzyme activities, many enzyme systems have been assayed using this technique. By means of the kinetic method, rate measurements are made as a function of enzyme concentration in the presence of excess substrate. Stationary-solution methods and automated flow-through systems have been described. A list of some of the enzymes that have been determined with ion-selective electrodes is found in Table 2. Several of these systems will be described here.

TABLE 2. Enzymes Determined with Ion-Selective Electrodes

Enzyme	Substrate	Measured species	Electrode	Mode of operation	Reference
Urease (E.C. 3.5.1.5)	Urea	NH_4^+	Monovalent cation	Stationary solution	21
		NH_4^+	Monovalent cation	Stationary solution	9
		NH_3	Air gap	Stationary solution	74
		H^+	pH	Stationary solution	81
Glutaminase (E.C. 3.5.1.2)	Glutamine	NH_4^+	Monovalent cation	Stationary solution	9
		NH_3	Gas sensing	Stationary solution	73
Asparaginase (E.C. 3.5.1.1)	Asparagine	NH_4^+	Monovalent cation	Stationary solution	9

continued overleaf

TABLE 2 *(Continued)*

Enzyme	Substrate	Measured species	Electrode	Mode of operation	Reference
		NH_4^+	Monovalent cation	Stationary solution	72
D-Amino acid oxidase (E.C. 1.4.3.3)	D-Amino acids	NH_4^+	Monovalent cation	Stationary solution	9
L-Amino acid oxidase (E.C. 1.4.3.2)	L-Amino acids	NH_4^+	Monovalent cation	Stationary solution	9
Arginase (E.C. 3.5.3.1)	Arginine	NH_3	Air gap	Stationary solution	74
Creatinine deiminase (E.C. 3.5.4.21)	Creatinine	NH_3	Gas sensing	Flow stream	75
Cyanoalanine synthase (E.C. 4.4.1.9)	CN^- and cysteine	S^{-2}	Sulfide	Stationary solution	61
Cysteine synthase (E.C. 4.2.99.8)	S^{-2} and *O*-acetyl L-serine	S^{-2}	Sulfide	Stationary solution	79
Rhodanese (E.C. 2.8.1.1)	$S_2O_3^{-2} + CN^-$	SCN^- or CN^-	Thiocyanide cyanide	Stationary solution	63
		CN^-	Cyanide	Flow stream	64
		CN^-	Cyanide	Flow stream	65
β-Glucosidase (E.C. 3.2.1.21)	Amygdalin	CN^-	Cyanide	Stationary solution	62
		CN^-	Cyanide	Flow stream	65
Glucose oxidase (E.C. 1.1.3.4)	β-D-glucose	I^-	Iodide	Flow stream	65
Nitrate reductase (E.C. 1.7.99.4)	Nitrate	NO_3^-	Nitrate	Stationary solution	71
		NO_3^-	Nitrate	Flow stream	25
Nitrite reductase (E.C. 1.6.6.4)	Nitrite	NH_4^+	Ammonium ion	Flow stream	25
Acetyl-cholinesterase (E.C. 3.1.1.7)	Acetyl-choline	Acetyl-choline	Acetyl-choline	Stationary solution	67
Cholinesterase (E.C. 3.1.1.8)	Acetylcholine or butyryl-choline	Acetyl-choline	Acetyl-choline	Stationary solution	66, 68
	Acetylthio-choline	Thiocholine	Sulfide	Flow stream	69
	Acetylcholine	H^+	pH	Stationary and flow stream	70

TABLE 2 (*Continued*)

Enzyme	Substrate	Measured species	Electrode	Mode of operation	Reference
Adenosine deaminase (E.C. 3.5.4.4)	Adenosine	NH_3	Gas sensing	Flow stream	76
Monoamine oxidase (E.C. 1.4.3.4)	Primary monoamines	NH_3	Gas sensing	Stationary solution	77
Protease (E.C. 3.4.21.14)	Casein	Leucine	Leucine	Stationary solution	78
Glutamate decarboxylase (E.C. 4.1.1.15)	Glutamic acid	CO_2	Gas sensing	Flow stream	80
α-Chymotrypsin (E.C. 3.4.21.1)	DPCF	F^-	Fluoride	Stationary solution	60

An interesting method was described by Erlanger and Sack[60] for the determination of α-chymotrypsin. Diphenylcarbamyl fluoride (DPCF), an inactivator of the enzyme, combines with active α-chymotrypsin according to the reaction

$$\text{DPCF} + \alpha\text{-chymotrypsin} \rightarrow \text{DPC-chymotrypsin} + F^- \quad (18)$$

The fluoride ion produced was measured with a fluoride electrode enabling the determination of the operational normality of the enzyme, i.e., the number of active centers per liter.

Guilbault *et al.*[61] developed a method for the assay of β-cyanoalanine synthase using a precipitate-based sulfide electrode. The enzyme catalyzes the reaction

$$CN^- + \text{cysteine} \xrightarrow[\text{synthase}]{\beta\text{-cyanoalanine}} HS^- + \beta\text{-cyanoalanine} \quad (19)$$

The sulfide electrode was used to monitor the sulfide ion produced. However, since CN^- interferes with the electrode, standard "mixed-response" calibration curves had to be used.

A procedure for the determination of β-glucosidase activity was described by Llenado and Rechnitz.[62] A precipitate-based cyanide electrode was used to monitor the rate of the enzymatic hydrolysis of amygdalin:

$$\text{amygdalin} + H_2O \xrightarrow{\beta\text{-glucosidase}} \text{HCN} + 2\ \text{glucose} + \text{benzaldehyde} \quad (20)$$

By using the initial-slope kinetic method, a relative precision of 2% was obtained with an analysis time of approximately 1 min.

For the assay of the enzyme rhodanese, which catalyzes the following reaction,

$$S_2O_3^{2-} + CN^- \xrightarrow{\text{rhodanese}} SCN^- + SO_3^{2-} \tag{21}$$

Llenado and Rechnitz[63] described two possible methods. The rate of disappearance of cyanide was measured by using a precipitate-based cyanide electrode. Alternatively, the rate of appearance of thiocyanate was measured with a precipitate-based thiocyanate electrode. It was found that the method based on the measurement of the disappearance of cyanide was preferred, since cyanide interfered with the thiocyanate electrode. In addition the response time of the thiocyanate electrode was slower. A similar approach was used by Hussein *et al.*[64] for the development of a flow-through procedure for rhodanese. The cyanide electrode was used to monitor the disappearance of cyanide. A second cyanide electrode served as the reference electrode to compensate for any possible response to thiosulfate. Results obtained were comparable to those of spectrophotometric procedures.

Automated systems for enzyme assay using an autoanalyzer setup were described by Llenado and Rechnitz.[65] The system is similar to that shown in Fig. 1 except that the substrate solution replaces the enzyme solution and the sample contains the enzyme to be determined. Methods were developed for glucose oxidase based on reactions (13) and (12), rhodanese based on reaction (21), and β-glucosidase based on reaction (20). The analyses were performed at sampling rates of more than 40 samples per hour.

Several different approaches have been taken for the assay of cholinesterase. Baum and Ward[66] used a liquid-membrane acetylcholine-selective electrode to measure the rate of disappearance of acetylcholine due to the cholinesterase-catalyzed reaction:

$$\text{acetylcholine} + H_2O \xrightarrow{\text{cholinesterase}} \text{choline} + \text{acetic acid} \tag{22}$$

This same approach had previously been used in the determination of acetylcholinesterase, which catalyzes the same reaction.[67] Baum *et al.*[68] further applied this method to the determination of cholinesterase in whole blood, serum, and red blood cells.

Von Storp and Guilbault[69] developed a method for cholinesterase using acetylthiocholine as the substrate for reaction (22), i.e.,

$$\text{acetylthiocholine} + H_2O \xrightarrow{\text{cholinesterase}} \text{thiocholine} + \text{acetic acid} \tag{23}$$

The thiocholine produced was monitored with a sulfide-selective electrode.

Measurements were made under flow-stream conditions with a second sulfide electrode as the reference electrode.

Gibson and Guilbault[70] described still another method for the determination of cholinesterase activity. A pH electrode was used to monitor the linear rate of change of pH of a weak tris buffer resulting from reaction (22). Two techniques were used, a stationary-solution method and an automated method capable of sampling rates of up to 40 samples per hour based on an autoanalyzer system. Pooled human serum samples were assayed with comparable speed and greater precision than a spectrophotometric method, although there is a possible interference problem from other acid–base equilibria.

Potentiometric-based assays of the enzymes nitrate and nitrite reductases were reported by Hussein and Guilbault.[25,71] Nitrate reductase, which catalyses reaction (8), was determined by measuring the rate of disappearance of nitrate with a liquid-membrane nitrate-selective electrode. In the assay of nitrite reductase, reaction (9), the appearance of ammonium ion was monitored with an ammonium-ion-selective electrode. Measurements were made under stationary and flow-stream conditions. The results obtained compared favorably with standard spectrophotometric techniques.

The measurement of the rate of ammonia production, with either a monovalent cation electrode, a liquid-membrane ammonium ion electrode, or more recently the gas-sensing ammonia electrode, has been used as the basis for many enzyme assays. The enzyme asparaginase, which catalyzes the reaction

$$\text{L-asparagine} + H_2O \xrightarrow{\text{asparaginase}} \text{L-aspartate} + NH_4^+ \qquad (24)$$

was assayed in stationary solutions by Ferguson *et al.*[72] using a monovalent cation electrode. Huang,[73] using a gas-sensing ammonia electrode, developed a solution method for the determination of glutaminase:

$$\text{L-glutamine} + H_2O \xrightarrow{\text{glutaminase}} \text{L-glutamate} + NH_3 \qquad (25)$$

Three glutaminase isoenzymes from rat tissues were assayed with results comparable to those obtained with conventional methods. The enzymes urease, reaction (5), and arginase, using reactions (7) and (5), were determined by Larsen *et al.*[74] using an ammonia air gap electrode. The analyses were done in stationary solutions using initial rate measurements.

Meyerhoff and Rechnitz[75] described an automated assay procedure for creatininase, reaction (10), using a flow-through gas-sensing ammonia electrode with an autoanalyzer system. A unique advantage of this system is that it is capable of monitoring creatininase activity via the ammonia produced. Colorimetric assays, being indirect based on creatinine consumption, cannot distinguish between creatininase and creatinine amidohydrolase activity. Using a similar automated system, Hjemdahl-Monsen *et al.*[76] developed an assay procedure for the enzyme adenosine deaminase:

$$\text{adenosine} + H_2O \xrightarrow[\text{deaminase}]{\text{adenosine}} \text{inosine} + NH_3 \tag{26}$$

The automated system was modified somewhat by the addition of a second probe to the sample probe arm of the automatic sampler. The twin-probe system permits simultaneous determination of ammonia and enzyme activities to be made, thus simplifying blank determinations. Studies on aqueous standards, protein-containing solutions, and horse serum gave good results over the clinically important range of enzyme levels.

Meyerson *et al.*[77] described a method for the determination of monoamine oxidase activity. The rate of ammonia production from the reaction

$$RCHNH_2 + H_2O + O_2 \xrightarrow[\text{oxidase}]{\text{monoamine}} RCHO + NH_3 + H_2O_2 \tag{27}$$

was measured with a gas-sensing electrode. Activities were determined in homogenates and mitochondria from rat brain with comparable sensitivity to other methods. The enzyme activity could be measured in crude homogenates with minimal interference from endogenous ammonia, which was measured in a blank and subtracted out.

Other ion-selective-based enzyme assays include the following: the assay of proteolytic enzyme using a leucine-selective liquid-membrane electrode,[78] the determination of cysteine synthase using a sulfide electrode,[79] and the assay of glutamate decarboxylase with a carbon dioxide gas-sensing electrode.[80]

2.3. Inhibitor Determinations

Inhibitors of enzymes can be determined by their effect on reaction rates. Since inhibitors often produce effects in the nanogram range, a highly sensitive determination of these substances is possible. However, this approach is not very selective since other substances which inhibit the enzyme reaction will interfere. These other inhibitors, if present, must be removed or masked before the determination is attempted. This technique, based on potentiometric detection, has not been widely used. Several examples of inhibitor determinations using soluble enzymes will be discussed here.

Toren and Burger[19] described a pH stat method for the determination of the ions Ag(I), Hg(II), Cu(II), Cd(II), Mn(II), Co(II), Pb(II), and Ni(II) based on the inhibition of the urease reaction (5). Perchloric acid was added to keep the pH at 7.05 using a commercially available pH stat instrument. Silver ion was determined in the 2×10^{-8} to 1×10^{-7} M range with a maximum error of 11.6%. The other ions were determined in the 10^{-6} M range.

Weisz and Rothmaier[81] used a new kinetic-catalytic method to determine cadmium by its inhibition of urease. A variable-time kinetic method was used in which the pH was monitored during the course of the reaction using a pH electrode. Standard sulfuric acid was repeatedly added to give replicate measurements in the same system. Cadmium could be determined in the range 0.1–1.0 μg/ml.

The cholinesterase inhibitors, Paraoxon and Tetram, were determined by Baum and Ward.[261] The rate of the reaction was determined by following the disappearance of the substrate acetylcholine with a liquid-membrane acetylcholine-selective electrode. These organophosphate pesticides could be determined in the ng/ml range.

Meyerson *et al.*[77] determined the monoamine oxidase inhibitors, Chlorgyline and Deprenyl, by their effect on the rate of reaction (27), which was followed with an ammonia gas-sensing electrode. Chlorgyline could be determined in the range 1×10^{-10} to 1×10^{-7} M, while Deprenyl could be determined from 1×10^{-8} to 1×10^{-6} M.

3. IMMOBILIZED ENZYME SYSTEMS

The use of immobilized enzymes in analytical systems offers many advantages over soluble enzymes. A large economic advantage is realized since the enzyme can be readily recovered and therefore used for many analyses. In addition the long-term stability and the temperature stability of the immobilized enzyme are often greater than that of the isolated enzyme because the enzyme is kept in a more natural environment. Furthermore, the immobilized enzyme may be less susceptible to inhibition. The possible reasons for the improved stability of immobilized enzymes have recently been discussed.[82]

3.1. Methods of Immobilization

There have been many methods developed for the immobilization of enzymes. These have been the subject of numerous books and review articles.[82–90] A brief survey of various immobilization techniques will be undertaken here. For a more thorough discussion the reader is referred to

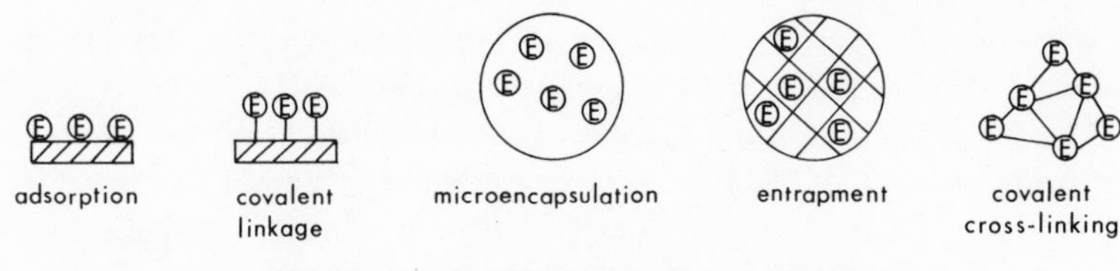

Fig. 3. Enzyme immobilization methods.

the cited review articles, many of which contain detailed immobilization procedures.

Methods for enzyme immobilization can be divided into the following categories[87]: adsorption, gel entrapment, microencapsulation, and covalent linkage. These are illustrated in Fig. 3.

3.1.1. Adsorption

The simplest and most economically attractive method of immobilization involves the adsorption of the enzyme onto a surface-active material. Adsorbents which have been employed include alumina, ion exchange resins, carbon, clay, glass, collagen, and collodian (cellulose nitrate).[84,91,92] The immobilization is carried out simply by contacting an aqueous solution of the enzyme with the surface-active adsorbent. The interactions involved include hydrogen bonding, ionic, and hydrophobic interactions. This method offers advantages in terms of its extreme simplicity, the mild conditions used, the large choice of carriers available, and the possibility of simultaneous purification and immobilization. Furthermore, regeneration is easily achieved by exposing the inactivated carrier to fresh enzyme. The ease with which the enzyme can be desorbed by a change of conditions, e.g., pH, ionic strength, and temperature, is a serious disadvantage.

3.1.2. Gel Entrapment

This method involves the entrapment of the enzyme within a cross-linked insoluble polymer. In the immobilization process the monomer, cross-linking agent, and the material to be trapped are mixed in a buffer solution and then the catalyst is added. The polymer structure is such that the large enzyme molecules cannot diffuse out of the smaller pores, but small substrate molecules can freely diffuse in. The pore size can be controlled somewhat by the ratio of monomer and cross-linking agent used. The polymers most often used include polyacrylamides, silicon rubber, fibrin, and collagen.[84,93,94]

The principal advantage of this method is the mild reaction conditions used in which few significant changes occur in the enzyme structure. A disadvantage is that due to variations in pore size, enzyme molecules slowly leak out of the gel matrix. In addition, components of the reaction mixture can cause denaturation of the enzyme.

3.1.3. Microencapsulation

In the microencapsulation method the enzyme is enveloped within various forms of semipermeable membranes.[84,95,96] The large enzyme molecules are trapped in the microcapsules, which generally have mean diameters ranging from 5 to 300 μm, while substrate and product molecules are free to diffuse across the membrane. This type of immobilization differs from gel entrapment in that instead of having individual molecules of enzyme trapped in polymer lattices, any volume and any amount of enzyme can be enclosed in such a way as to produce an intracellular environment. Microencapsulation is accomplished by depositing polymer around emulsified aqueous droplets either by interfacial polycondensation or coacervation. Membrane materials which have been used include collodian, nylon, silicon rubber, polystyrene, cellulose derivatives, lipid bilayers, and liposomes.

An advantage of this method is that it is applicable to many enzymes with a high recovery of enzyme activity. Also, it allows for the simultaneous immobilization of a number of enzymes in one step. Possible leakage of the enzyme and the inactivation of some enzymes due to the high protein concentration necessary for microcapsule formation constitute some disadvantages. Furthermore, substrates are restricted to low-molecular-weight substances.

3.1.4. Covalent Linkage

Two types of covalent linkages are possible, i.e., bonding to an insoluble support and intermolecular cross-linking. In the first method a covalent bond is formed between nonessential amino acid residues of the enzyme and a reactive group attached to the surface of a solid carrier phase. Possible carriers include synthetic polymers (e.g., polyacrylamides, polystyrene, nylon, and silicon rubber), natural polymers (e.g., cellulose, agarose, sepharose, and derivatives thereof, starch, dextran, and sephadex derivatives), and inorganic materials (e.g., silica glass beads and metal oxides). The most commonly used method involves the cyanogen bromide activation of cellulose or sepharose derivatives. Other possibilities include linkage via azide, isocyanate, carbodiimide derivatives of the support, and coupling using glutaraldehyde.[84,86,97]

An advantage of this method is that an increase in stability of the enzyme often results. However, some knowledge of the chemical makeup or requirements of the enzyme, particularly the amino acid residues that are essential for catalysis, is necessary. If these amino acids are used in the binding, inactivation of the enzyme will result. Also, the enzyme must be attached in a configuration in which the active site is accessible to the substrate. Therefore, if a "shot in the dark" approach is used, inactivation of the enzyme or a blocking of the active site can result.

The second type of covalent cross-linkage, intermolecular cross-linking, involves the formation of intermolecular covalent bonds between molecules of the enzyme and a low-molecular-weight multifunctional cross-linking agent. Glutaraldehyde, a bifunctional reagent with two identical functional groups to react with the enzyme, is one of the most frequently used cross-linking agents. Reactions involving glutaraldehyde can be carried out under mild conditions.[98] Other cross-linking reagents which have been used include diazobenzidine, dimethyl adipimidate, diisothiocyanate, chlorotriazines, and 1,5-difluoro-2,4-dinitrobenzene.[84,98,99]

There are several ways in which this approach can be used to produce insoluble enzyme conjugates. The pure enzyme can be reacted with the multifunctional reagent. Also, crystals of the enzyme can be treated with the cross-linking reagent thereby enhancing the mechanical strength of the matrix. The enzyme can first be adsorbed onto a surface-active support followed by cross-linking. Co-cross-linking is another approach in which the enzyme is reacted with a second protein, e.g., bovine serum albumin, and the multifunctional reagent.

Advantages of intermolecular cross-linking include its simplicity, chemical binding of the enzyme, and the fact that one reagent can be used to prepare different kinds of conjugates. A disadvantage is the need to control the experimental conditions fairly rigidly to obtain maximum activity. In addition, many enzymes are sensitive to cross-linking reagents and therefore lose activity in the process.

3.1.5. Combined Methods

Several combinations of immobilization techniques can be used to alleviate some of the problems involved in the individual methods. One example was mentioned in the previous section, i.e., adsorption of the enzyme onto a solid support followed by cross-linking to provide a more permanent, insoluble conjugate.[100,101] Similarly, the enzyme can be entrapped in nylon or collodian microcapsules and then treated with a cross-linking reagent.[102] Cross-linked enzyme conjugates can be trapped in a polymer matrix to prevent leakage of the enzyme.[103,104] A combination of the gel entrapment procedure with simultaneous covalent coupling of the enzyme to the polymer to decrease linkage has also been described.[105,106] Moreover, covalently bound, adsorbed, or gel-matrix-entrapped enzymes can be used in conjunction with microencapsulation techniques.[107]

3.2. Characteristics of Immobilized Enzymes

Upon immobilization of an enzyme, several changes can occur in enzyme behavior. The activity of the bound enzyme is generally less than that of the corresponding soluble enzyme. This can be due to the reaction of necessary amino acid residues in the immobilization process, steric hindrance, changes in protein structure and diffusional limitations.

The pH profile of the enzyme may shift depending on the carrier used.[108,109] When the enzyme is bound to a positively charged matrix, the pH optimum shifts to a lower pH region. Conversely, when a negatively charged support is used, the pH optimum is shifted to higher pH values. This behavior can be explained in terms of a microenvironment effect in which the local hydrogen ion concentration is affected by the electrostatic charge of the carrier. The local hydrogen ion concentration can be either higher, in the case of a negatively charged matrix, or lower, in the case of a positively charged matrix, than the bulk solution.

The kinetics of the enzyme reaction can also be affected by immobilization. Kinetic measurements for immobilized systems are generally made in terms of "apparent values." In many cases the rate of diffusion of the substrate toward the immobilized enzyme is the rate-controlling step. Upon immobilization the K_m value of the enzyme often increases. This can be caused by charge effects, diffusion effects, or changes in the enzyme configuration. There have been several excellent discussions dealing with the kinetics of immobilized enzymes.[110–114]

The long-term stability as well as the thermal stability of the enzyme can often be increased by the proper immobilization procedure.[82] Alterations in the specificity of the enzyme upon immobilization can even occur due to changes in the net charge of the enzyme and conformational changes.

More detailed discussions on the effect of immobilization on enzyme properties can be found elsewhere.[110,112,115]

3.3. Analytical Applications with Ion-Selective Electrodes

Like soluble enzymes, immobilized enzymes can be used with ion-selective electrodes under both stationary and flow-stream conditions. In stationary solutions, the enzyme, which is insolubilized, is added to the substrate to be determined. The product formed or a second substrate consumed is measured with an appropriate membrane electrode. The enzyme can be recovered for reuse by filtration or centrifugation. Alternatively, the substrate can be passed through an enzyme reactor column and the product formed measured in the effluent. For use in automated flow systems the enzyme is immobilized in a reactor column which can be used over and over again, thereby enabling a "reagentless" determination.

Ion-selective electrodes are particularly advantageous for the determination of the activity of immobilized enzymes. In addition, enzyme inhibitors can be determined.

3.3.1. Substrate Determinations

A list of substances determined using immobilized enzymes with potentiometric detection is found in Table 3.

Kiang *et al.*[116] developed a stationary-solution method for nitrite using

TABLE 3. Substrates Determined Using Immobilized Enzymes in Conjunction with Ion-Selective Electrodes

Substrate	Enzyme	Electrode	Mode of operation	Concentration range (M)	Reference
Nitrite	Nitrite reductase (E.C. 1.6.6.4)	NH_3 air gap	Stationary[a] solution	1×10^{-4}–5×10^{-2}	116
Nitrate	Nitrate reductase (E.C. 1.9.6.1) and nitrite reductase (E.C. 1.6.6.4)	NH_3 air gap	Reactor[b] column	5×10^{-5}–1×10^{-2}	130
Urea	Urease (E.C. 3.5.1.5)	NH_3 air gap	Immobilized[c] enzyme stirrer	1×10^{-4}–5×10^{-2}	117
		Monovalent cation	Reactor[b] column	1×10^{-2}–1×10^{-1}	129
		NH_3 gas sensing	Flow through[b] reactor column	3×10^{-3}–2×10^{-1}	131
		NH_3 gas sensing	Flow-through reactor column	3×10^{-3}–2×10^{-1}	134
		NH_3 gas sensing	Flow-through[b] reactor column	5×10^{-5}–3×10^{-2}	135
		pH	Flow-through[b] pH stat	1×10^{-4}–1×10^{-1}	140
L-Amino acids	L-Amino acid oxidase (E.C. 1.4.3.2)	NH_3 gas sensing	Flow-through[b] reactor column	3×10^{-5}–1×10^{-3}	136
Penicillin	Penicillinase (E.C. 3.5.2.6)	pH	Flow-through[b] reactor column	8×10^{-5}–5×10^{-4}	137
L-Asparagine	Asparaginase (E.C. 3.5.1.1)	Ammonium ion	Flow-through[d] reactor column	1×10^{-4}–1×10^{-2}	138

TABLE 3 (*Continued*)

Substrate	Enzyme	Electrode	Mode of operation	Concentration range (M)	Reference
Arginine	Arginase (E.C. 3.5.3.1) and urease (E.C. 3.5.1.5)	Ammonium ion	Flow-through[d] reactor column	1×10^{-4}–1×10^{-2}	138
Natural lipids	Lipase (E.C. 3.1.1.3)	pH	Flow-through[e] reactor column	5×10^{-6}–5×10^{-5}	139

Methods of immobilization:
[a] Covalent co-cross-linking.
[b] Covalently coupled to glass beads.
[c] Commercial preparation.
[d] Covalently bound to nylon tube.
[e] Collagen membrane.

immobilized nitrite reductase partially purified from spinach leaves. The ammonium ion produced from reaction (9) was measured with an air gap electrode. The nitrite reductase was immobilized by co-cross-linking using bovine serum albumin (BSA) and glutaraldehyde. The spongelike copolymer was placed in a beaker containing the nitrite sample. After incubation for 5 min at 30 °C, sodium hydroxide was added and the ammonia level determined. The immobilized enzyme was recovered by filtration, and washed with distilled water between assays. Nitrite levels in the range 1×10^{-4} to 5×10^{-2} M could be determined. The immobilized enzyme was stable for at least three weeks and could be used repeatedly for about 100 determinations.

A more convenient method of utilizing immobilized enzymes in stationary solutions was described by Guilbault and Stokbro.[117] Commercially available immobilized urease was placed onto a Teflon-coated magnetic stirring bar and was held in place by means of a nylon net. The "immobilized enzyme stirrer," shown in Fig. 4, both stirs the solution and

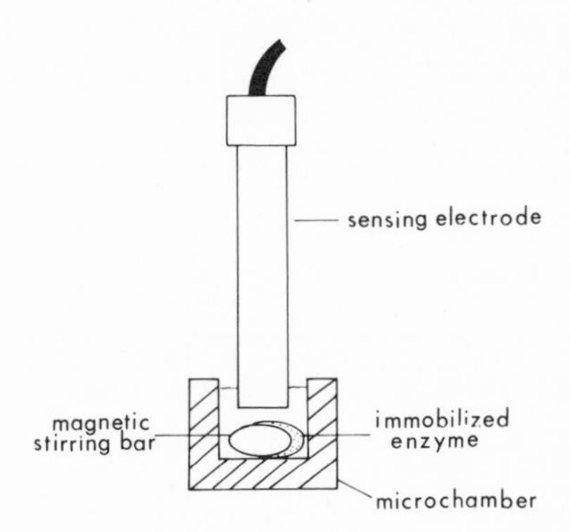

Fig. 4. Immobilized enzyme stirrer. Reprinted with permission from reference 117.

affects the enzymatic transformation. Urea in blood serum was determined with an accuracy and precision of approximately 2% using an air gap electrode to measure the ammonia formed. After the analysis was completed the immobilized enzyme stirrer was removed, washed with distilled water, and patted dry. It was then ready for the next assay. Linear calibration curves for urea solutions were obtained from 1×10^{-4} to 5×10^{-2} *M*. The stirrer could be used for 450 assays or a period of four weeks.

Another possible configuration for analytical systems using immobilized enzymes is the immobilized enzyme reactor. This arrangement is readily combined with a flow system for automated analyses, although it can also be used for manual determinations.

The enzyme reactor can be designed in several different ways.[(118)] The most popular type is the fixed-bed reactor in which the enzyme is attached to solid support particles, generally glass beads. The beads are then packed into a column. Another approach is the tubular enzyme reactor. In this configuration the enzyme is immobilized onto the walls of nylon tubing. This type of reactor is especially useful in air-segmented flow systems.[(119)]

Immobilized enzyme reactors have been used in analytical systems with many types of detection including spectrophotometric,[(120–122)] thermal,[(122–126)] and electrochemical[(127,128)] detectors.

One of the first reports of the use of an immobilized enzyme reactor with potentiometric detection was that of Weetall and Hersh.[(129)] Urease was covalently coupled to glass beads which were used in a reactor column. The ammonium ion produced by the enzymatic reaction was measured with a monovalent cation electrode after the substrate urea was passed through the column.

Kiang *et al.*[(130)] developed an enzymatic method for nitrate using a dual-enzyme reactor column. The enzymes MVH (methyl viologen, reduced form) nitrate reductase, isolated from *Escherichia coli* K12, and MVH-nitrite reductase, isolated from spinach leaves, were immobilized onto glass beads via glutaraldehyde coupling. The reduced form of methyl viologen serves as the electron donor in the enzymatic reduction of nitrate to ammonia, reactions (8) and (9). The ammonia formed was measured, after the addition of strong base, in the collected column effluent using an air gap electrode. Calibration curves for both nitrate and nitrite were linear in the range 5×10^{-5} to 1×10^{-2} *M*. The immobilized enzyme reactor could be used for up to 100 determinations with no loss of activity. Of the ions tested, only Hg(II) and Cu(II) were interferences due to the inhibition of the enzymatic reactions.

The use of immobilized enzyme reactors in flow systems represents the most economical configuration for automated enzyme-based analysis. A diagram of such a system is shown in Fig. 5. The components are similar to

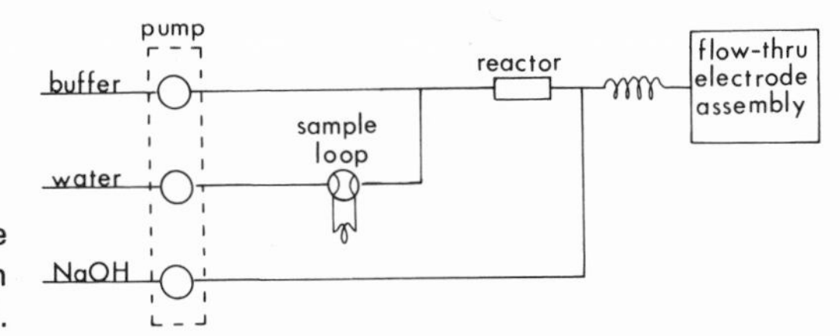

Fig. 5. Flow-through enzyme reactor system. Reprinted with permission from reference 135.

those described for soluble enzyme automated analysis (Fig. 1), however the enzyme is contained in the reactor.

Watson and Keyes[131] developed an enzyme-reactor-based automated system for the determination of urea. Urease was immobilized on an inert alumina support. The ammonia produced by the enzymatic reaction was measured after the addition of base with a flow-through gas-sensing electrode. The system uses the equilibrium method in which the urea is completely converted to ammonia. The immobilized enzyme column was stable for three months and 1000 determinations. The analyzer was capable of performing 30–45 tests per hour. The method proved to be more accurate and precise than the diacetyl monoxime autoanalyzer method.[132] The only known interference in the analysis of plasma and serum samples was blood ammonia. However, since normal blood ammonia levels are much lower than normal blood urea nitrogen (BUN) levels, there was no significant interference. This system is the basis for the commercially available Kimble BUN analyzer.[133] A similar automated system was described by Klein and Montalvo[134] for the continuous monitoring of urea levels during hemodialysis.

Johansson and co-workers described flow-through reactor systems for urea[135] and L-amino acids.[136] For the determination of urea, urease was immobilized onto glass beads by coupling with glutaraldehyde. A flow-through gas-sensing ammonia electrode was used to measure the ammonia produced by the enzymatic reaction. The response to urea was linear from 5×10^{-5} to 3×10^{-2} *M*. Using the equilibrium method, up to eight samples could be run per hour. The enzyme column showed no appreciable loss of activity after a period of one month.

A similar system was used for the determination of L-amino acids.[136] The enzymes L-amino acid oxidase and catalase were both immobilized in the reactor column. L-amino acid oxidase catalyzes the formation of ammonia, which was measured with a flow-through gas sensor, from L-amino acids, reaction (16). Hydrogen peroxide was added to the sample and was converted to oxygen, by the action of catalase, i.e.,

$$2H_2O_2 \xrightarrow{\text{catalase}} O_2 + 2H_2O \tag{28}$$

thereby increasing the amount of oxygen available for reaction (16). The response to the amino acid L-leucine was linear from 3×10^{-5} to 1×10^{-3} *M*.

Rusling *et al.*[137] reported a flow system for the determination of penicillin. Pencillinase, which catalyzes the following reaction,

$$\text{penicillin} + H_2O \xrightarrow{\text{penicillinase}} \text{penicilloic acid} \quad (29)$$

was covalently linked to glass beads and used in a reactor column. A pH glass electrode was used to monitor the change in pH produced by the enzymatic reaction. By maintaining a constant buffer capacity, a linear relationship was obtained between the measured electrode potential and the penicillin concentration at low concentration levels, i.e., 0.8×10^{-4} to 5×10^{-4} *M*. A simplified mathematical model was used to explain this linear relationship as opposed to the normal logarithmic response. The system was applied to the determination of penicillin in fermentation broths. When a dialysis system was used to clean up the sample, the results obtained were in good agreement with an established colorimetric method.

Ngo[138] described the use of tubular enzyme reactor systems for the determination of L-arginine and L-asparagine. For the determination of asparagine the enzyme asparaginase, reaction (24), was attached to the walls of a nylon tube using glutaraldehyde. A dual-enzyme system was used for the determination of L-arginine. Both arginase and urease were immobilized in the nylon tube reactor. Ammonium ion was produced from arginine via reactions (7) and (5). For both determinations an ammonium-ion-selective electrode was used to measure the ammonium ion produced by the enzymatic reactions.

A method for the determination of neutral lipids using an immobilized collagen membrane reactor was described by Satoh *et al.*[139] Lipase, which catalyzes the following reaction,

$$\text{neutral lipids} + H_2O \xrightarrow{\text{lipase}} \text{fatty acids} + \text{glycerol} \quad (30)$$

was immobilized in a collagen membrane which was inserted into a biocatalytic-type reactor. A pH electrode was used to measure the pH change resulting from the enzymatic reaction. The column was stable for a period of 15 days. Analysis of neutral lipids in blood serum gave results which were comparable to those obtained with a conventional colorimetric assay procedure.

Adams and Carr[140] described a coulometric flow analyzer for use with immobilized enzyme reactors. A totally electrochemical pH stat was used in which pH changes were measured with a pH glass electrode and coulometrically restored to a preset value. The system studied involved the determination of urea using urease covalently linked to glass beads. Urea could be determined with good precision in simulated sera and in quality control reference sera. However, problems were encountered in the analysis of actual human sera. In these samples very large and irreproducible blanks limited the precision attainable. Other problems encountered due to the low buffer capacity of the carrier fluid necessitated by the nature of the measurement were also discussed.

3.3.2. Assay of Immobilized Enzymes

Ion-selective electrodes offer an especially attractive method for the assay of immobilized enzymes. Conventional spectrophotometric techniques must be modified to determine immobilized enzymes, which are usually particulate in nature.[141] However, since ion-selective electrodes are unaffected by sample turbidity, assays can be easily and directly made in heterogeneous suspensions.

A novel application of ion-selective electrodes, which involves the activity determination of immobilized enzymes, has been in potentiometric-based enzyme immunoassay. This technique will be discussed here.

Potentiometric-Based Enzyme Immunoassay. Radioimmunoassay (RIA), a competitive binding technique based on the use of radiolabeled ligands, has found widespread use in biochemical analysis. Recently, enzyme labels have been used in place of radiolabels for the measurement of antigens, antibodies, and haptens. Enzyme immunoassay (EIA)[142,143] offers advantages in terms of speed, convenience, and cost.

The general features of EIA involve the reaction of a test substance (P), the same substance labeled with an enzyme (P*) and the antibody for P (Q) which can specifically bind both P and P*, i.e.,[142]

$$(P + P^*) + Q \rightleftharpoons PQ + P^*Q \tag{31}$$

Since P is the only variable in the test system, the quantity of P*Q formed will be inversely proportional to the initial concentration of P due to the competitive binding of P and P* for Q. It is necessary to separate P and P* from PQ and P*Q and then measure the activity of the enzyme-labeled P*Q. In EIA there are several ways in which P*Q can be isolated. The bound fraction can be precipitated using an insolubilized second antibody directed against the first antibody. Alternatively, the enzyme-linked immunosorbent (ELISA) method can be used in which the antibody is linked to an insoluble support, i.e.,

$$\vdash P + P + Q^* \rightleftharpoons \vdash PQ^* + PQ^* \tag{32}$$

where $\vdash$ is the insoluble support, and Q* is the enzyme-labeled antibody.

The amount of P governs the distribution of Q^* between $\vdash PQ^*$ and PQ^* due to competitive binding. The $\vdash PQ^*$ can be easily separated from PQ^*, P, and Q^*, and the immobilized enzyme activity assayed. This is preferred to measuring the decrease in activity in the solution phase which requires complete purification of the labeled substance to remove any unbound enzyme.

Another approach to the ELISA method is the "sandwich technique" in which the immobilized antibody is exposed to P and then to the enzyme-labeled antibody, i.e.,

$$\vdash Q + P \rightleftharpoons \vdash QP + Q^* \rightleftharpoons \vdash QPQ^* \tag{33}$$

where $\vdash Q$ and Q^* are in excess. The amount of $\vdash QPQ^*$ is proportional to the initial amount of P.

In all these cases the actual measurement involves the determination of the activity of the enzyme label, preferably made on the immobilized fraction. This determination can be made advantageously with ion-selective electrodes.

The first report of a potentiometric-based enzyme immunoassay was reported by Boitieux *et al.*[144,145] A method for the determination of hepatitis B surface antigen in biological fluids was developed using the enzyme-linked immunoassay sandwich procedure. Antibodies to hepatitis B surface antigen were immobilized onto a gelatin membrane. This membrane was then immersed for 30 min in a dilute solution of the antigen. The membrane was thoroughly washed with distilled water. Next, the membrane was incubated for 2 hr in a solution containing the peroxidase-labeled antibody, according to the sandwich principle. After washing with buffer the membrane was placed onto the surface of a precipitate-based iodide electrode. The activity of the peroxidase label was determined via reaction (12) in the presence of hydrogen peroxide and iodide. A fixed-time kinetic method, in which the potential was taken after 1 min, was used to determine the rate of decrease of iodide and hence the enzyme activity. Optimal conditions of incubation times, temperature, pH, and substrate concentrations were studied. The electrode response (ΔE) was linearly related to the antigen concentration in the range 0.5–50 μg/liter.

This technique offers several advantages over radioimmunoassay. It is less time consuming, more convenient, and more sensitive. In addition small quantities of antibodies, and no purified antigens are required. Moreover, most nonspecific interferences present in spectrophotometric-based procedures are eliminated.

Meyerhoff and Rechnitz[146] developed a potentiometric-based enzyme immunoassay procedure for bovine serum albumin (BSA) and cyclic AMP using the double-antibody technique. For the determinations rabbit antibody, a urease-labeled conjugate and a standard amount of BSA or cyclic

AMP were mixed for 1 hr according to reaction (31). An addition of insolubilized antibody (goat antirabbit γ-globulin) to the primary antibody, was added followed by incubation for two more hours. The insolubilized fraction was collected by centrifugation. The activity of the bound enzyme label was determined by measuring the rate of the urease-catalyzed reaction by means of an ammonia gas-sensing electrode using initial rate measurements. The optimum amounts of antibody, urease conjugate, and insolubilized antibody were determined by a series of titration experiments. Determinations of BSA in the range 10–1000 ng/ml and cyclic AMP in the range 10^{-9}–10^{-6} *M* were made with good reproducibility.

This approach appears to offer exciting new possibilities for potentiometric-based analyses. By coupling immunological reactions with existing enzymatic methods the number of substances which can be determined using ion-selective electrodes can be further increased.

Several polarographic-based enzyme immunoassay procedures have also been reported.[147–149]

3.3.3. Inhibitor Determinations

Inhibitors can be determined in immobilized enzyme systems if the inhibition is reversible. This was demonstrated by Ogren and Johansson[150] in the determination of Hg(II) using an immobilized urease reactor. An ammonia gas-sensing electrode was used to monitor the ammonia produced by the urease-catalyzed reaction. Theoretical considerations indicated that the amount of inhibition should be a linear function of the mercury concentration and this was shown to be true experimentally. A linear relationship was obtained between the percent inhibition and the Hg(II) concentration from 5×10^{-9} to 1.5×10^{-7} *M*. The column was regenerated with a solution containing thioacetamide and EDTA between determinations. Selectivity was obtained over Ni(II), Cd(II), Pb(II), and Zn(II). However, Ag(I) and Cu(II) were interferences.

4. ENZYME ELECTRODES

In a provisional nomenclature suggestion of the IUPAC an enzyme electrode has been defined as "a sensor in which an ion-selective electrode is covered with a coating that contains an enzyme which causes the reaction of an organic or inorganic substance (substrate) to produce a species to which the electrode responds."[151] The basic principle of operation of an enzyme electrode is very simple. The substrate to be determined diffuses into the enzyme layer where the enzymatic reaction occurs producing a product or consuming a reactant which is sensed by the ion-selective electrode. A

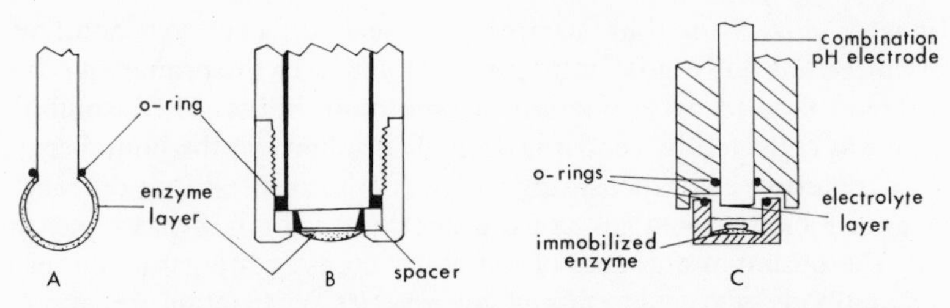

Fig. 6. Enzyme electrode configurations: (A) Glass membrane electrode type. (B) Gas-sensing electrode type. (C) Air gap type. Reprinted with permission from reference 161.

steady state potential is reached after some time due to the attainment of equal rates of product formation and diffusion out of the enzyme layer. The kinetic method based on the rate of potential change can also be used. Since only small amounts of substrate are consumed in the process, the method is essentially nondestructive. Several enzyme electrode configurations are shown in Fig. 6.

The first report of an enzyme-coupled electrode was given by Clark and Lyons in 1962.[40] Two possible electrodes for glucose were proposed. The first involved the use of the enzyme glucose oxidase, held on a pH electrode by means of a cuprophane membrane. The decrease in pH due to the production of gluconic acid via reaction (13) was proportional to the glucose concentration. The second approach involved the polarographic determination of glucose using glucose oxidase held between a hydrophobic membrane (e.g., polyethylene) and a dialysis membrane on a p_{O_2} electrode. The oxygen uptake was proportional to the glucose concentration.

The term "enzyme electrode" was introduced by Updike and Hicks.[152] They described a polarographic enzyme electrode for glucose using glucose oxidase immobilized in a polyacrylamide gel coated on an oxygen electrode. The decrease in oxygen pressure was proportional to the glucose concentration. Measurements were made in blood and plasma with a response time of less than 1 min.

The first potentiometric enzyme electrode was reported by Guilbault and Montalvo[153,154] for the determination of urea. Since then many potentiometric enzyme electrodes have been described (Table 4). In the

TABLE 4. Potentiometric Enzyme Electrodes

Substrate	Enzyme	Electrode	Method of immobilization	Concentration range $(M)^a$	Reference
Urea	Urease (E.C. 3.5.1.5)	Monovalent cation	Gel entrapment	5×10^{-5}–1×10^{-1}	153–155

TABLE 4 *(Continued)*

Substrate	Enzyme	Electrode	Method of immobilization	Concentration range $(M)^a$	Reference
		Monovalent cation	Covalently bound	1×10^{-4}–1×10^{-2}	156
		Ammonium ion	Gel entrapment	1×10^{-4}–1×10^{-2}	158
		NH_3 gas sensing	Covalently bound	1×10^{-4}–1×10^{-2}	160
		NH_3 air gap	Commercial preparation	1×10^{-4}–1×10^{-2}	161
		NH_3 gas sensing	Dialysis membrane entrapment	1×10^{-4}–1×10^{-1}	162
		NH_3 gas sensing	Covalently bound	1×10^{-4}–5×10^{-3}	163
		CO_2 gas sensing	Dialysis membrane entrapment	1×10^{-4}–1×10^{-1}	164
		CO_2 gas sensing	Covalently bound	1×10^{-4}–1×10^{-2}	165
		pH	Gel or dialysis membrane entrapment	5×10^{-5}–5×10^{-3}	166
Glucose	Glucose oxidase (E.C. 1.1.3.4) and peroxidase (E.C. 1.11.1.7)	Iodide	Covalently bound	3×10^{-4}–1×10^{-3}	37
	Glucose oxidase (E.C. 1.1.3.4)	pH	Gel entrapment	1×10^{-3}–1×10^{-1}	166
Amygdalin	β-glucosidase (E.C. 3.2.1.21)	Cyanide	Gel entrapment	1×10^{-5}-1×10^{-3}	167, 168
		Cyanide	Dialysis membrane entrapment	1×10^{-4}–1×10^{-1}	170
Penicillin	Penicillinase (E.C. 3.5.2.6)	pH	Gel entrapment	1×10^{-4}–5×10^{-2}	171
		pH	Dialysis membrane entrapment	1×10^{-4}–1×10^{-2}	166
		pH	Adsorption on glass disk	1×10^{-5}–3×10^{-3}	172
L-Amino acids	L-Amino acid oxidase (E.C. 1.4.3.2)	Monovalent cation	Dialysis membrane entrapment	1×10^{-4}–1×10^{-2}	174, 175
D-Amino acids	D-Amino acid oxidase (E.C. 1.4.3.3)	Monovalent cation	Dialysis membrane entrapment	1×10^{-4}–5×10^{-2}	176

continued overleaf

TABLE 4 *(Continued)*

Substrate	Enzyme	Electrode	Method of immobilization	Concentration range (M)[a]	Reference
L-Tyrosine	Tyrosine decarboxylase (E.C. 4.1.1.25)	CO_2 gas sensing	Dialysis membrane entrapment	1×10^{-4}–5×10^{-2}	164
		CO_2 gas sensing	Covalently bound	1×10^{-4}–1.5×10^{-3}	177
		CO_2 gas sensing	Magnetic membrane	2.5×10^{-4}–1.5×10^{-3}	178
L-Glutamine	Glutaminase (E.C. 3.5.1.2)	Monovalent cation	Dialysis membrane entrapment	1×10^{-4}–1×10^{-1}	179
L-Phenyl-alanine	L-Amino acid oxidase (E.C. 1.4.3.2)	Ammonium ion	Covalently bound	1×10^{-4}–1×10^{-2}	180
	L-Amino acid oxidase (E.C. 1.4.3.2) and peroxidase (E.C. 1.11.1.7)	Iodide	Covalently bound	5×10^{-5}–1×10^{-3}	180
	Phenylalanine ammonia lyase (E.C. 4.3.1.5)	NH_3 air gap	Soluble enzyme	6×10^{-5}–1×10^{-3}	181
	Phenylalanine decarboxylase (E.C. 4.1.1.53)	CO_2 gas sensing	Covalently bound	2.5×10^{-3}–1.5×10^{-2}	177
		CO_2 gas sensing	Magnetic membrane	2.5×10^{-3}–1.5×10^{-2}	178
L-Asparagine	Asparaginase (E.C. 3.5.1.1)	Monovalent cation	Gel entrapment	5×10^{-5}–1×10^{-2}	176
		NH_3 gas sensing	Covalently bound	8×10^{-5}–8×10^{-3}	182
L-Lysine	Lysine decarboxylase (E.C. 4.1.1.18)	CO_2 gas sensing	Covalently bound	1×10^{-4}–2×10^{-3}	177
		CO_2 gas sensing	Magnetic membrane	1×10^{-4}–1×10^{-3}	178
		CO_2 gas sensing	Covalently bound	6×10^{-5}–3×10^{-2}	183
L-Glutamate	Glutamate dehydrogenase (E.C. 1.4.1.2)	Monovalent cation	Dialysis membrane entrapment	1×10^{-4}–1×10^{-3}	184
	Glutamate decarboxylase (E.C. 4.1.1.15)	CO_2 gas sensing	Covalently bound	7×10^{-4}–7×10^{-3}	185
5′-AMP	AMP deaminase (E.C. 3.5.4.6)	NH_3 gas sensing	Dialysis membrane entrapment	8×10^{-5}–1.5×10^{-2}	186

TABLE 4 (*Continued*)

Substrate	Enzyme	Electrode	Method of immobilization	Concentration range (*M*)[a]	Reference
Cyclic AMP	Phospho-diesterase (E.C. 3.1.4.17) and AMP deaminase (E.C. 3.5.4.6)	NH_3 gas sensing	Dialysis membrane entrapment	1×10^{-5}–1×10^{-2}	187
Uric acid	Uricase (E.C. 1.7.3.3)	CO_2 gas sensing	Adsorption and dialysis membrane entrapment	1×10^{-4}–2.5×10^{-3}	188
Creatinine	Creatinine deiminase (E.C. 3.5.4.21)	NH_3 gas sensing	Dialysis membrane entrapment	7×10^{-5}–1×10^{-2}	75
Acetylcholine	Acetyl-cholinesterase	pH	Covalently bound	1×10^{-4}–1×10^{-2}	189
	(E.C. 3.1.1.7)	pH	Covalently bound	1×10^{-5}–1×10^{-4}	190
D-Gluconate	Gluconate kinase (E.C. 2.7.1.12) and 6-phospho-D-gluconate dehydrogenase (E.C. 1.1.1.44)	CO_2 gas sensing	Dialysis membrane entrapment	1.2×10^{-4}–2.4×10^{-3}	191
Lactate	Lactate dehydrogenase (E.C. 1.1.2.3)	Redox	Gel entrapment	1×10^{-4}–1×10^{-3}	192

[a] Analytically useful range.

subsequent discussion a survey of reported enzyme electrodes will be undertaken starting with the urea electrode. The many improvements and refinements made in this particular electrode illustrate the developments made in enzyme electrode technology from its conception to the current state of the art.

4.1. Urea Electrodes

The first urea electrode[153,154] consisted of physically entrapped urease [reaction (5)] in a polyacrylamide gel held over the surface of a monovalent cation electrode (Fig. 6). The electrode responded to changes in urea concentration in the range 5×10^{-5} to 1×10^{-1} *M* with a response time of about 35 sec. There was little loss of activity over a 14-day period. By adding a thin cellophane film over the polyacrylamide gel, the lifetime of the electrode could be extended to three weeks.[154,155] The cellophane membrane is permeable to the small substrate molecules but prevents the much larger enzyme molecules from leaching out of the gel layer.

Tran-Minh *et al.*[156] also developed a urea enzyme electrode based on the monovalent cation electrode. Urease was immobilized by covalent coupling using bovine serum albumin (BSA) and glutaraldehyde.

Since the monovalent cation electrode was used in these sensors, sodium and potassium ions interfered with the measurement. Therefore, these enzyme electrodes were of limited usefulness in the analysis of biological samples. In order to circumvent this problem, Guilbault and Hrabankova[157] used an uncoated cation electrode as the reference electrode and treated blood and urine samples with an ion exchange resin to reduce interferences. This procedure, although cumbersome, proved useful for the assay of urea in biological fluids providing good precision and accuracy (2–3%).

In a further attempt to improve the selectivity of the urea electrode, Guilbault and Nagy[158] employed an ammonium-ion-selective electrode[23] as the base sensor. This electrode has improved selectivity characteristics over sodium and potassium ions. The enzyme electrode was prepared by immobilizing urease in a polyacrylamide gel over the ammonium ion electrode. The interference problems were diminished somewhat in that sodium ion response was eliminated. However, potassium ion concentrations at the high end of the normal range in blood still caused errors.

Guilbault *et al.*[159] described another method for the measurement of blood urea. A three-electrode system was used which consisted of a urea enzyme electrode, an uncoated ammonium ion electrode (used to measure the background interference level), and a reference electrode. This system allowed for the dilution of the sample to a constant interference level before the urea measurement was made. Although still somewhat tedious, the analysis of blood serum was in good agreement with a standard spectrophotometric assay.

Anfalt *et al.*[160] designed an enzyme electrode for urea based on an ammonia gas-sensing electrode. Urease was covalently linked directly to the gas-permeable membrane of the ammonia sensor using glutaraldehyde. It was found that even at a pH as low as 7 enough ammonia was produced in the enzyme layer to be detected by the gas-sensing electrode. Since the gas sensor does not respond to ions in solution, this urea electrode was free from the ionic interferences which plagued earlier electrodes. The response of the urea sensor was linear in the concentration range of 1×10^{-4} to 1×10^{-2} with a response slope of 69.5 mV/decade at pH 7.0. The response time was 5 min for a urea concentration of 1×10^{-4} M and 2–3 min for concentrations of 1×10^{-3} to 1×10^{-2} M.

The discovery that the ammonia gas-sensing electrode could be used as the base sensor for an enzyme electrode at low pH, necessary for most enzymes, had a very important impact on enzyme electrode technology. It has led to the development of many enzyme electrode systems based on this sensor. At pH 7 less than 1% of the total ammonia is present as NH_3, which is the measured species. Apparently, the reason for the good response of these enzyme electrodes is due to the concentrating effect of the very small volume of the enzyme layer.

Guilbault and Tarp[161] described an enzyme electrode for urea based on the ammonia air gap electrode.[29] Commercially available immobilized urease was placed at the bottom of a microchamber and covered with a piece of nylon net. The sample solution was placed in the microchamber which was then sealed by the insertion of the air gap electrode (Fig. 6). The solution was stirred by means of a Teflon-coated magnetic stirring bar. This configuration is not an enzyme electrode in the true sense, since the enzyme is not immobilized at the electrode surface and total conversion of the sample results. This method would more properly be classified as a stationary-solution method using immobilized enzymes (Section 3.3.1). The response slope of the electrode was 0.95 pH units/decade in the range 5×10^{-5} to 1×10^{-1} *M* at pH 8.5. Response times were 2 min at a concentration of 1×10^{-2} *M* and 3–4 min at lower concentrations. In between measurements the assembly was taken apart and the microchamber and the immobilized enzyme were rinsed with distilled water. The air gap electrode was renewed after each measurement by replacing the electrolyte layer. This resulted in a very rapid return to base line (about 20 sec).

The use of the air gap electrode offers several possible advantages over the gas membrane type of sensor in that the response time and the recovery time are faster. In addition, the sensing electrode never touches the sample solution so that proteins, blood cells, etc., which have been shown to adversely affect the conventional ammonia gas membrane, have no effect on the air gap sensor. However, the electrode assembly must be taken apart and reassembled and the electrolyte solution renewed between measurements.

A urea electrode suitable for whole blood measurements was reported by Papastathopoulos and Rechnitz.[162] The sensor was based on an ammonia gas-sensing electrode. The enzyme urease was held on the surface of the gas-permeable membrane by means of a dialysis membrane (Fig. 6). The use of the dialysis membrane also prevents proteins from coming in contact with and clogging the gas-permeable membrane. This method of physically entrapping the enzyme is simple and convenient to use when enzyme stability is not a problem. The response of the urea electrode was linear in the concentration range 5×10^{-4} to 7×10^{-2} *M* with a slope of 90.7 mV/decade. The slope remained constant for a period of 20 days. The response time was about 5 min in the linear concentration range, slightly longer than that reported by Anfalt *et al.*[160] and Guilbault and Tarp[161] for gas-sensor-based urea electrodes. Good agreement was obtained with a spectrophotometric method in the analysis of whole blood and serum.

A still further improved urea electrode was described by Mascini and Guilbault.[163] The electrode consisted of an ammonia gas sensor with a special Teflon membrane. Urease was immobilized onto the Teflon membrane via covalent linkage with BSA and glutaraldehyde. The response time of this electrode, about 2 min, was greatly improved over that of previously reported urea electrodes based on gas sensors. It was therefore possible to run up to 20 assays/hour. The response slope of the electrode was 55 mV/decade in the linear range 1×10^{-4} to 5×10^{-3} *M*. No loss of activity was noted over 200 assays. Better precision was obtained in the analysis of serum samples than with currently used spectrophotometric methods.

Another important aspect of this work was the demonstration of the effect of the buffer capacity on the response of gas-sensor-based urea electrodes. Previous investigators[160,162] had reported super-Nernstian responses for these urea sensors. This unusual response was explained in this study as being due to an increase in pH in the enzyme layer, which affects the NH_3–NH_4^+ equilibrium, as a result of the enzymatic reaction. From reaction (5) for the urease catalyzed reaction at pH 7.4, it can be seen that hydroxide ions are produced. The use of a buffer with increased buffer capacity (0.2 *M* Tris-HCl) was effective in eliminating the super-Nernstian response up to a urea concentration of 5×10^{-3} *M*.

A different approach to the development of an enzyme electrode for urea was taken by Guilbault and Shu.[164] The sensor was based on a carbon dioxide gas-sensing electrode to measure the CO_2 produced by the enzymatic reaction. The enzyme electrode was prepared by dispersing an enzyme suspension over the electrode surface covered with a nylon netting. A dialysis membrane was placed over the enzyme layer to hold the enzyme in place. A linear response was obtained in the range 1×10^{-4} to 1×10^{-1} *M* with a slope of 57 mV/decade. Response times varied with substrate concentration being 5 min at 1×10^{-4} *M* and 1 min at 0.1 *M*. The response was reproducible over a three-day period.

Tran-Minh and Broun[165] also reported a urea electrode based on the carbon dioxide gas sensor. Urease immobilized by co-cross-linking with human serum albumin (HSA) and glutaraldehyde was used. The response time was from 2 to 3 min in the linear concentration range, 1×10^{-4} to 1×10^{-2} *M*.

These carbon-dioxide-based urea electrodes, although free from ionic interferences, are not as convenient as the ammonia-based sensors for the determination of blood urea levels due to the presence of interfering levels of carbon dioxide in serum.

Still another approach to the development of a urea enzyme electrode was described by Nilsson *et al.*[166] A pH electrode was used to measure the pH change as a result of the enzymatic reaction. The enzyme was held on the glass electrode either by entrapment in a polyacrylamide gel or a cellophane dialysis membrane. The response to urea was linear in the range 5×10^{-5} to 5×10^{-3} *M* in a weak 1×10^{-3} *M* Tris buffer containing 0.1 *M* NaCl as ionic strength adjustor. Response times were from 5 to 7 min. The electrode remained stable for a period of over two weeks. Although the electrode exhibited no interference problems due to the high selectivity of the pH sensor, the response was very dependent on the buffer capacity of the buffer used and would likewise depend on the buffer capacity of the sample. In addition the response times were relatively long.

4.2. Amygdalin Electrodes

The amygdalin electrode reported by Llenado and Rechnitz[167,168] was the first successful example of a potentiometric enzyme sensor which utilized a nonglass membrane. The enzyme β-glucosidase, which catalyzes reaction (20), was coupled with a solid state cyanide electrode. The enzyme was entrapped in a polyacrylamide gel which was polymerized in a Plexiglas–Teflon syringe-type chamber. In this way, membrane slices were cut and stored on a porcelain spot plate under refrigeration until use. The cyanide electrode was used in an upside-down configuration. The polymer enzyme disk was held in place on the electrode by a Plexiglas cap which also served as the sample cell. Samples as small as 0.2 ml could be used. The response of the electrode was linear from 1×10^{-5} to 1×10^{-2} *M* with a slope of 48 mV/decade. Response times were approximately 10 min above 1×10^{-4} *M* and about 30 min from 1×10^{-5} to 1×10^{-4} *M*. The electrode was stable for a period of four days. The pH used in this study was 10.4, which is not in the optimum pH range for the enzyme. This high pH was chosen since the cyanide electrode responds only to CN^- and not to HCN, which predominates at lower pH. It was believed that the use of a lower pH, which would be better suited for the enzyme, would decrease the sensitivity of the enzyme electrode.

Mascini[169] later demonstrated that the cyanide electrode could be used at a pH as low as seven with little loss in sensitivity. Using this knowledge, Mascini and Liberti[170] were able to improve the response characteristics of the amygdalin enzyme electrode by using a pH of 7. The enzyme electrode was prepared by trapping the enzyme β-glucosidase in a dialysis membrane over the surface of the cyanide electrode. The response of the electrode was linear over the concentration range 1×10^{-4} to 1×10^{-1} *M* with a slope of 53 mV/decade. The response times were greatly improved, ranging from 1 to 3 min in the 1×10^{-3} to 1×10^{-1} *M* range and about 6 min at 1×10^{-4} *M*. The lifetime of the electrode was also improved, being increased to a period of one week.

4.3. Glucose Electrodes

Nagy *et al.*[37] described a dual-enzyme electrode for glucose based on a solid state iodide electrode. The enzymes glucose oxidase and peroxidase were immobilized onto the electrode either by entrapment with a dialysis membrane, or in a polyacrylamide gel. Covalent bonding to a polyacrylic acid derivative was also used. The iodide electrode measures the decrease in iodide concentration, due to reactions (13) and (12), which is proportional to the glucose concentration. A linear response was obtained only over a very narrow range, i.e., 3×10^{-4} to 1×10^{-3} *M*. The immobilized electrode was stable for a period of 30 days. Interferences were a serious problem with uric acid, tyrosine, ascorbic acid, and Fe(II) being the principal interfering species due to competition with the reduction of hydrogen peroxide in the indicator reaction. There were several problems with this sensor, including the narrow linear range, long response times (2–8 min), and many interferences.

Nilsson *et al.*[166] developed a glucose sensor based on a pH electrode. Glucose oxidase was immobilized on the glass electrode by entrapment in a polyacrylamide gel. The pH change due to the production of gluconic acid via reaction (13), was measured by the pH electrode. The response was linear in the glucose concentration range 1×10^{-3} to 1×10^{-1} *M*. This glucose electrode was relatively free from interferences but the response was very dependent on the buffer capacity of the buffer used.

These potentiometric glucose enzyme electrodes are inferior to those based on polarographic detection[40,42,152] in terms of linear range, response time, and selectivity. Furthermore, the polarographic technique gives a linear relationship between the measured current and the concentration. Of course, the relationship between the potential and the concentration in potentiometric analysis is logarithmic.

4.4. Penicillin Electrodes

An enzyme electrode for penicillin was reported by Papariello *et al.*[171] The electrode was prepared by immobilizing the enzyme penicillinase, which catalyzes reaction (29), onto a pH electrode using a polyacrylamide gel. The decrease in pH due to the production of penicilloic acid was sensed by the glass electrode and the potential change was proportional to the logarithm of the penicillin concentration. The response slope varied with the type of penicillin tested, ranging from 38 to 52 mV/decade. The useful concentration range was between 1×10^{-4} and 5×10^{-2} *M* with a response time of 15–30 sec. The electrode lifetime was reported to be up to two weeks. The reproducibility of the electrode, however, was poor possibly because neither the ionic strength nor the pH were controlled. Also, response was observed for all monovalent cations.

Nilsson *et al.*[166] prepared a penicillin electrode in essentially the same manner using dialysis membrane entrapment of the enzyme. The response was linear between 1×10^{-3} and 1×10^{-2} *M* and useful down to 1×10^{-4} *M*. The electrode was stable for three weeks and had a response time of about 2 min. The ionic strength and the pH were controlled with a 5×10^{-3} *M* phosphate buffer, pH = 6.8, containing 0.1 *M* NaCl. The reproducibility was much better than that obtained by Papariello *et al.*[171] Average deviations for repeated measurements were about 2%.

An improved penicillin electrode was described by Cullen *et al.*[172] A new electrode configuration was used in which the enzyme penicillinase was immobilized by adsorption onto a fritted glass disk. The disk was then affixed to a flat-surface pH electrode. This design eliminated the cation interferences observed with the earlier polyacrylamide-entrapped enzyme electrode. The change of selectivity of the pH sensor when coated with the enzyme layer was believed to be due to the adsorption of the penicillinase onto the glass surface of the electrode. The response to monovalent cations was attributed to a mechanism of cation adsorption to the negative charges from the ionized glutamic and aspartic acid residues of the enzyme. A similar effect on the selectivity of a urease-coated monovalent cation electrode has also been reported.[173]

The response of the penicillin enzyme electrode was linear from 1×10^{-5} to 3×10^{-3} *M* with a slope of 56–58 mV/decade. Measurements were made in a 0.01 *M* KCl ionic strength adjustor. Stability was reported for a period of up to six weeks. The electrode was used to determine penicillin in commercial capsules and filtered fermentation broths. Good agreement was obtained with the established iodometric titration procedure. One disadvantage of the configuration used was the long wash times necessary to remove the buildup of product from the fritted glass disk.

4.5. Amino Acid Electrodes

4.5.1. General Amino Acid Electrodes

A general L-amino acid electrode was reported by Guilbault and Hrabankova.[174,175] The enzyme L-amino acid oxidase was held on the surface of a monovalent cation electrode by means of a dialysis membrane. The ammonium ion produced via reaction (16) was detected by the cation electrode. Response was obtained between 1×10^{-4} and 5×10^{-2} *M* for the following amino acids: L-cysteine, L-leucine, L-tyrosine, L-tryptophan, L-phenylalanine, and L-methionine. The degree of response varied with the amino acid tested. The electrode was stable for two weeks and had a response time of 2 min.

A general D-amino acid electrode was reported by Guilbault and Hrabankova.[176] The electrode was constructed using the enzyme D-amino acid oxidase in conjunction with a monovalent cation electrode. The enzyme catalyzes the formation of ammonium ion from D-amino acids by the following reaction:

$$\text{D-amino acids} + H_2O + O_2 \rightarrow NH_4^+ + \text{keto acid} + H_2O_2 \qquad (34)$$

Response was obtained to the following amino acids: D-phenylalanine, D-alanine, D-valine, D-methionine, D-leucine, D-norleucine, and D-isoleucine in the range 1×10^{-4} to 5×10^{-2} *M*. The electrode was stable for 21 days when stored in buffered flavine adenine dinucleutide (FAD) solution. The FAD is a necessary cofactor and is weakly bound to the active site of the enzyme.

These general amino acid electrodes have little practical application due to their poor selectivity. More selective amino acid electrodes have been developed by utilizing highly specific decarboxylase and deaminase enzymes as described below.

4.5.2. L-Tyrosine Electrodes

The enzyme tyrosine decarboxylase in conjunction with a carbon dioxide gas-sensing electrode was used by Guilbault and Shu[164] to construct a highly selective enzyme electrode for L-tyrosine. The enzyme, which catalyzes the reaction

$$\text{L-tyrosine} \xrightarrow[\text{decarboxylase}]{\text{tyrosine}} \text{tyramine} + CO_2 \qquad (35)$$

was held on the carbon dioxide sensor with a dialysis membrane. The enzyme electrode had a slope of 55 mV/decade in the linear range 2.5×10^{-4} to 1×10^{-2} *M*. Response times of under 5 min were obtained at concentrations above 1×10^{-4} *M*. This enzyme electrode had good selectivity due to the specificity of the enzyme and the carbon dioxide sensor.

Berjonneau *et al.*[177] reported a similar enzyme electrode for L-tyrosine with the decarboxylase enzyme immobilized on a Teflon membrane via co-cross-linking with BSA and glutaraldehyde.

Calvot *et al.*[178] also reported an enzyme electrode for L-tyrosine based on the decarboxylase enzyme and the CO_2 sensor. A magnetic enzyme membrane was used which was prepared by cross-linking the enzyme with BSA and glutaraldehyde in the presence of magnetic ferrite particles. The membrane was held on the CO_2 electrode by means of a cylinder magnet.

The response of the tyrosine electrode was linear over the range 2.5×10^{-4} to 1.5×10^{-3} *M*.

4.5.3 L-Glutamine Electrode

Guilbault and Shu[179] described an enzyme electrode for L-glutamine. Glutaminase was held on the surface of a monovalent cation electrode using a dialysis membrane. Ammonium ion produced by reaction (25) was measured with the cation electrode. The response of the enzyme sensor was linear from 1×10^{-3} to 1×10^{-1} *M* with a slope of 45 mV/decade. The response time was about 2 min at 5×10^{-3} *M*. The stability of the electrode was very poor, i.e., it was usable for only one day. A response to asparagine was also observed.

4.5.4. L-Phenylalanine Electrodes

Two kinds of enzyme electrodes for L-phenylalanine were reported by Guilbault and Nagy.[180] One was based on an ammonium-ion-selective electrode used to detect the ammonium ion produced by reaction (16) catalyzed by L-amino acid oxidase. The enzyme was chemically bound to a polyacrylic acid derivative and held on the electrode by means of a dialysis membrane. The response time of the electrode was from 2 to 3 min in the linear concentration range 1×10^{-4} to 1×10^{-2} *M*.

The second approach was to use a dual-enzyme layer containing L-amino acid oxidase and peroxidase chemically bound to a polyacrylic gel over an iodide electrode. The amount of hydrogen peroxide formed by reaction (16) was measured using reaction (12), where the iodide consumed was monitored with the iodide electrode. A linear response was obtained in the range 5×10^{-5} to 1×10^{-3} *M* using initial rate measurements.

Neither of these electrodes was very selective due to the nonspecific enzyme used.

A more selective phenylalanine electrode was reported by Hsiung *et al.*[181] The enzyme L-phenylalanine ammonia-lyase from potato tuber was used in conjunction with an ammonia air gap electrode. The enzyme catalyzes the following reaction:

$$\text{L-phenylalanine} \xrightarrow[\text{ammonia-lyase}]{\text{phenylalanine}} \textit{trans}\text{-cinnamate} + NH_3 \qquad (36)$$

The ammonia produced was measured with the air gap electrode in a microchamber as was done in the case of the urea electrode based on the air gap sensor.[161] After incubation at pH 7 and 30 °C for 10 min, the reaction was quenched with the addition of sodium hydroxide and the ammonia level measured. The response was linear over a fairly limited range, 1×10^{-4} to

6×10^{-4} *M*, with a slope of 1.03 pH units/decade, due to the incomplete conversion of the substrate during the incubation period. However, the response was usable between 6×10^{-5} and 1×10^{-3} *M*. Due to the specificity of the enzyme used, there was no response to other amino acids.

Enzyme electrodes for phenylalanine using the decarboxylase enzyme in conjunction with a CO_2 gas sensor have also been reported.[177,178] The enzyme catalyzes the formation of carbon dioxide from phenylalanine by the reaction

$$\text{L-phenylalanine} \xrightarrow[\text{decarboxylase}]{\text{phenylalanine}} \text{phenylethylamine} + CO_2 \qquad (37)$$

Berjonneau *et al.*[177] prepared a phenylalanine electrode using phenylalanine decarboxylase immobilized on a Teflon membrane via co-cross-linking with BSA and glutaraldehyde. The electrode had a linear response range of 2.5×10^{-3} to 1.5×10^{-2} *M*. Calvot *et al.*[178] immobilized the enzyme in a magnetic membrane held to the CO_2 electrode with a cylinder magnet. A linear response was obtained between 2.5×10^{-3} and 1.5×10^{-2} *M*.

4.5.5. L-Asparagine Electrodes

An enzyme electrode for L-asparagine was constructed by Guilbault and Hrabankova[176] using the enzyme asparaginase and a monovalent cation electrode to measure the ammonium ion produced by reaction (24). The enzyme was stable for a period of three weeks when immobilized in a polyacrylamide gel. The response was linear in the range 5×10^{-5} to 1×10^{-2} *M* with a response time of 1 to 2 min.

Wawro and Rechnitz[182] described an electrode for asparagine based on the ammonia gas-sensing electrode. The enzyme asparaginase was linked to a 210 mesh nylon net via glutaraldehyde coupling. The net was then placed over the gas membrane of the ammonia sensor. The response of the electrode was linear in the range 8×10^{-5} to 8×10^{-3} *M* with a slope of 50 mV/decade. There was little loss of activity over a 15-day period. Due to the large open area (~35%) of the nylon net there is little impediment of movement of the substrate to the enzyme. Consequently, the response time (1.3 min) was very rapid for this type of electrode.

4.5.6. L-Lysine Electrodes

A lysine-specific enzyme electrode produced by coupling the enzyme lysine decarboxylase with a CO_2 gas-sensing electrode was reported by Berjonneau *et al.*[177] The enzyme, which catalyzes the following reaction,

$$\text{L-lysine} \xrightarrow[\text{decarboxylase}]{\text{lysine}} \text{cadaverine} + CO_2 \qquad (38)$$

was bound to a Teflon membrane using BSA and glutaraldehyde. Linear response was obtained in the range 1×10^{-4} to 2×10^{-3} *M*.

Calvot *et al.*[178] also reported a lysine electrode using a CO_2 gas-sensing electrode. Lysine decarboxylase was immobilized in a magnetic membrane. Response was linear from 1×10^{-4} to 1×10^{-3} *M*.

White and Guilbault[183] described an enzyme electrode for L-lysine based on the same system. Lysine decarboxylase was immobilized directly onto the surface of the gas membrane of the CO_2 sensor. The response of the electrode was linear between 1×10^{-4} and 3×10^{-2} *M* with a slope between 0.80 and 0.90 pH units/decade. The electrode was usable for a period of up to 50 days when stored in a buffer containing pyridoxal phosphate, a necessary enzyme cofactor. Due to the high specificity of the enzyme, no response was obtained for other common amino acids. The only limitation of the electrode was a somewhat long response time, i.e., 5–10 min. This enzyme electrode was shown to be effective in determining the lysine content of grains.

4.5.7. L-Glutamate Electrodes

Davies and Mosbach[184] described an enzyme electrode for L-glutamate which utilized an immobilized enzyme cofactor. The enzymes glutamate dehydrogenase and lactate dehydrogenase along with dextran-bound NAD^+ (nicotinamide adenine dinucleotide) were held on the surface of a monovalent cation electrode with a dialysis membrane. The cation electrode responds to the ammonium ion produced by the reaction:

$$\text{L-glutamate} + H_2O + NAD^+ \xrightarrow[\text{dehydrogenase}]{\text{glutamate}} \alpha\text{-ketoglutarate} + NH_4^+ + NADH \qquad (39)$$

The second enzyme is needed to recycle the cofactor, i.e.,

$$\text{pyruvate} + NADH \underset{\text{dehydrogenase}}{\overset{\text{lactate}}{\rightleftharpoons}} \text{L-lactate} + NAD^+ \qquad (40)$$

For the determination of glutamate, pyruvate was added to the buffer solution at a concentration of 2 m*M*. The same system was used to indirectly measure pyruvate concentrations in the presence of excess glutamate, i.e., 10 m*M*. The response to glutamate was linear in the range 1×10^{-4} to 1×10^{-3} *M* with a response time from 3 to 4 min. The response slope was from 15 to 20 mV/decade, and this response decreased over a 15-day period. The electrode responded to pyruvate in a nonlinear manner in the range 1×10^{-5} to 1×10^{-3} *M*.

This was the first attempt to utilize an immobilized enzyme cofactor in a potentiometric enzyme electrode. Most previous electrodes used simple hydrolytic enzymes which did not require cofactors. In cases where cofactors were required, they were included in the buffer solution. This is a very uneconomical approach due to the general high cost of these substances. The low slope obtained in this study, i.e., 15–20 mV/decade, was probably due to the failure to optimize conditions. This sytem is much more complex than a simple enzyme electrode, requiring two enzymes in addition to the immobilized cofactor. Parameters that need to be optimized include the enzyme concentrations, cofactor concentration, and the enzyme-to-cofactor ratio.

Ahn *et al.*[185] reported a sensor for glutamate in which a CO_2 gas sensor was used in conjunction with glutamate decarboxylase which catalyzes the reaction

$$\text{L-glutamate} \xrightarrow[\text{decarboxylase}]{\text{glutamate}} \text{4-aminobutyrate} + CO_2 \tag{41}$$

The enzyme was cross-linked with glutaraldehyde and held on the electrode with a dialysis membrane or a silicon adhesive. The enzyme electrode had a very long response time, i.e., 15–20 min, and responded only at high glutamate levels, 7×10^{-4} to 7×10^{-3} *M*. This poor response is probably due to the low initial activity of the enzyme used and the large loss of activity during the immobilization process, i.e., only 1.4% of the original activity remained. Better results were obtained using the soluble enzyme in the stationary-solution mode.

4.6. Nucleotide Electrodes

4.6.1. AMP Electrode

Papastathopoulos and Rechnitz[186] described a highly selective enzyme electrode for 5′-adenosine monophosphate (AMP). The enzyme AMP deaminase selectively cleaves the substrate to produce ammonia according to the reaction

$$5'\text{-AMP} + H_2O \xrightarrow[\text{deaminase}]{\text{AMP}} 5'\text{-IMP} + NH_3 \tag{42}$$

An ammonia gas-sensing electrode was used to measure the ammonia produced. A commercial enzyme suspension was concentrated by molecular filtration and coupled to the gas sensor by means of a dialysis membrane. A linear response was obtained between 8×10^{-5} and 1.5×10^{-2} *M* with a slope of 46 mV/decade. The response time varied with concentration being about 6 min between 1×10^{-4} and $1 \times 10^{-3}\,M$ and about 2 min at $1 \times 10^{-2}\,M$. No response was observed to other nucleotides tested including 5′-ATP, 5′-ADP, 3′,5′-cyclic AMP, adenine, and adenosine.

4.6.2. Cyclic AMP Electrode

A dual-enzyme electrode responsive to 3′,5′-cyclic AMP was described by Rechnitz *et al.*[187] The enzymes phosphodiesterase, which converts cyclic AMP to 5′-AMP, i.e.,

$$\text{3',5'-cyclic AMP} + H_2O \xrightarrow[\text{phosphodiesterase}]{\text{cyclic AMP}} \text{5'-AMP} \tag{43}$$

and AMP deaminase, reaction (42), were used to produce ammonia from cyclic AMP. The enzymes were held on the surface of an ammonia gas-sensing electrode with a dialysis membrane. Magnesium ion, a necessary cofactor for the phosphodiesterase enzyme, was included in the buffer solution. The response of the enzyme electrode was linear in the concentration range 1×10^{-5} to $1 \times 10^{-2}\,M$ with a slope of 56 mV/decade. In this range the response times varied from 3 to 6 min. The electrode, as would be expected, responded to AMP. In addition, response was observed to ADP (adenosine diphosphate), possibly due to the contamination of phosphodiesterase with myokinase which converts ADP to AMP, i.e.,

$$2\,\text{ADP} \xrightleftharpoons{\text{myokinase}} \text{ATP} + \text{AMP} \tag{44}$$

Further work needs to be done to purify the enzyme to improve the selectivity.

4.7. Uric Acid Electrode

Kawashima and Rechnitz[188] described an electrode for uric acid based on the carbon dioxide gas sensor. The enzyme uricase was used to catalyze the production of CO_2 from uric acid via the following reaction:

$$\text{uric acid} + O_2 \xrightarrow{\text{uricase}} \text{allantoin} + H_2O_2 + CO_2 \tag{45}$$

The enzyme was adsorbed on hydroxyethylcellulose and held on the surface of the gas sensor by means of a dialysis membrane. Optimum conditions for the system were determined in stationary solutions. An automated flow system using the soluble enzyme was also investigated. The enzyme sensor gave a linear response from 1×10^{-4} to 2.5×10^{-3} with a slope of 57 mV/decade. Response times were from 5 to 10 min depending on the concentration level. The electrode was stable for a period of ten days. The analysis of urine control samples gave results in good agreement with conventional procedures.

4.8. Creatinine Electrode

Meyerhoff and Rechnitz[75] reported an enzyme electrode responsive to creatinine. An ammonia gas-sensing electrode was used to measure the ammonia produced by the enzyme creatinine deiminase [reaction (10)]. The commercial enzyme preparation was purified on DEAE anion exchange cellulose and concentrated by molecular filtration. It was discovered that tripolyphosphate, an activator of the enzyme, significantly improved the response characteristics of the electrode, and consequently was added to the buffer solution. The activated enzyme electrode exhibited a linear response in the range 7×10^{-5} to 1×10^{-2} M with a slope of 49 mV/decade. The response time was dependent on the concentration, being 6–10 min below 5×10^{-3} M and 2–5 min above this concentration level. The electrode was usable for a period of eight days with some loss of activity. No response was observed for arginine, urea, or creatine. After the removal of ammonia by ion exchange, the response of the electrode in urine control samples was similar to that obtained in buffered standards, thereby making possible the determination of creatinine in urine samples. In control serum however, the response slope decreased to 29 mV/decade and the response was not linear in the normal range of serum creatinine. Therefore, the electrode could only be used as a semiquantitative screening procedure for abnormally high serum creatinine values.

4.9. Acetylcholine Electrodes

Acetylcholine enzyme electrodes using the enzyme acetylcholine esterase in conjunction with a pH electrode to sense the change in pH due to reaction (22) have been reported.[189,190] Tran-Minh *et al.*[189] described an electrode in which the enzyme was immobilized directly onto the surface of the pH sensor by cross-linking with BSA and glutaraldehyde. The measured potential was proportional to the logarithm of the acetylcholine concentration in the range 1×10^{-4} to 1×10^{-2} M with a response slope of 35 mV/decade. The electrode had a response time of 3 min and was very stable with only a 10% loss of activity over a two-month period.

Durand *et al.*[190] reported a similar electrode in which acetylcholinesterase was immobilized in a gelatin membrane on the surface of a pH electrode. By maintaining a constant buffer capacity a linear relationship between the measured potential and the acetylcholine concentration was obtained at low concentration levels, i.e., 1×10^{-5} to 1×10^{-4} M. The kinetic method, using initial slope measurements, was employed. The electrode was stable for a period of 40 days.

4.10. D-Gluconate Electrode

Jensen and Rechnitz[191] recently reported an enzyme sequence electrode for D-gluconate using a dual-enzyme reaction layer. The substrate was decarboxylated using the following reaction sequence:

$$\text{D-gluconate} + \text{ATP} \xrightarrow[\text{Mg(II)}]{\text{gluconate kinase}} \text{6-phospho-D-gluconate} + \text{ADP} \tag{46}$$

$$\text{6-phospho-D-gluconate} + \text{NADP}^+ \xrightarrow[\text{dehydrogenase, Mg(II)}]{\text{6-phospho-D-gluconate}} \text{D-ribulose-5-phosphate} + \text{NADPH} + \text{H}^+ + \text{CO}_2 \tag{47}$$

The two enzymes were coupled to the CO_2 sensor either by a dialysis membrane or BSA-glutaraldehyde cross-linking directly onto the gas-permeable membrane. This is the most complicated enzyme system used to date in a potentiometric enzyme electrode. In addition to the two enzymes, it was necessary to have present ATP, $NADP^+$, and Mg^{+2}, which were added to the buffer solution. In Tris-HCl buffer the potentiometric response of the dialysis membrane electrode was linear in the range 8×10^{-5} to 1.2×10^{-3} *M* with a slope of 51 mV/decade. The response time was found to be dependent on the concentration level, being about 6 min at 1×10^{-4} *M* and 1–2 min at 1×10^{-3} *M*. However, the electrode lifetime was only one day. Both the response slope and the stability of the electrode were improved in glycylglycine buffer. Under these conditions the response was linear from 1.2×10^{-4} to 2.4×10^{-3} *M* with a slope of 56 mV/decade and a usable lifetime of four days. The electrode showed no response to D-glucose, D-mannose, D-mannose-6-phosphate, glucaric acid, or glucuronic acid. There was a slight response to D-glucose-6-phosphate possibly due to a trace impurity in the enzymes used. Of course, the electrode also responded to 6-phospho-D-gluconate. It was found that electrodes prepared by cross-linking the enzymes yielded a poorer response, i.e., slope of 28 mV/decade. This can be attributed to either the restricted mobility of the substrates and the two necessary cofactors in the gel layer or the inactivation of the enzymes during the immobilization process.

4.11. Lactate Electrode

A different approach to constructing a potentiometric enzyme electrode was recently reported by Shinbo *et al.*[192] A redox electrode was used to construct a sensor for lactate in conjunction with the enzyme lactate dehydrogenase, i.e.,

$$\text{lactate} + 2\text{Fe(CN)}_6^{-3} \xrightarrow[\text{dehydrogenase}]{\text{lactate}} \text{pyruvate} + 2\text{Fe(CN)}_6^{-4} + 2\text{H}^+ \tag{48}$$

The change in the ratio of ferricyanide to ferrocyanide due to the enzymatic reaction was sensed by the redox electrode. The sensor membrane of the redox electrode consisted of a plasticized polyvinyl chloride membrane containing dibutylferrocene. A silver–silver-chloride wire was used as the internal reference electrode. The enzyme lactate dehydrogenase along with a small amount of catalase, added to diminish the inactivation of the enzyme, were immobilized in a gelatin membrane and held on the electrode by means of a dialysis membrane. The response time of the electrode was very long, i.e., 20–30 min. An S-shaped response curve was obtained which possessed a narrow linear range, 1×10^{-4} to 1×10^{-3} *M*, due to the low K_m value of the enzyme ($K_m = 1.2 \times 10^{-3}$ *M*). The response was usable, however, from 5×10^{-5} to 1×10^{-2} *M*. The disadvantages of this sensor include the unusually long response time and possible interference by reducing agents.

4.12. Inhibitor Determination

Enzyme electrodes can also be used for the determination of reversible enzyme inhibitors, as was shown by Tran-Minh and Beaux.[193,194] A urea enzyme electrode, based on a CO_2 gas sensor, was used to determine fluoride. Urease was immobilized directly onto the gas-permeable membrane by cross-linking with HSA and glutaraldehyde. It was found that the sensitivity to fluoride increased with increasing cross-linking time.[193] Fluoride response was determined in the presence of 1×10^{-2} *M* urea. Steady state potentials were obtained after a period of 5–6 min. The potentiometric response was linear in the fluoride concentration range 3×10^{-4} to 1×10^{-2} *M*. Conditions for optimum inhibitor response were found to be the opposite of those for substrate determination. An enzyme electrode used for substrate determinations, where the effect of inhibitors must be minimized, must contain a large amount of immobilized enzyme with high activity. For inhibitor determinations, a low immobilized enzyme concentration was necessary. The behavior of these electrodes as a function of various parameters was explained by a theoretical analysis of the concentration profiles of substrate and product in the enzyme layer. This approach can be extended to other enzyme–inhibitor couples.

4.13. Substrate Electrodes

Substrate electrodes for the determination of several enzymes have been reported.[195–198] In this type of sensor the substrate is coupled to an appropriate ion-selective electrode which detects the product of the enzymatic reaction. Since the substrate is consumed by the reaction, the substrate layer must be continually renewed by a flow-through arrangement. The system can be quite complicated requiring pumps and regulators to control the substrate flow. This approach is not very practical and therefore has not been widely used.

Montalvo[195,196] reported the first substrate electrode for the determination of urease. The surface of a monovalent cation electrode was covered with a dialysis membrane trapping a thin layer of concentrated urea solution between the glass bulb and the membrane. The urea solution was continually renewed by gravity flow from a reservoir or by a means of a roller pump. When the electrode is dipped in a solution containing urease, urea diffuses out of the substrate layer and reacts with the enzyme at the surface of the membrane. A fraction of the ammonium ion formed diffuses into the membrane layer, due to the concentration gradient, and is detected by the cation electrode. A steady state potential, proportional to the enzyme activity, was obtained in about 5 min.

Crochet and Montalvo[197] described a substrate electrode for the determination of serum cholinesterase activity. A pH electrode was used to measure pH changes in a complex micro electrochemical cell. The cell contained two thin layers separated by a dialysis membrane. Cholinesterase solution was passed through the layer closest to the electrode while the substrate acetylcholine was pumped through the other layer. The acetic acid produced when the substrate diffuses into the enzyme layer and reacts with the enzyme was sensed by the pH electrode. A kinetic method was used enabling the assay to be completed in 2–5 min. The rate of potential change was proportional to the enzyme concentration. As was mentioned, the experimental setup used was fairly complicated, requiring two pumps, one for enzyme injection and one for substrate injection, and a special micro electrochemical cell.

A different approach to constructing a substrate electrode for cholinesterase was described by Gibson and Guilbault.[198] An insoluble reineckate salt of acetylcholine was held on the tip of a pH electrode with a nylon net. The enzyme diffuses into the substrate layer and the consequent enzymatic reaction produces acetic acid, which is sensed by the pH electrode. The rate of potential change was directly related to the enzyme concentration. The limitation of this approach is obvious, i.e., the immobilized substrate has to be periodically replaced.

4.14. Current Trends

There have been many advances made in enzyme electrode technology in recent years. Many of these were mentioned in an earlier section dealing with the different types of enzyme electrodes. In this discussion the current trends and recent innovations in enzyme electrode methodology will be considered in more detail.

Probably the most significant improvement that has been made in enzyme electrodes has been the use of gas-sensing electrodes as the sensing element. The first gas-sensor-based enzyme electrodes, reported by Guilbault and Shu[164] and Tran-Minh and Broun,[165] utilized a CO_2 sensor. A significant contribution was made by Anfalt *et al.*,[160] who were the first to report an ammonia gas-sensor-based enzyme electrode. This study demonstrated the feasibility of using the ammonia gas sensor in the construction of enzyme electrodes despite the difference in pH optima for the electrode (pH = 12) and most enzyme reactions (pH = 6–8). The urea electrode constructed was found to have good sensitivity at a pH as low as 7.0. Both the CO_2 and the NH_3 gas sensors have since been widely used in the fabrication of enzyme electrodes (Table 4) and currently are the most frequently employed detectors in potentiometric-based enzymatic analysis.

The popularity of the gas-sensing electrodes is due to their high selectivity, since they are unaffected by ions in solution. The resulting enzyme sensors are highly selective and can be used in many cases without prior separation in complex biological fluids such as blood serum,[161–163] whole blood,[162] and urine.[188] No other potentiometric electrode including the highly selective pH electrode can match this selectivity. The response of the pH-electrode-based enzyme sensors is strongly dependent on the buffer capacity of the solution. This can cause problems in the analysis of biological fluids and other complex media, i.e., fermentation broths,[199] where the buffer capacity can vary between samples.

A disadvantage of these gas-sensor-based enzyme electrodes is an inherently longer response time. Response times of 5 min or more are not unusual. Generally, this is not a serious handicap in light of the specificity and simplicity of the method. Moreover, it has been possible to reduce these response times considerably by the immobilization of the enzyme directly onto a new Teflon gas membrane,[163] or onto a porous nylon net.[182]

Considerable improvements have also been made in the enzyme immobilization procedures used in the construction of enzyme electrodes. Initially enzyme electrodes were formed by physically entrapping the enzyme either in a polyacrylamide gel[153–155] or by means of a dialysis membrane.[162,166,170] Although the latter method is still frequently used due to its simplicity, the use of chemically bound enzymes has become increasingly popular because of their improved stability. The most frequently used chemical methods have been the insolubilization onto a porous organic polymer[37,180] or co-cross-linking with BSA and glutaraldehyde.[185] The insolubilized enzyme is held on the electrode by means of a dialysis membrane. Covalent attachment directly to the gas membrane of gas-sensing electrodes has also been used.[160,163,177,183] These methods yield electrodes with greater stability, although the procedures can be time consuming and reproducibility from electrode to electrode is difficult to obtain.

Tran-Minh and Broun[200] described two methods of constructing enzyme electrodes using cross-linked enzymes which have been shown to be superior to other methods for forming simple, cheap, and reproducible electrodes.[201] The first method, called the direct binding method, was used for the binding of an enzyme to a glass electrode using BSA and glutaraldehyde. The electrode was dipped into a solution containing the enzyme, BSA, and glutaraldehyde and then rotated around its axis for 15 min to obtain an even coating. The electrode was then rinsed with water, a glycine solution, and once again with water. The second method involves the preparation of an "active membrane" to be used with gas-sensing electrodes. The active membrane was prepared by cross-linking the enzyme and BSA with glutaraldehyde until complete solidification was achieved. This membrane was then placed onto the gas-sensing electrode along with the gas-permeable membrane. These techniques result in the simple and reproducible formation of stable homogeneous layers of covalently bound enzymes.

Wawro and Rechnitz[182] described a procedure for the immobilization of enzymes which greatly reduced the response time of an ammonia gas-sensor-based asparagine electrode. Asparaginase was immobilized onto a porous nylon net via glutaraldehyde coupling. A circular segment of the enzyme-containing net was placed on the gas membrane of the ammonia sensor. Although the method was fairly time consuming, a response time of under 2 min was obtained due to the porosity of the nylon net.

Another area of current investigation has been the development of enzyme sequence electrodes, involving more than one enzyme, to increase the applicability of enzyme sensors. However, there are several problems with this approach. If the enzymes used have different pH optima, then a compromise of conditions is necessary. Also, selectivity problems can arise since the electrode will respond to each intermediate substrate. Another complicating factor is the requirement of enzyme cofactors. If the enzymes require a cofactor, this substance must either be added to the buffer solution, which in many cases is uneconomical, or immobilized cofactors can be used. In this case an additional enzyme reaction is needed to continually regenerate the cofactor. Davies and Mosbach[184] described an electrode for glutamate which involved a single enzyme along with an immobilized cofactor, regenerated by means of a second enzymatic reaction. Due to the complexity of the system, poor response slopes were obtained, i.e., 15–20 mV/decade. A system involving more than one enzyme along with cofactors would be even more complicated.

There have been relatively few potentiometric enzyme electrodes using this multistep approach, none of which has involved the use of more than two enzymes. Early enzyme sequence electrodes include a glucose electrode using the enzymes glucose oxidase and peroxidase along with an iodide sensor[37] and a phenylalanine electrode using L-amino oxidase and peroxidase with an iodide electrode.[180] More recently a dual-enzyme cyclic AMP-selective electrode has been reported.[187] This electrode employs an ammonia gas-sensing electrode in conjunction with the enzymes AMP deaminase and phosphodiesterase. These systems are relatively simple, requiring two enzymatic reactions which do not need cofactors. A more complicated system has been reported for a D-gluconate electrode.[191] Two enzymes, gluconate kinase and 6-phospho-D-gluconate dehydrogenase, both of which require cofactors, were used in conjunction with a CO_2 gas-sensing electrode. The enzymes were held on the electrode with a dialysis membrane and the cofactors were added to the buffer solution. The electrode functioned very well using this configuration. However, a considerable decrease in response was observed when the enzymes were covalently bound to the gas membrane using BSA and glutaraldehyde. This could indicate a possible problem of using immobilized enzymes in this complicated system, i.e., the restricted mobility of all the various substrates and cofactors in the enzyme layer. However, the decrease in response could simply be due to a loss of enzyme activity during the immobilization process.

Despite the problems mentioned, multistep enzyme electrodes will undoubtedly become more popular due to the many recent advances in multiple-enzyme immobilization,[202,203] and cofactor immobilization and regeneration[184,204–209] procedures. With the use of these enzyme sequence electrodes, the versatility and applicability of enzyme electrodes can be markedly increased. Another approach to the development of these multistep enzyme electrodes is described in Section 5, which deals with the construction of sensors that use intact bacterial cells.

The major emphasis, until recently, has been directed toward the development of macroscale enzyme electrodes. With recent advances in the design of miniature and micro size ion-selective electrodes,[210–215] including gas-sensing electrodes for CO_2 and NH_3,[216–220] the construction of micro enzyme sensors for *in situ* and *in vivo* analysis has become possible.[220]

Pui *et al.*[220] described a micro gas probe for either NH_3 or CO_2 and have demonstrated the feasibility of using this sensor in the construction of a micro enzyme electrode. The design of this micro enzyme electrode is shown in Fig. 7. The electrode was based on the air gap design and had a tip diameter of less than 10 μm. A spear-type glass micro pH electrode was contained in an insulating pipet. Two additional pipets were used in the construction of the micro gas sensor. The insulated pH electrode was placed in an inner pipet, which contains the internal filling solution (0.1 *M* NaCl and 0.005 *M* NH_4Cl or $NaHCO_3$ depending on the electrode desired). This whole assembly was enclosed in an outer pipet. A silver–silver-chloride wire

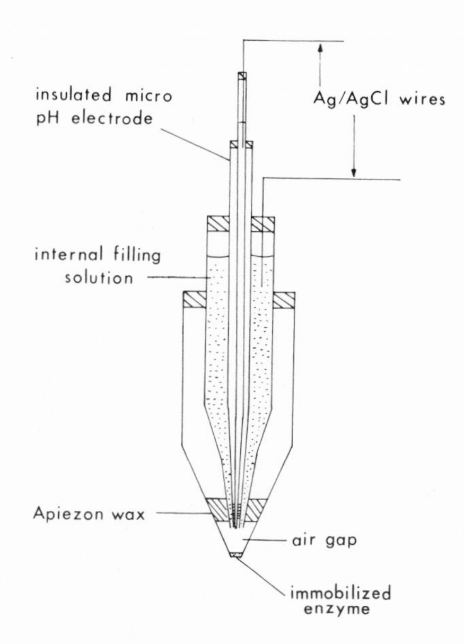

Fig. 7. Micro enzyme electrode. Reprinted with permission from reference 220.

served as the reference electrode. This design offers the advantage of a conventional air gap electrode in that the absence of a gas membrane facilitates free diffusion of the gaseous species into the internal solution. However, the electrolyte layer does not need to be replaced after every measurement in order to obtain the best reproducibility. The inner pipet serves to retain the bulk of the internal electrolyte, which stabilizes the composition of the thin layer on the pH electrode by a slow, continuous exchange process. Response characteristics of the micro gas sensor were comparable to those of the commercial macroelectrodes.

A micro urea electrode was constructed by introducing a solution containing urease and BSA into the air gap by capillary action. After the solution dried by evaporation, a drop of glutaraldehyde solution was added to cross-link the enzyme, forming a sturdy layer at the pipet tip. The potentiometric response of the micro urea electrode was linear from 1×10^{-4} to 1×10^{-2} *M*, similar to the macro enzyme electrodes. However, the slope was considerably lower than previously observed, i.e., 30 mV/decade. This was probably due to having insufficient enzyme activity present. Although the conditions were not optimized, this study demonstrates the feasibility of fabricating micro enzyme sensors. More developments will certainly be made in this area, particularly with the number of possible micro electrodes which have been developed.

Another recent development of interest has been the use of an enzyme activator to improve the response characteristics of an enzyme electrode.[75] Previously, the effects of pH, temperature, ionic strength, and amount of enzyme (see Section 4.17) were studied as a means of optimizing electrode response. The activation of the enzyme creatinine deiminase by polyphosphate was utilized to construct an activated enzyme electrode. The activator was added to the buffer solution at a concentration of 3×10^{-2} *M*. The unactivated electrode had a response slope of 44 mV/decade in the linear range 4.2×10^{-4} to 8.9×10^{-3} *M* and showed a considerable loss of activity after a four-day period. The activated electrode had an extended linear range, i.e., 7×10^{-5} to 8.9×10^{-3} *M*, with an improved slope of 49 mV/decade. The stability of the electrode was also significantly improved. It was usable for a period of eight days. Kinetic studies were done to elucidate the origin of the activation effect. Eadie–Hoftsee plots for the activated and nonactivated enzymes indicated that the increased sensitivity and response slope of the activated electrode were due to an increase of V_m of the enzyme. The K_m values for both cases were the same. This approach offers another possible way to improve enzyme electrode response characteristics.

Still another interesting development has been the use of an enzyme electrode to determine an enzyme inhibitor.[194] Generally, immobilized enzyme electrodes prepared for optimum substrate determination are not significantly affected by most inhibitors. This study demonstrates that optimum conditions for inhibitor determinations are opposite of those necessary for substrate determinations, i.e., low enzyme concentrations are needed. Fluoride was determined in the range 3×10^{-4} to 1×10^{-2} *M* by its inhibition of a urea electrode. This approach can be extended to other enzyme–inhibitor systems, making possible a very sensitivite determination of many more substances.

4.15. Applications of Enzyme Electrodes

The main applications of enzyme electrodes have been in the clinical analysis of biological fluids. These applications were already discussed in the enzyme electrode survey section (Section 4). Here, several other interesting uses of enzyme electrodes will be considered.

The urea enzyme electrode has been used to monitor the dissolution rate of industrial retarded urea animal feeds.[221] Urea is used as a source of nitrogen for protein synthesis in feedstuffs for domestic ruminant animals, e.g., cattle and sheep. The release of the urea has to be slow enough so that the ammonia produced by hydrolysis does not reach toxic levels in the blood. Therefore, it is necessary to administer the urea in the form of a special preparation having a relatively slow dissolution rate. The rate of dissolution from different industrial feeds is an important property of the product which can be simply measured using the urea enzyme electrode. For the determination, an appropriate amount of the sample was added to a buffer solution and the rate of urea production was monitored with an ammonium-ion-based urea electrode. The times required to dissolve 50% (T_{50}) and 90% (T_{90}) of the urea were used to compare dissolution rates of different products. The method provides a simple and convenient technique for *in vitro* dissolution studies as long as the rate of dissolution is slow compared to the response time of the electrode.

A lysine enzyme electrode has been used for the determination of lysine in grains and foodstuffs.[183] The electrode was highly specific, exhibiting no response to other amino acids, so that it could be used in mixtures without the need for extensive separations. Grain samples were analyzed after acid hydrolysis, giving results in good agreement with an amino acid analyzer method.

The penicillin enzyme electrode has been applied to the measurement of penicillin levels in fermentation broths.[199,222–224] Filtered samples from penicillin fermentation broths were analyzed using an enzyme electrode formed by entrapping penicillinase in a dialysis membrane, over a combination pH electrode.[199] In the buffered solutions studied, i.e., phosphate buffer of different buffer capacities and culture media, the measured potential was linearly related to the penicillin concentration in the range 1×10^{-3} to 1×10^{-2} *M*. This linear relationship between the potential and the concentration has been explained as being due to the constant buffer capacity of the solution which makes the change in pH directly proportional to the concentration of the acid-producing species.[56,137] Since the electrode is based on measuring pH changes, the sensitivity is dependent on the buffer capacity of the medium. Due to the differences in buffer capacities between standard and sample solutions, best agreement with a standard method (the hydroxylamine method) was obtained using a standard addition procedure. The penicillin electrode method is applicable to discrete sample analysis and offers a simple procedure for the determination of penicillin in fermentation broths.

A sterilizable and regenerable penicillin electrode for the direct monitoring of penicillin levels in a fermenter for purposes of fermentation control has been reported.[222–224] The enzyme electrode was constructed using a flat pH electrode mounted in a stainless steel housing. The enzyme solution was held in a chamber, one wall of which being the surface of the glass electrode, the other being a semipermeable membrane. The electrode could be filled with enzyme solution through stainless steel tubes connected to the enzyme chamber. This made it possible to sterilize the electrode body in an autoclave and then introduce the enzyme solution, sterilized by filtration. In addition, the electrode could be regenerated simply by replacing the enzyme solution. In 0.025 *M* phosphate buffer the potential was linearly related to the concentration up to 3.0×10^{-2} *M*. The response time was about 1 min. This electrode design meets many of the necessary requirements for continuous monitoring fermentation control. It is autoclavable, rechargeable, has a sufficient lifetime, and responds in the required concentration range. The major problem with the electrode is that it cannot be precalibrated in an uninoculated broth because the buffer capacity of the broth is not constant during the fermentation process. This change in buffer capacity, of course, affects the response of the electrode.

Enzyme electrodes have also been shown to be useful in determining enzyme activities, thereby enabling the potentiometric assay of many more enzymes. The first application of this kind was the determination of arginase activity using a urea electrode reported by Booker and Haslam.[225] A monovalent-cation-based urea electrode was used to measure the rate of urea production from reaction (7). The enzyme electrode and a reference electrode were placed in a buffer solution containing the substrate, arginine. The enzyme solution was added and the kinetic curve of potential versus time was recorded. Three methods were evaluated for the determination, i.e., the initial slope method in which a tangent was drawn to the kinetic curve at the initial stage of the reaction, the lapsed-time–slope method in which a tangent was drawn to the curve at $t = 30$ sec, and a timed-potential method (fixed-time method) where the potential was obtained after a constant time. The initial slope method was found to be the most sensitive, but a response was obtained even in the absence of substrate, apparently from impurities in the enzymes used. This also caused problems using the fixed-time method. However, by using the lapsed-time–slope method the response caused by the contaminants could be eliminated since 98% of this response occurred in the first 30 sec. Measurements were made in less than 10 min with a precision of about 3%. The simple potentiometric determination of arginase in serum could provide a convenient means of diagnosis of liver disorders. Although this method is limited by the interference of sodium and other monovalent cations, it demonstrated the feasibility of using enzyme electrodes to determine enzyme activities.

Gebauer *et al.*[226] have recently extended this approach using ammonia gas-sensor-based enzyme electrodes, thus eliminating cation interference problems. It was found that despite the relatively slow response of these enzyme electrodes, it was possible to measure enzyme activities over a considerable range. The enzymes determined were arginase [reaction (7)], with a urea electrode; 3′,5′-cyclicnucleotide phosphodiesterase [reaction (43)], with an AMP electrode; and 5′-nucleotidase, which catalyzes the following reaction:

$$5'\text{-AMP} + H_2O \xrightarrow{5'\text{-nucleotidase}} \text{adenosine} + \text{Pi} \qquad (49)$$

using an adenosine electrode.[227] The initial rate of the reactions, determined using the slope method, were directly proportional to the enzyme activities. Arginase activity was also determined in beef liver homogenates.

The results obtained were in good agreement with an established colorimetric method. Advantages of this potentiometric technique over other coupled enzyme methods include the ability to reuse the enzyme, which is immobilized on the electrode. In addition, the need for long incubation and color development times, as well as colorimetric reagents and centrifugation steps are eliminated. However, if the enzymes used have different pH optima, a compromise of conditions is necessary.

Another interesting application of enzyme electrodes has been the determination of formation constants. Riechel and Rechnitz[228] used an AMP-selective enzyme electrode to study the interaction of AMP with the enzyme D-fructose-1-6-diphosphatase to demonstrate the feasibility of the technique. AMP is an allosteric inhibitor of this enzyme. The binding study is possible since the enzyme electrode responds only to the free AMP since the protein-bound AMP is screened out of the enzyme layer by the dialysis membrane. Binding measurements were made using a potential comparison method to determine the free AMP concentration. The bound concentration was then easily determined since the total AMP concentration was known. Binding constants and site concentrations were evaluated graphically using a Lineweaver–Burk plot or a direct linear plot. Both the binding constant and the stoichiometry determined were in fairly good agreement with earlier studies. The convenience, simplicity, and low cost of this method make it an attractive alternative to traditional gel filtration techniques for such binding studies.

4.16. Theoretical Studies

There are many factors which affect the response of enzyme electrodes. The various steps which need to be considered include the following: diffusion of the substrate through the solution to the surface of the electrode; diffusion through the enzyme membrane; phase equilibrium between the enzyme layer and the solution; kinetics of the enzymatic reaction; and diffusion of the product to the electrode surface where it is detected. There have been several theoretical studies on potentiometric enzyme electrode response dealing with the steady state[200,229,230] and transient behavior.[200,231] Frequently, a simplified model is used in which the response of the electrode is taken to be governed by the rate of the enzymatic reaction and the diffusion of the substrate and product in the enzyme layer.[200,231] It is assumed that the enzyme reaction obeys Michaelis–Menten kinetics with the rate given by equation (2). The concentrations of substrate (S) and product (P) in the enzyme layer are then given by

$$\frac{\partial[S]}{\partial t} = D_s \frac{\partial^2[S]}{\partial X^2} - \frac{V_m[S]}{K_m + [S]} \tag{50}$$

$$\frac{\partial[P]}{\partial t} = D_p \frac{\partial^2[P]}{\partial X^2} + \frac{V_m[S]}{K_m + [S]} \tag{51}$$

where t is the time of the reaction, and D_s and D_p are the diffusion coefficients of the substrate and product in the enzyme layer. Two limiting cases are usually considered: (1) $[S] \ll K_m$, which is the most important analytical situation since the reaction is first order with respect to substrate, and (2) $[S] \gg K_m$, i.e., zero-order kinetics with respect to substrate. Other factors which have been considered include the mass transfer of substrate from the bulk of the solution to the electrode surface, diffusion through a semipermeable membrane, and partitioning of the substrate between the enzyme layer and the solution.

Blaedel *et al.*[229] considered the steady state behavior of potentiometric enzyme electrodes taking into consideration the diffusion within the enzyme layer, external mass transfer, phase equilibrium, and the kinetics of the enzyme reaction. Experimental support for the derived equations was obtained in three systems: an enzyme-membrane-covered monovalent cation electrode; an enzyme membrane separating two solutions; and an enzyme membrane immersed in solution. For the potentiometric sensor the calibration curve predicted by the theoretical equations agreed with that observed experimentally. As predicted, curvature of the plot at low substrate concentrations ($<10^{-4}$ M) was observed, which is governed by the lower detection limit of the cation electrode. A linear portion was obtained where the measured potential was directly proportional to the log of the substrate concentration. At high concentrations the response became independent of substrate due to the saturation of the enzyme. The effect of enzyme concentration could be predicted quantitatively. The response increased as the enzyme concentration in the membrane increased until an upper limit was reached, after which no further improvement was realized.

Racine and Mindt[230] also considered the steady state behavior of enzyme electrodes. The main consideration was the effect of diffusion of the substrate through a semipermeable membrane. Diffusion within the membrane was assumed to be rapid. The derived equations indicated that if the diffusion through the membrane is made the rate-limiting step by using a membrane with a low permeability, the response of the sensor would be linear up to concentrations considerably higher than the K_m value of the enzyme. In addition the effect of enzyme denaturization on the time stability of the electrode would be minimized. A negative consideration would be a longer response time. These predictions were verified experimentally.

Tran-Minh and Broun[200] solved the diffusion equations for a potentiometric sensor by computer calculation. Both the steady state and transient behavior were estimated. The theoretical equations obtained indicated that the concentration of substrate and product near the surface of the electrode depend on several constant parameters, i.e., the K_m of the enzyme, the enzyme activity, the thickness of the enzyme layer and the diffusion coefficients of the substrate and product. It was predicted that the response time could be improved by either reducing the thickness of the enzyme layer or increasing its permeability. The most favorable condition would be to have a high enzyme activity in a very thin layer.

Carr[231] studied the transient response of potentiometric enzyme electrodes using Fourier analysis of the partial differential equations which govern electrode response [equations (50) and (51)]. The assumptions that were made include the following: mass transfer from the bulk of the solution to the electrode is very fast; the partition coefficients for the substrate and product between the enzyme layer and the solution are equal to unity; and the diffusion coefficients of the substrate and product are equal, i.e., $D_s \approx D_p = D$. It was determined that when the enzyme reaction is first order with respect to substrate, the response is dictated by two dimensionless terms Dt^2/L and $K_m D/V_m L^2$, where D is the diffusion coefficient of both the substrate and product, t is the time, and L is the thickness of the enzyme layer. When the enzyme activity is reasonable, i.e., $K_m D/V_m L^2 \leqslant 1$, the response time is not a strong function of the specific activity of the enzyme. In the case of zero-order kinetics neither the concentration nor the K_m value have any influence on the transient behavior, which is governed only by diffusion. However, these factors do affect the steady state potential.

4.17. Response Characteristics

4.17.1. Response Curve

The general shape of an enzyme electrode calibration curve is shown in Fig. 8. As predicted theoretically,[200,229] the curve contains a linear portion, generally between 1×10^{-4} and 1×10^{-2} *M*, where the potential is directly proportional to the logarithm of the substrate concentration. The curve levels off at low concentrations ($<10^{-4}$ *M*) due to the detection limit of the sensing electrode used. The leveling off at high concentrations is due to the saturation of the enzyme at concentrations above the K_m value of the enzyme, as predicted by the Michaelis–Menten equation.

In the case of electrodes based on the pH sensor, it has been shown that by maintaining a constant buffer capacity, a linear relationship can be obtained between the measured potential and the substrate concentration (not the logarithm of the substrate concentration as predicted by the Nernst equation).[190,199] Since the buffer capacity is given by

$$\beta = \frac{dC_B}{d\,\mathrm{pH}} \tag{52}$$

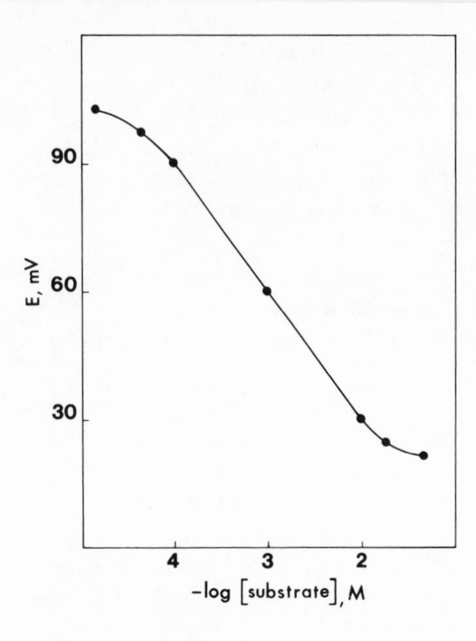

Fig. 8. Typical enzyme electrode response curve.

if β is kept constant the change in pH becomes a linear function of the amount of protons produced or consumed in the enzymatic reaction.[56,137]

The shape and the slope of the response curve are dependent on several factors. One of these is the amount of enzyme in the enzyme layer. The slope increases with increasing amounts of enzyme until a level is reached where further increases produce only small changes in electrode response. In addition, increases in enzyme activity shift the response curves toward potentials which indicate a higher concentration of product in the enzyme layer.[154,161,172,176,179] These observations are illustrated in Fig. 9 for a urea enzyme electrode. Theoretical studies have also predicted these effects.[200,229] Enzyme activators have been shown to have the same effect on the response curve as increasing the enzyme concentration.[75] Figure 10 shows the result of an activator, polyphosphate, on the response of a creatinine enzyme electrode. A shift in potential and an increase in the response slope can be seen.

The pH is another important factor which affects enzyme electrode response. Most enzymes have a pH range where maximum activity is observed. On either side of this range, the activity decreases rapidly. As was noted in Section 3.2, the pH optimum of an immobilized enzyme is not necessarily the same as that of the soluble enzyme. Furthermore, the pH requirements of the sensing electrode may not be compatible with the pH optimum of the enzyme. In this case a compromise in the pH conditions must be made, e.g., enzyme electrodes based on the ammonia gas sensor or the

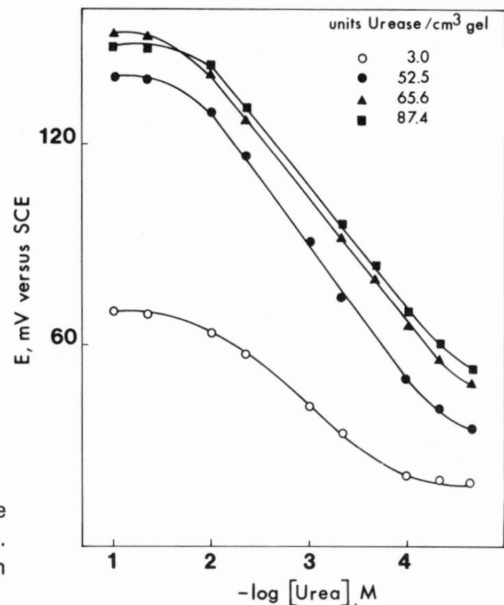

Fig. 9. Effect of the amount of enzyme on the response of a urea electrode. Reprinted with permission from reference 154.

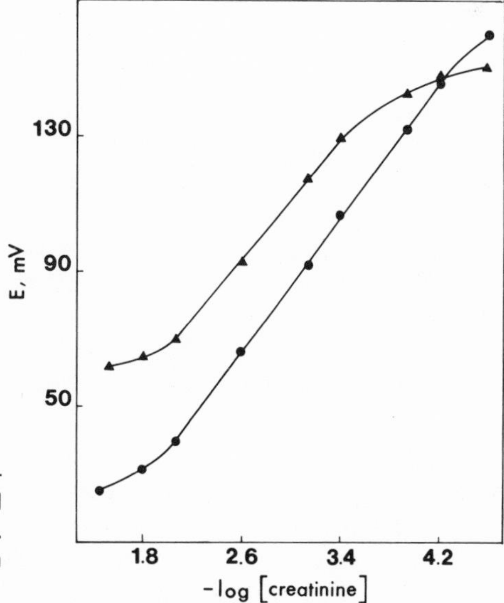

Fig. 10. Response curves of a creatinine electrode under nonactivated (▲) and activated (●) conditions. Reprinted with permission from reference 75.

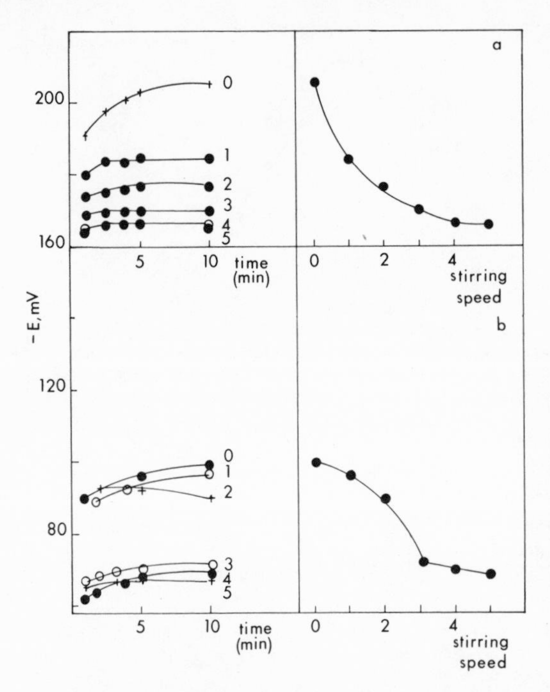

Fig. 11. Effect of stirring rate on response time and steady state potential of an amygdalin electrode. The number 0 corresponds to an unstirred solution and numbers 1–5 to increasing stirring speeds in arbitrary units. [amygdalin] = (a) 10^{-2} *M*, (b) 10^{-4} *M*. Reprinted with permission from reference 170.

solid state cyanide electrode. Generally, it is better to adjust the pH to suit the enzyme rather than the electrode.[160,170]

The absolute value of the steady state potentials depends on several additional factors. The rate of stirring has been shown to affect the value of the potential. The stirring rate affects both the rate of mass transfer of substrate to the electrode and the rate of mass transfer of the product away from the electrode. The net effect, however, seems to be to lower the concentration of product in the enzyme layer, therefore shifting the measured potential until a limiting value is obtained (Fig. 11). Higher stirring rates have also been shown to produce more reproducible potentials and shorter response times.[170,172]

The thickness of the enzyme layer also influences the steady state potential,[170] as shown in Fig. 12. As the membrane is made thicker, the concentration of substrate in the enzyme layer decreases, due to the effect of the membrane on the diffusion process. The concentration of measured product is therefore lowered and a shift in potential results.

Papastathopoulos and Rechnitz[162] found that the potentials obtained with a urea electrode in buffer solutions differed from those obtained in whole blood and serum samples (Fig. 13). A possible explanation for this effect is that the increased viscosity of the biological fluids alters the diffusion coefficients of the species involved. This result indicates that calibration curves should be done in a medium similar to that of the sample solution.

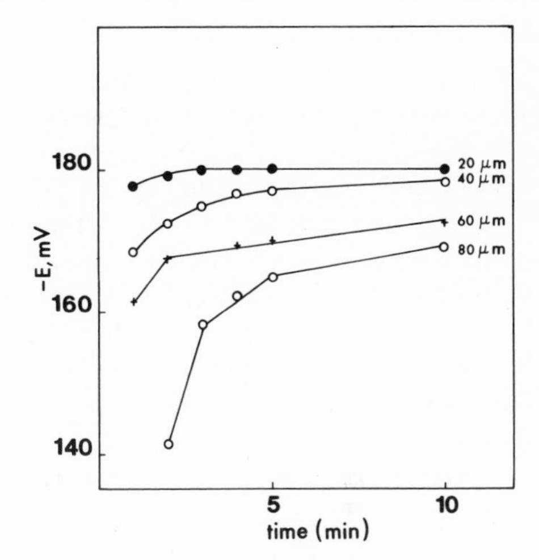

Fig. 12. Effect of thickness of dialysis membrane on response time and steady state potentials. [amygdalin] = 10^{-2} *M*. Reprinted with permission from reference 170.

Alternatively, a standard addition procedure can be used since no change in the response slope results.

In general, the temperature has been found to have little effect on the steady state potentials or the shape of the response curves of enzyme electrodes.[172,175,176] However, in the case of gas-sensing electrodes, the potentials will vary significantly with temperature. These electrodes respond to the partial pressure of dissolved gases in solution, which of course is dependent on the temperature. Accurate temperature control is therefore necessary to obtain reproducible potentials when these sensors are used.

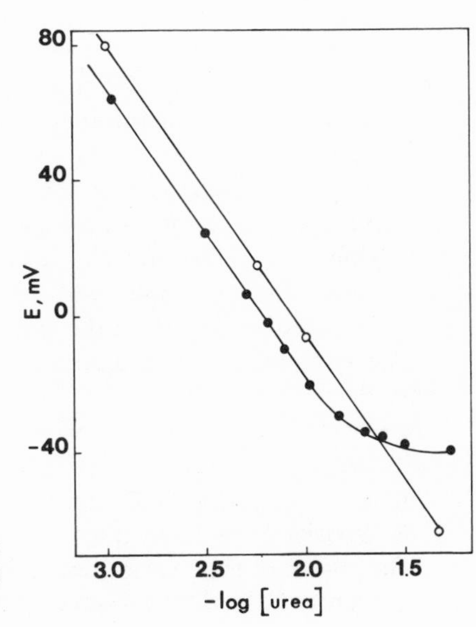

Fig. 13. Response of urea electrode in buffer (○) and in serum (●). Reprinted with permission from reference 162.

4.17.2. Response Times

The response times of enzyme electrodes are determined by several factors. Response times are shorter at high substrate concentrations than at lower concentrations due to the increased rate of both the enzymatic reaction and the diffusion of substrate into the enzyme layer.[154,168,170]

The activity of the enzyme also affects the response time of the electrode. As the activity of the enzyme is increased, the response time decreases up to a certain point, after which no further improvement is realized.[170,231] Further increases in the amount of enzyme can actually lead to longer response times due to the resulting increase in the thickness of the enzyme layer.[170] Since the pH affects the activity of the enzyme, it also affects the response time. Response times will be at a minimum in the optimum pH range.[170,172,179] Similarly, the response time of enzyme electrodes generally increases with age due to the loss of enzyme activity.[161,172]

According to theory[200,230,231] and in practice,[154,160,161,170,172,179] the response time is very dependent on the thickness and the permeability of the enzyme layer. As is shown in Fig. 12 taken from the work of Mascini and Liberti,[170] the response time increases with increasing thickness of the enzyme membrane. For optimum response times the enzyme layer should be kept as thin as possible while using as high an enzyme activity as possible.[231]

Another way to improve response times is to increase the permeability of the enzyme membrane. This was demonstrated by Cullen *et al.*,[172] who prepared penicillin electrodes by immobilizing penicillinase on fritted glass disks of different porosities. Wawro and Rechnitz[182] were able to achieve response times of less than 2 min for an ammonia-sensor-based asparagine electrode by immobilizing the enzyme on a nylon net. Due to the large open area of the net, response times were considerably faster than previously reported for this type of sensor.

The response time is also strongly influenced by the rate of stirring of the solution.[170,172] This effect is illustrated in Fig. 11. As would be expected the response time decreases with increases in stirring rate owing to the increased rate of transport of the substrate into the enzyme layer.

The composition of the solution can also affect the response time. Papastathopoulos and Rechnitz[162] reported response times of 6–8 min for a urea electrode when used in serum and whole blood. While under similar conditions, response times in buffer solutions were found to be about 5 min. The increased response times in the biological fluids can be attributed to the higher viscosity of these samples.

The response time varies depending upon the sensing electrode used. Enzyme sensors based on gas-sensing membrane electrodes have considerably longer response times than those based on other ion-selective electrodes.[160,162–164] This is due to the inherently long response time of the gas sensors, because of the diffusion of the gaseous species through the gas-permeable membrane.

Temperature has been found to have only a slight effect on the response time.[171,172,176]

Another important characteristic of enzyme electrodes is the wash time or the recovery time. This is the time required for the electrode to regain its potential in buffer solution after being exposed to substrate. This time depends, in a similar manner as the response time, on the rate of stirring, and the membrane thickness and permeability. The wash time, however, increases with increasing concentration.

4.17.3. Electrode Lifetime

The usable lifetime of an enzyme electrode is determined primarily by the stability of the enzyme used. The stability of the ion-selective electrode is rarely the limiting factor. An aging electrode exhibits characteristics that would be expected for a decrease in enzyme activity in the membrane. As the electrode ages, there is a shift in the calibration curve, although in most cases the slope remains fairly constant. The electrode can still be used, but frequent recalibration is necessary. There is also an increase in the response times. Eventually the slope begins to decrease, after which the loss in response is usually very rapid.

The lifetime of an enzyme electrode in many cases can be improved by immobilizing the enzyme.[37,176] In the case of glucose oxidase, Guilbault and Lubrano[43] demonstrated that the lifetime of the glucose electrode decreased in the order chemically bound > physically entrapped (polyacrylamide gel entrapment) > soluble enzyme (dialysis membrane entrapped). The stability obtained will depend on the nature of the enzyme. The immobilized glucose oxidase lost only 30% of its activity over a period of 13 months! Although immobilized enzymes are generally more stable than soluble enzymes, the method of immobilization chosen must be compatible with the enzyme or a large loss of activity will result. An example of this is the L-amino acid electrode.[175] The electrode prepared using polyacrylamide-gel-entrapped L-amino acid oxidase was found to be less stable and had a poorer response than the electrode prepared by entrapping the enzyme in a dialysis membrane. This was due to the inactivation and inhibition of the enzyme during the immobilization process.

With the early enzyme electrodes, in which the enzyme was entrapped in a polyacrylamide gel, it was found that leaching of the enzyme from the gel limited the lifetime of the electrode.[153–155] The urea electrode so constructed had a lifetime of about 14 days, after which it began to lose activity.

By placing a thin cellophane dialysis membrane over the gel layer to prevent leaching, the lifetime was extended to over 21 days. The use of a dialysis membrane also protects the enzyme from bacterial attack which can decrease the lifetime of the electrode.

As was mentioned earlier, there is a critical amount of enzyme necessary to produce a Nernstian response. Further increases in the amount of enzyme have little effect on the response characteristics of the electrode. However, the use of an excess amount of enzyme can increase the lifetime of the electrode.[171,175,176,179] In this way enzyme activity can be lost while still maintaining a Nernstian response.

Use of an enzyme electrode at a pH far from the pH optimum will adversely affect the lifetime of the electrode due to the rapid loss of enzyme activity. This was the case with the amygdalin electrode, which was initially used at pH 10.4.[168] The electrode was usable for only a period of four days. When the electrode was used at pH 7.0, the electrode was stable for one week.[170]

Enzyme activators which increase enzyme activity can also increase the lifetime of the electrode.[75]

The lifetime of enzyme electrodes can be affected by storage conditions. Generally, the electrodes are stored under refrigeration in distilled water or buffer solution. However, in some cases, the storage solution can be important. Guilbault and Hrabankova[176] found that the D-amino acid electrode lost activity rapidly even though the crude enzyme was stable in solution for weeks. It was discovered that the reason for this was the loss of a weakly bound cofactor, flavine adenine dinucleotide (FAD). Storage of the electrode in buffer containing FAD extended the lifetime to at least 21 days. The storage buffer was also found to be important for the stability of the L-amino acid electrode.[175] Storage of the electrode in pH 5.5 phosphate buffer containing 1×10^{-2} *M* chloride improved the stability. This effect was due to the fact that the enzyme undergoes an unusual type of reversible inactivation, which was prevented by the presence of chloride at a pH less than seven.

Other factors which can affect the lifetime of an enzyme electrode include the inhibition by trace amounts of heavy metal ions, e.g., Hg^{+2}. This can be prevented by masking, using cysteine or EDTA.[162] The temperature is another important factor. The lifetime of an enzyme electrode is generally longer at low temperature due to the higher rate of enzyme deactivation at elevated temperatures.[172]

As pointed out by White and Guilbault,[183] when talking about enzyme electrode stability, it is necessary to distinguish between storage stability, where the electrode is stored and tested from time to time, and operational stability, in which the electrode is continually in use during the period. The operational stability will obviously be less than the storage stability.

4.17.4. Electrode Interferences

Enzyme electrode interferences can arise from two sources, i.e., interferences with the ion-selective electrode and the lack of specificity of the enzyme. It is obviously desirable to use as selective a sensing electrode as is possible. In this regard, the gas-sensing electrodes are superior since they do not respond to ions in solution. Early enzyme electrodes based on the monovalent cation electrode had serious interference problems from monovalent cations, e.g., Na^+, K^+.[153–156,174–176,179] Enzyme sensors based on the solid state iodide[37,180] or cyanide electrode[167–170] also have considerable selectivity problems. Ions capable of forming insoluble silver salts will interfere along with substances that reduce or complex silver ions.

With pH-electrode-based enzyme sensors, any acidic or basic substances present will interfere. This effect can be eliminated by adjusting the pH of the sample prior to the measurement. A more serious problem with this type of electrode is the effect of the buffer capacity of the sample. The response is strongly dependent on the buffer capacity since the measurement involves the determination of changes in pH. The penicillin enzyme electrode, based on a pH sensor, was also found to become sensitive to monovalent cations owing to the adsorption of the enzyme on the glass electrode.[171,172] This change in selectivity has been attributed to cation adsorption on the fixed negative sites of the enzyme.[172] This effect was eliminated by separating the enzyme from the surface of the glass electrode either by an additional dialysis membrane or by immobilizing the enzyme on a fritted glass disk which is held on the surface of the electrode.

The specificity of the enzyme will vary depending on the enzyme used. Some enzymes are highly specific, e.g., urease,[154,160,162] and AMP deaminase,[186] while other enzymes will react with several substrates, causing additional selectivity problems. For example, L-amino acid oxidase,[179,175] D-amino acid oxidase,[176] and penicillinase[171] are much less selective. Consequently, enzyme electrodes using these enzymes respond in varying degrees to a wide variety of substances. The presence of impurities in the enzyme will also make the electrode less selective.

The response of enzyme electrodes can be affected by the presence of various inhibitors, e.g., heavy metal ions. This problem can be eliminated by masking. Furthermore, the effects of many inhibitors can be markedly decreased by proper enzyme immobilization procedures.[82,193]

5. BACTERIAL ELECTRODES

A new approach to the development of potentiometric biosensors has been the use of intact bacterial cells instead of isolated enzymes. These bacterial electrodes function in an analogous manner to conventional enzyme electrodes, but offer several unique advantages. Enzyme extraction and purification steps, necessary when the desired enzyme is not commercially available, are eliminated. The bacterial cells are more economical to use than commercially available enzymes. Since the loss of enzyme activity in intact cells is often less than that of isolated enzymes a longer electrode lifetime is possible. In addition, the bacterial electrodes can be regenerated by storage in growth medium to replenish any inactive cells, thereby further increasing electrode lifetime. Enzyme activity is also higher in the bacterial cells, which provide an environment optimized through evolution. Furthermore, the use of bacterial cells offers an interesting alternative in the construction of the multistep enzyme electrodes discussed earlier (Section 4.14). The bacterial cells can be chosen so as to contain the necessary enzymes and cofactors as well as a means for cofactor regeneration, thereby greatly simplifying the system needed.

The immobilization of biologically active bacterial cells has been accomplished by several methods[232–239] including entrapment in polyacrylamide gels,[232,233] immobilization in a collagen membrane,[234] and immobilization on insoluble metal hydroxides using a chelation process.[235] In many cases, activity is retained for a period of months.

Immobilized bacterial cells have been used for some time as biocatalysts, particularly in industrial applications.[240–242] The primary use has been in bioconversions and synthesis. Some examples include the production of foodstuffs, e.g., yogurt, cheese; preparation of primary metabolites, e.g., amino acids, vitamins; and the preparation of secondary metabolites, e.g., penicillin, dextran. Bacterial cells have also been used in waste disposal. It has not been until fairly recently, however, that bacterial cells have been used as catalysts for analytical determinations.

Apparently, the first application of bacterial cells in conjunction with electrochemical detection for analytical purposes was reported by Divies.[243] A "microbial" electrode for ethanol was produced by coupling the bacterium *Acetobacter xylinium* with a polarographic oxygen electrode. Ethanol was oxidized to acetic acid by the reaction

$$\text{ethanol} + O_2 \rightarrow \text{acetic acid} + H_2O \tag{53}$$

and the oxygen consumed was sensed by the oxygen electrode. There have been several other bacterial or microbial electrodes based on polarographic measurement, including electrodes for the determination of biochemical oxygen demand (BOD).[244–246] In addition, a solution method for the microbioassay of vitamin B1 using electrochemical detection has been reported.[247] A microbe thermistor for the determination of various sugars has also been described.[248] The primary concern here, however, will be with potentiometric-based methods.

5.1. L-Arginine Electrode

The first potentiometric-based bacterial sensor was reported by Rechnitz *et al.*[249] The bacterium *Streptococcus faecium* was coupled with an ammonia gas-sensing electrode to yield a biosensor for the amino acid L-arginine (see Table 5). This microorganism metabolizes the substrate to produce ammonia by the following reaction:

$$\text{L-arginine} + H_2O \xrightarrow[\text{deiminase}]{\text{arginine}} \text{citrulline} + NH_3 \quad (54)$$

The citrulline produced is further metabolized by a series of steps to produce another mole of ammonia, although only the first reaction is necessary to produce the desired product. The enzyme arginine deiminase is not commercially available, consequently an enzyme electrode for L-arginine has not been reported.

The bacterial electrode was prepared by spreading the cells, collected by centrifugation, onto the gas-permeable membrane of the ammonia sensor. A cellophane dialysis membrane was used to hold the cells in place (Fig. 14). Since the stability of enzymes contained in bacterial cells is generally much greater than that of isolated enzymes, there is little to be gained by more elaborate immobilization procedures. The electrode exhibited a linear response to L-arginine in the concentration range 5×10^{-5} to 1×10^{-3} *M* with a slope from 40 to 45 mV/decade. After a period of 40 days, only a slight decrease in the upper limit of the linear range was

TABLE 5. Potentiometric Bacterial Electrodes

Substrate	Bacterium	Electrode	Concentration range (*M*)[a]	Reference
L-Arginine	*Streptococcus faecium* ATCC # 9790	NH_3 gas sensing	5×10^{-5}–1×10^{-3}	249
L-Aspartate	*Bacterium cadaveris* ATCC # 9760	NH_3 gas sensing	3×10^{-4}–7×10^{-3}	250
L-Glutamine	*Sarcina flava* ATCC # 147	NH_3 gas sensing	2×10^{-5}–1×10^{-2}	251
L-Cysteine	*Proteus morganii* ATCC # 8019	H_2S gas sensing	5×10^{-5}–9×10^{-4}	252
NAD	NADase and *Escherichia coli* ATCC # 27195	NH_3 gas sensing	5×10^{-5}–8×10^{-4}	253
Nitrate	*Azotobacter vinelandii* ATCC # 9104	NH_3 gas sensing	1×10^{-5}–8×10^{-4}	254

[a] Analytically useful range.

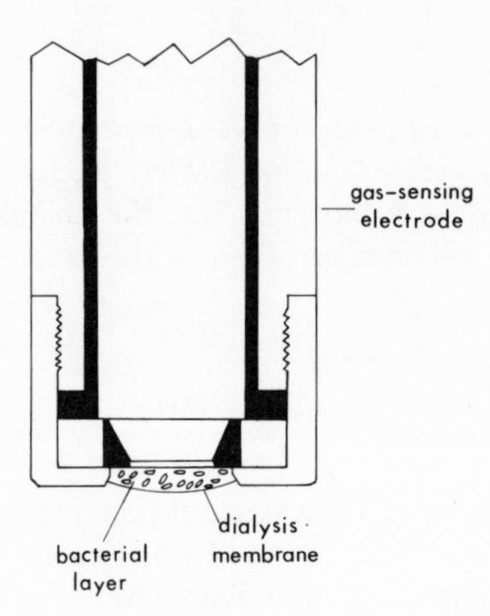

Fig. 14. Bacterial electrode.

observed when the electrode was stored in buffer. Response times were unusually long, i.e., about 20 min. Apparently, the long response time was due to the additional diffusion step through the cell membrane, since electrodes prepared with ruptured cells had response times of about 10 min. The bacterial electrode showed a moderate response to some other amino acids, i.e., L-glutamine and L-asparagine. Since the enzyme involved is known to be specific for arginine, this response must be due to the presence of other deaminating enzymes present in the bacterial cells. Deaminase enzymes are quite common in bacteria, making selectivity a frequent problem with ammonia-electrode-based bacterial sensors.

5.2. L-Aspartate Electrode

The increased stability and the possibility of regeneration of electrode response of bacterial electrodes was demonstrated by Kobos and Rechnitz.[250] The response characteristics of an enzyme electrode and a bacterial electrode for L-aspartate were compared. The enzyme L-aspartase, which catalyzes the following reaction:

$$\text{L-aspartate} \xrightarrow{\text{aspartase}} \text{fumarate} + NH_3 \tag{55}$$

is known to be very unstable. The bacterium used to construct the bacterial electrode, *Bacterium cadaveris*, is the source of the commercially available enzyme. The ammonia produced using either the commercial enzyme or the bacterial cells, was detected with an ammonia gas-sensing electrode.

The linear range, response slope, and even the response time for the two electrodes were very similar. The linear range for both electrodes was between 3×10^{-4} and 7×10^{-3} *M*. The bacterial electrode had a response slope from 45 to 50 mV/decade, while the slope of the enzyme electrode was 50–54 mV/decade. Response times for both electrodes were approximately 5 min. The enzyme electrode was usable for only one day. The bacterial electrode showed somewhat better stability, being usable for two days when stored in buffer. However, when the bacterial sensor was stored in sterile growth medium, there was no decrease in response for a period of over 14 days. In fact, there was an increase in the linear range at high concentrations. Apparently, the bacterial cells continually reproduce and replenish the supply of the necessary enzyme. With age the bacterial layer became thicker, causing longer response times. Ultimately, this was the limiting factor for practical purposes. Once again, selectivity was a problem with the bacterial sensor, which responded to adenosine and L-asparagine in a similar manner to L-aspartate.

5.3. L-Glutamine Electrode

A truly specific bacterial electrode was reported by Rechnitz *et al.*[251] for the determination of L-glutamine. The bacterium *Sarcina flava* was used to deaminate L-glutamine [reaction (25)], and the ammonia produced was sensed with an ammonia gas-sensing electrode. The response curve was linear in the range of 1×10^{-4} to 1×10^{-2} *M* with a slope of 48 mV/decade. A useful response was obtained down to 2×10^{-5} *M*. Response times were approximately 5 min over the linear range. In contrast to previously reported bacterial electrodes, the L-glutamine sensor showed only negligible response to other amino acids. The calibration curve obtained in pooled human serum had the same slope as that obtained in buffer, indicating that the electrode could be used in serum with no interferences from other constituents. The electrode was stable for at least two weeks, further demonstrating the improved stability of the bacterial cells. A previously reported enzyme electrode for glutamine had a useful lifetime of only one day.[179]

5.4. L-Cysteine Electrode

A bacterial electrode for L-cysteine was described by Jensen and Rechnitz.[252] A strain of *Proteus morganii* was coupled, by means of a dialysis membrane, to a hydrogen sulfide gas-sensing electrode which measured the H_2S produced by the following reaction:

$$\text{L-cysteine} + H_2O \xrightarrow[\text{desulfhydrase}]{\text{L-cysteine}} \text{pyruvate} + NH_3 + H_2S \qquad (56)$$

This enzyme has not been isolated, so it is only through the use of bacterial cells possessing the enzyme that a sensor for cysteine can be made.

The response of the bacterial sensor was linear from 5×10^{-5} to 9×10^{-4} *M* with a slope of 25 mV/decade. Response times ranged from 5 to 8 min. The electrode was usable for a period of six days when stored in buffer. However, the response could be regenerated by storage in growth medium. Selectivity was a problem, not only due to the presence of other enzymes in the cells, but also due to the sensing electrode. The bacterial sensor showed a moderate response to homocysteine, and a slight response to cystathionine and methionine. The H_2S-sensing electrode responded equally well to CO_2, which caused two problems. Due to the presence of urease in the bacterium, which produces CO_2 and NH_3 from urea [reaction (5)], the electrode exhibited a significant response to urea. Another problem caused by the CO_2 response of the hydrogen sulfide electrode is that even though residual levels of CO_2 in the solution could easily be removed by degassing, the CO_2 produced by bacterial respiration was detected. A more selective H_2S sensor is needed to make this electrode more practical. Some of these selectivity problems could possibly be eliminated by using either an ammonia gas-sensing electrode to measure the ammonia produced by reaction (56) or a solid state sulfide electrode to measure the sulfide produced.

5.5. Nicotinamide Adenine Dinucleotide Electrode

Riechel and Rechnitz[253] described a hybrid bacterial and enzyme electrode for the determination of nicotinamide adenine dinucleotide (NAD). The enzyme NADase from *Neurospora crassa* extracts was used in conjunction with whole cells of *Escherichia coli* which contain nicotinamide deamidase. An ammonia gas-sensing electrode was used to measure the ammonia produced by the following reaction sequence:

$$NAD^+ + H_2O \xrightarrow{\text{NADase}} \text{nicotinamide} + \text{ADP-ribose} \qquad (57)$$

$$\text{nicotinamide} + H_2O \xrightarrow[\text{deamidase}]{\text{nicotinamide}} \text{nicotinic acid} + NH_3 \qquad (58)$$

As would be expected, the electrode responds to both NAD^+ and nicotinamide. The response slope was between 45 and 50 mV/decade in the linear range 1×10^{-4} to 1×10^{-3} *M*. The electrode was usable for at least one week with response times from 5 to 10 min. It was shown that both components, the enzyme and the bacterial cells, were needed for response to NAD^+. However, the electrode also responded to glutamine and NADH in a similar manner. This study demonstrated the feasibility of coupling an enzyme with bacterial cells to further increase the applicability of potentiometric biosensors.

5.6. Nitrate Electrode

The advantages of using intact bacterial cells to mediate a multistep enzymatic process requiring cofactors was demonstrated by Kobos *et al.*[254] A strain of *Azotobacter vinelandii*, coupled to an ammonia gas sensor, was used to carry out the enzymatic reduction of nitrate to ammonia:

$$NO_3^- + NADH \xrightarrow[\text{reductase}]{\text{nitrate}} NO_2^- + H_2O + NAD^+ \tag{59}$$

$$NO_2^- + 3NADH \xrightarrow[\text{reductase}]{\text{nitrite}} NH_3 + 2H_2O + 3NAD^+ \tag{60}$$

The two enzymes involved, the cofactor NADH, as well as a means of cofactor regeneration, are all present in the bacterial cells. The system is much less complicated than would be necessary if the isolated enzymes were used.

It was found that it was necessary to have present in the buffer solution both sucrose, which serves as an energy source for the cells, and isonicotinic acid hydrazide, an inhibitor to prevent the cells from consuming the ammonia produced, to obtain a Nernstian response. The bacterial sensor had a response slope of 45–50 mV/decade in the linear range 1×10^{-5} to 8×10^{-4} *M*. Response times ranged from 5 to 8 min depending on the concentration. The electrode showed no decrease in response over a two-week period when stored in the growth medium, although the response times gradually increased. The sensor responded equally well to nitrite and hydroxylamine and showed varying degrees of response to L-aspartate, L-glutamate, L-glutamine, L-asparagine, and urea. Of the common anion interferences found in other methods of nitrate determination, only chlorate was found to inhibit the electrode response. The effect was only to shorten the linear range at high nitrate levels. The interferences observed would not pose a serious problem in most nitrate determinations, e.g., natural and potable waters, since the concentration of these substances would be below the detection limit of the electrode. However, application in biological samples would be limited. It was possible to analyze potable and wastewater reference standards with a precision and accuracy of 3–4%.

It should be emphasized that these bacterial electrodes are still in the developmental stage. Although some of these electrodes have the required selectivity to be practical sensors in real samples,[251,254] selectivity in most cases remains a problem. This is a similar situation as existed with the early enzyme electrodes. Further research is needed to improve the selectivity characteristics. Possible ways of accomplishing this include the use of enzyme inhibitors to prevent undesired reaction sequences, variations in growth conditions to enhance desired pathways and retard the development of undesired ones, and possible genetic manipulation to eliminate the development of unwanted enzymes.

5.7. Potentiometric-Based Analysis with Bacterial Cells

In addition to the bacterial electrodes described above, there have been studies which utilized bacterial cells for analytical determinations in other ways. Matsunaga *et al.*[255] described a method for nicotinic acid using immobilized *Lactobacillus arabinosus*. This bacterium produces lactic acid during growth, which requires the presence of the vitamin nicotinic acid. The method consisted of adding an equal amount of the bacterial cells, immobilized in an agar gel, to two test tubes containing double-strength assay medium. The nicotinic acid sample was added to one of the tubes, the other tube served as a blank. The test tubes were incubated for 1 hr, after which the amount of lactic acid produced was measured with a pH electrode. The potential difference (ΔE_2) between the incubated blank and the initial medium was also determined. The potential difference $\Delta E_1 - \Delta E_2$ was found to be proportional to the logarithm of the nicotinic acid concentration. The relationship was linear in the range 4.2×10^{-7} to 4.2×10^{-5} *M*. The immobilized bacteria were active for one month.

Still another use of bacterial cells was reported by D'Orazio *et al.*[256] for the measurement of lysozyme enzyme activity. Cells of *Micrococcus lysodeikticus* were loaded with an ion marker, trimethylphenylammonium ion ($TMPA^+$). Lysozyme ruptures the bacterial cells in hypotonic solutions, releasing the marker which is sensed by a $TMPA^+$-selective electrode.[257] The initial rate of potential change is a function of the rate of lysis and therefore the amount of lysozyme present. The marker used was chosen because of the high selectivity of the sensing electrode over other monovalent cations, making possible the application in biological fluids such as blood serum. The calibration curve of the initial reaction rate versus lysozyme concentration was linear above 0.36 μg/ml. Preliminary determinations of lysozyme in serum indicated that measurements could be made with no interference from serum constituents.

6. TISSUE-BASED ELECTRODE

Other possible biological mediators, besides enzymes and bacterial cells, are being studied for use in biosensors. Rechnitz *et al.*[258] recently reported an electrode for glutamine which used tissue slices. A thin layer of porcine kidney tissue was fixed to an ammonia gas-sensing electrode using either a dialysis membrane or a nylon net. Pork kidneys were obtained from freshly slaughtered animals and showed a high glutaminase activity. The resulting sensor was stable for a period of 28 days. The potentiometric response was linear in the range 1×10^{-4} to 5×10^{-3} M with a response time of 5–7 min. However, the response was useful in the range 5×10^{-5} to 1×10^{-2} M. Only a negligible response was observed to other amino acids and urea. This approach yielded an electrode which had excellent selectivity and sensitivity, while being extremely simple and inexpensive.

7. FUTURE OF POTENTIOMETRIC ENZYME METHODS

Advances in potentiometric-based enzyme technology show no signs of diminishing. The method continues to provide an attractive assay procedure for an ever increasing number of substances. Due to the increasing number of enzyme available and the continual developments in ion-selective electrode methodology, many possibilities exist for the development of new biosensors. The development of micro size enzyme electrodes for *in situ* measurements, for example, is an area worthy of considerable research.[220]

Besides the determination of substrates, inhibitors of enzymes can also be determined,[19,66,77,81,150,194] thereby enabling the determination of inorganic ions, e.g., Hg^{+2}, and other substances at very low concentration levels. There are many inhibitor–enzyme systems which can be utilized in this manner.

Another very recent application of ion-selective electrodes which merits further study is their use in enzyme immunoassay.[144–146] In this manner, potentiometric methods can be extended to the determination of immunological reagents, i.e., antibodies, antigens, and haptens.

The combination of two or more enzymes to produce enzyme sequence electrodes[187,191] should further increase the number of important substances that can be determined potentiometrically. Recent advances in enzyme and cofactor immobilization techniques should stimulate developments in this area.

The use of bacterial cells, once selectivity can be more effectively controlled, can further increase the applicability of potentiometric-based sensors.[249–256] Electrodes can be developed in this manner even if the enzymes involved have not been isolated or if the exact reaction sequence is not known. This approach also offers an attractive alternative for the development of enzyme sequence electrodes, especially when enzyme cofactors are involved.[254]

Recent investigations have shown that biosensors need not be limited by the use of enzymes or bacterial cells. A whole range of mediators have been proposed[259] including various organelles, tissues,[258] and immunoagents.[260] The use of these other mediators is still at an early stage and much more can be expected in the future.

In summary, there have been many interesting developments in the field of potentiometric enzyme analysis in recent years. These advances have opened up whole new areas of investigation which should prove very fruitful for future study.

REFERENCES

1. Fishman, M. M., *Anal. Chem.* **50**, 261R (1978).
2. Bergmeyer, H. U., *Methods in Enzymatic Analysis*, 2nd ed., Academic Press, New York (1975).
3. Guilbault, G. G., *Handbook of Enzymatic Methods of Analysis*, Marcel Dekker, New York (1976).
4. Michaelis, L., and Menten, M., *Biochem. Z.* **49**, 333 (1913).
5. Laidler, K. J., and Bunting, P. S., *The Chemical Kinetics of Enzyme Action*, 2nd ed., Clarendon Press, Oxford (1973).
6. Ingle, J. D., and Crouch, S. R., *Anal. Chem.* **45**, 333 (1973).
7. Crouch, S. R., in *Computers in Chemistry and Instrumentation* (J. S. Mattson, H. B. Mark, and H. C. MacDonald, eds.), Marcel Dekker, New York (1973).
8. Carr, P. W., *Anal. Chem.* **50**, 1602 (1978).
9. Guilbault, G. G., Smith, R. K., and Montalvo, J. G., *Anal. Chem.* **41**, 600 (1969).
10. Cammann, K., *Fresenius' Z. Anal. Chem.* **287**, 1 (1977).
11. Guilbault, G. G., in *Immobilized Enzymes, Antigens, Antibodies and Peptides* (H. H. Weetall, ed.), p. 294, Marcel Dekker, New York (1975).
12. Guilbault, G. G., in *Methods in Enzymology* (K. Mosbach, ed.), Vol. 44, p. 579, Academic Press, New York (1976).
13. Rechnitz, G. A., *Chem. Eng. News* **53**(4), 29 (1975).
14. Moody, G. J., and Thomas, J. D. R., Analyst **100**, 609 (1975).
15. Elving, P. J., *Bioelectrochem. Bioenerg.* **2**, 251 (1975).
16. Chance, B., and Nishimura, M., in *Methods in Enzymology* (R. W. Estabrook and M. E. Pullman, eds.), Vol. 10, p. 641, Academic Press, New York (1967).
17. Jacobsen, C. F., Leonis, J., Linderstrom-Lang, K., and Ottesen, M., in *Methods of Biochemical Analysis* (D. Glick, ed.), Vol. 4, p. 171, Interscience, New York (1957).
18. Malmstadt, H. V., and Piepmeier, E. H., *Anal. Chem.* **37**, 34 (1965).
19. Toren, E. C., and Burger, F. J., *Mikrochim. Acta* **1968**, 1049.
20. Katz, S. A., and Rechnitz, G. A., *Z. Anal. Chem.* **196**, 248 (1963).
21. Katz, S. A., *Anal. Chem.* **36**, 2500 (1964).
22. Katz, S. A., and Cowans, J. A., *Biochim. Biophys. Acta* **107**, 605 (1965).
23. Scholer, R. P., and Simon, W., *Chimia* **24**, 372 (1970).
24. Neubecker, T. A., and Rechnitz, G. A., *Anal. Lett.* **5**, 653 (1972).
25. Hussein, W. R., and Guilbault, G. G., *Anal. Chim. Acta* **76**, 183 (1975).
26. Severinghaus, J. W., and Bradley, A. F., *J. Appl. Physiol.* **13**, 515 (1958).
27. Severinghaus, J. W., *Ann. N.Y. Acad. Sci.* **148**, 115 (1968).
28. Ross, J. W., Riseman, J. H., and Krueger, J. A., *Pure Appl. Chem.* **36**, 473 (1973).
29. Ruzicka, J., and Hansen, E. H., *Anal. Chim. Acta* **69**, 129 (1974).
30. Hansen, E. H., and Ruzicka, J., *Anal. Chim. Acta* **72**, 353 (1974).
31. Thompson, H., and Rechnitz, G. A., *Anal. Chem.* **46**, 246 (1974).
32. Yao, S. J., Wolfson, S. K., and Tokarsky, J. M., *Bioelectrochem. Bioenerg.* **2**, 348 (1975).
33. Yao, S. J., Wolfson, S. K., Tokarsky, J. M., and Weiner, S. B., *Bioelectrochem. Bioenerg.* **3**, 106 (1976).

34. Berman, E. M., Yao, S. J., Wolfson, S. K., and Tokarsky, J. M., *Bioelectrochem. Bioenerg.* **5**, 63 (1978).
35. Llenado, R. A., and Rechnitz, G. A., *Anal. Chem.* **45**, 2165 (1973).
36. Thompson, H., and Rechnitz, G. A., *J. Chem. Instrum.* **4**, 239 (1973).
37. Nagy, G., Von Storp, L. H., and Guilbault, G. G., *Anal. Chim. Acta* **66**, 443 (1973).
38. Papastathopoulos, D. S., and Rechnitz, G. A., *Anal. Chem.* **47**, 1792 (1975).
39. Mascini, M., and Palleschi, G., *Anal. Chim. Acta* **100**, 215 (1978).
40. Clark, L. C. and Lyons, C., *Ann. N.Y. Acad. Sci.* **102**, 29 (1962).
41. Clark, L. C., U.S. Patent 3,539,455 (1970).
42. Guilbault, G. G., and Lubrano, G. J., *Anal. Chim. Acta* **60**, 254 (1972).
43. Guilbault, G. G., and Lubrano, G. J., *Anal. Chim. Acta* **64**, 439 (1973).
44. Clark, L. C., *Proc. Int. Union Physiol. Sci.* **9** (1971).
45. Nanjo, M., and Guilbault, G. G., *Anal. Chim. Acta* **73**, 367 (1974).
46. Guilbault, G. G., and Lubrano, G. J., *Anal. Chim. Acta* **69**, 183 (1974).
47. Clark, L. C., *Biotechnol. Bioeng. Symp.* No. 3, 377 (1972).
48. Suzuki, S., Takahashi, F., Satoh, I., and Sonobe, N., *Bull. Chem. Soc. Jpn.* **48**, 3246 (1975).
49. Nanjo, M., and Guilbault, G. G., *Anal. Chim. Acta* **75**, 169 (1975).
50. Clark, L. C., and Emory, C. R., in *Ion and Enzyme Electrodes in Biology and Medicine* (M. Kessler, L. C. Clark, D. W. Lubbers, I. A. Silver, and W. Simon, eds.), p. 161, University Park Press, Baltimore (1976).
51. Huang, H. S., Kuan, S. S., and Guilbault, G. G., *Clin. Chem.* **23**, 671 (1977).
52. Satoh, I., Karube, I., and Suzuki, S., *Biotechnol. Bioeng.* **17**, 1095 (1977).
53. Llenado, R. A., and Rechnitz, G. A., *Anal. Chem.* **46**, 1109 (1974).
54. Klein, E., Montalvo, J. G., Wawro, R., Holland, F. F., and Lebeouf, A., *Int. J. Artif. Organs* **1**, 116 (1978).
55. Schindler, J. G., Schindler, R. G., and Aziz, O., *J. Clin. Chem. Clin. Biochem.* **16**, 447 (1978).
56. Ruzicka, J., Hansen, E. H., Ghose, A. K., and Mottola, H. A., *Anal. Chem.* **51**, 199 (1979).
57. Tong, S. L., and Rechnitz, G. A., *Anal. Lett.* **9**, 1 (1976).
58. Herman, H. B., and Rechnitz, G. A., *Anal. Chem.* **184**, 1074 (1974).
59. Herman, H. B., and Rechnitz, G. A., *Anal. Chim. Acta* **76**, 155 (1975).
60. Erlanger, B. F., and Sack, R. A., *Anal. Biochem.* **33**, 318 (1970).
61. Guilbault, G. G., Gutknecht, W. F., Kuan, S. S., and Cochran, R., *Anal. Biochem.* **46**, 200 (1972).
62. Llenado, R. A., and Rechnitz, G. A., *Anal. Chem.* **44**, 468 (1972).
63. Llenado, R. A., and Rechnitz, G. A., *Anal. Chem.* **44**, 1366 (1972).
64. Hussein, W. R., Von Storp, L. H., and Guilbault, G. G., *Anal. Chim. Acta* **61**, 89 (1972).
65. Llenado, R. A., and Rechnitz, G. A., *Anal. Chem.* **45**, 826 (1973).
66. Baum, G., and Ward, F. B., *Anal. Biochem.* **42**, 487 (1971).
67. Baum, G., *Anal. Biochem.* **39**, 65 (1971).
68. Baum, G., Ward, F. B., and Yaverbaum, S., *Clin. Chim. Acta* **36**, 405 (1972).
69. Von Storp, L. H., and Guilbault, G. G., *Anal. Chim. Acta* **62**, 425 (1972).
70. Gibson, K., and Guilbault, G. G., *Anal. Chim. Acta* **76**, 245 (1975).
71. Hussein, W. R., and Guilbault, G. G., *Anal. Chim. Acta* **72**, 381 (1974).
72. Ferguson, D. A., Boyd, J. W., and Phillips, A. W., *Anal. Biochem.* **62**, 81 (1974).
73. Huang, Y. Z., *Anal. Biochem.* **61**, 464 (1974).
74. Larsen, N. R., Hansen, E. H., and Guilbault, G. G., *Anal. Chim. Acta* **79**, 9 (1975).
75. Meyerhoff, M., and Rechnitz, G. A., *Anal. Chim. Acta* **85**, 277 (1976).
76. Hjemdahl-Monsen, C. E., Papastathopoulos, D. S., and Rechnitz, G. A., *Anal. Chim. Acta* **88**, 253 (1977).

77. Meyerson, L. R., McMurtrey, K. D., and Davis, V. E., *Anal. Biochem.* **86**, 287 (1978).
78. Chien, P. T., and Michael, L. W., *Anal. Biochem.* **68**, 626 (1975).
79. Ngo, T. T., and Shargool, P. D., *Anal. Biochem.* **54**, 247 (1973).
80. Zeman, G. H., Sobocinski, P. Z., and Chaput, R. L., *Anal. Biochem.* **52**, 63 (1973).
81. Weisz, H., and Rothmaier, K., *Anal. Chim. Acta* **80**, 351 (1975).
82. Klibanov, A. M., *Anal. Biochem.* **93**, 1 (1979).
83. Weetall, H. H., *Anal. Chem.* **46**, 602A (1974).
84. Zaborsky, O. R., *Immobilized Enzymes*, CRC Press, Cleveland (1973).
85. Skinner, K. J., *Chem. Eng. News* **18**(8), 23 (1975).
86. Mosbach, K., ed., *Methods in Enzymology*, Vol. 44, Academic Press, New York (1976).
87. Chang, T. M. S., ed., *Biomedical Applications of Immobilized Enzymes and Proteins*, Vol. 1, Plenum Press, New York (1977).
88. Weetall, H. H., ed., *Immobilized Enzymes, Antigens, Antibodies and Peptides*, Marcel Dekker, New York (1975).
89. Guilbault, G. G., *Handbook of Enzymatic Methods of Analysis*, p. 447, Marcel Dekker, New York (1976).
90. Wingard, L. B., Katchalski-Katzir, E., and Goldstein, L., eds., *Applied Biochemistry and Bioengineering*, Vol. 1, Academic Press, New York (1976).
91. Messing, R. A., in *Methods of Enzymology* (K. Mosbach, ed.), Vol. 44, p. 148, Academic Press, New York (1976).
92. Zaborsky, O. R., in *Biomedical Applications of Immobilized Enzymes and Proteins* (T. M. S. Chang, ed.), Vol. 1, p. 37, Plenum Press, New York (1977).
93. O'Driscoll, K. F., in *Methods in Enzymology* (K. Mosbach, ed.), Vol. 44, p. 169, Academic Press, New York (1976).
94. Koch-Schmidt, A. C., in *Biomedical Applications of Immobilized Enzymes and Proteins* (T. M. S. Chang, ed.), Vol. 1, p. 47, Plenum Press, New York (1977).
95. Chang, T. M. S., in *Biomedical Applications of Immobilized Enzymes and Proteins* (T. M. S. Chang, ed.), Vol. 1, p. 69, Plenum Press, New York (1977).
96. Chang, T. M. S., in *Methods in Enzymology* (K. Mosbach, ed.), Vol. 44, p. 201, Academic Press, New York (1976).
97. Falb, R. D., in *Biomedical Applications of Immobilized Enzymes and Proteins* (T. M. S. Chang, ed.), Vol. 1, p. 7, Plenum Press, New York (1977).
98. Zaborsky, O. R., in *Biomedical Applications of Immobilized Enzymes and Proteins* (T. M. S. Chang, ed.), Vol. 1, p. 25, Plenum Press, New York (1977).
99. Broun, G. B., in *Methods in Enzymology* (K. Mosbach, ed.), Vol. 44, p. 263, Academic Press, New York (1976).
100. Haynes, R., and Walsh, K. A., *Biochem. Biophys. Res. Commun.* **36**, 325 (1969).
101. Broun, G., Selegny, E., Tran-Minh, C., and Thomas, D., *FEBS Lett.* **7**, 223 (1970).
102. Chang, T. M. S., *Biochem. Biophys. Res. Commun.* **44**, 1531 (1971).
103. Uriel, J., *Bull. Soc. Chim. Biol.* **48**, 969 (1966).
104. Paillot, B., Remy, M. H., Thomas, D., Broun, G., and Metayer, O., *Pathol. Biol.* **22**, 491 (1974).
105. Mosbach, K., *Acta. Chem. Scand.* **24**, 2082 (1970).
106. Mosbach, K., and Mattiason, B., *Acta. Chem. Scand.* **24**, 2093 (1970).
107. Chang, T. M. S., *Artificial Cells*, Charles C. Thomas Publisher, Springfield, Illinois (1972).
108. Levin, Y., Pecht, M., Goldstein, L., and Katchalski, E., *Biochemistry* **3**, 1905 (1964).
109. Goldstein, L., Levin, Y., and Katchalski, E., *Biochemistry* **3**, 1913 (1964).
110. Royer, G. P., in *Immobilized Enzymes, Antigens, Antibodies, and Peptides* (H. H. Weetall, ed.), p. 49, Marcel Dekker, New York (1975).
111. Stark, G. R., *Biochemical Aspects of Reactions on Solid Supports*, Academic Press, New York (1971).

112. Silman, I. H., and Katchalski, E., *Ann. Rev. Biochem.* **35**, 873 (1969).
113. Wingard, L. B., ed., *Enzyme Engineering Symposium No. 3*, Wiley, New York (1972).
114. Goldstein, L., in *Methods in Enzymology* (K. Mosbach, ed.), Vol. 44, p. 397, Academic Press, New York (1976).
115. Melrose, G. J. H., *Rev. Pure Appl. Chem.* **21**, 83 (1971).
116. Kiang, C. H., Kuan, S. S., and Guilbault, G. G., *Anal. Chim. Acta* **80**, 209 (1975).
117. Guilbault, G. G., and Stokbro, W., *Anal. Chim. Acta* **76**, 237 (1975).
118. Lilly, M. D., and Dunnill, P., in *Methods in Enzymology* (K. Mosbach, ed.), Vol. 44, p. 717, Academic Press, New York (1976).
119. Hornby, W. E., and Noy, G. A., in *Methods in Enzymology* (K. Mosbach, ed.), Vol. 44, p. 633, Academic Press, New York (1976).
120. Leon, L. P., Narayanan, S., Dellenbach, R., and Horvath, C., *Clin. Chem.* **22**, 1017 (1976).
121. Ngo, T. T., *Int. J. Biochem.* **1975**, 663.
122. Mattiasson, B. and Borrebaeck, C., *FEBS Lett.* **85**, 119 (1978).
123. Mattiasson, B., Danielson, B., and Mosbach, K., *Anal. Lett.* **9**, 317 (1976).
124. Danielsson, B., Gadd, K., Mattiasson, B., and Mosbach, K., *Anal. Lett.* **9**, 987 (1976).
125. Bowers, L. D., Canning, L. M., Sayers, C. N., and Carr, P. W., *Clin. Chem.* **22**, 1314 (1976).
126. Bowers, L. D., Carr, P. W., and Schifreen, R. S., *Clin. Chem.* **22**, 1427 (1976).
127. Weibel, M. K., Dritschilo, W., Bright, H. J., and Humphrey, A. E., *Anal. Biochem.* **52**, 402 (1973).
128. Updike, S. J., and Hicks, G. P., *Science* **158**, 270 (1967).
129. Weetall, H. H., and Hersh, L. S., *Biochim. Biophys. Acta* **185**, 464 (1969).
130. Kiang, C. H., Kuan, S. S., and Guilbault, G. G., *Anal. Chem.* **50**, 1319 (1978).
131. Watson, B., and Keyes, M. H., *Anal. Lett.* **9**, 713 (1976).
132. Hanson, D. J., and Bretz, N. S., *Clin. Chem.* **23**, 477 (1977).
133. Gray, D. N., Keyes, M. H., and Watson, B., *Anal. Chem.* **49**, 1067A (1977).
134. Klein, E., and Montalvo, J. G., *Int. J. Artif. Organs* **1**, 175 (1978).
135. Johansson, G., and Ogren, L., *Anal. Chim. Acta* **84**, 23 (1976).
136. Johansson, G., Edstrom, K., and Ogren, L., *Anal. Chim. Acta* **85**, 55 (1976).
137. Rusling, J. F., Luttrell, G. H., Cullen, L. F., and Papariello, G. J., *Anal. Chem.* **48**, 1211 (1976).
138. Ngo, T. T., *Can. J. Biochem.* **54**, 62 (1975).
139. Satoh, I., Karube, I., and Suzuki, S., *J. Solid Phase Biochem.* **2**, 1 (1977).
140. Adams, R. E., and Carr, P. W., *Anal. Chem.* **50**, 944 (1978).
141. Mattiasson, B., and Mosbach, K., in *Methods in Enzymology* (K. Mosbach, ed.), Vol. 44, p. 335, Academic Press, New York (1976).
142. Scharpe, S. L., Cooreman, W. M., Blomme, W. J., and Laekman, G. M., *Clin. Chem.* **22**, 733 (1976).
143. Schuurs, A. H. W. M., and Van Weeman, B. K., *Clin. Chim. Acta* **81**, 1 (1977).
144. Boitieux, J. L., Desmet, G., and Thomas, D., *Clin. Chim. Acta* **88**, 329 (1978).
145. Boitieux, J. L., Desmet, G., and Thomas, D., *Clin. Chem.* **25**, 318 (1979).
146. Meyerhoff, M. E., and Rechnitz, G. A., *Anal. Biochem.* **95**, 483 (1979).
147. Aizawa, M., Morioka, A., Matsuoka, H., Suzuki, S., Nagamura, Y., Shinohara, R., and Ishiguro, I., *J. Solid Phase Biochem.* **1**, 319 (1976).
148. Aizawa, M., Morioka, A., and Suzuki, S., *J. Membr. Sci.* **4**, 221 (1978).
149. Mattiasson, B., and Nilsson, H., *FEBS Lett.* **78**, 251 (1977).
150. Ogren, L., and Johannson, G., *Anal. Chim. Acta* **96**, 1 (1978).
151. International Union of Pure and Applied Chemistry, *Recommendations for Nomenclature of Ion-Selective Electrodes*, Information Bulletin: Appendices on Provisional Nomenclature, Symbols, Units and Standards, Number 43, Oxford (1975).

152. Updike, S. J., and Hicks, G. P., *Nature* **214**, 786 (1967).
153. Guilbault, G. G., and Montalvo, J. G., *J. Am. Chem. Soc.* **91**, 2164 (1969).
154. Guilbault, G. G., and Montalvo, J. G., *J. Am. Chem. Soc.* **92**, 2533 (1970).
155. Guilbault, G. G., and Montalvo, J. G., *Anal. Lett.* **2**, 283 (1969).
156. Tran-Minh, C., Selegny, E., and Broun, G., *C. R. Acad. Sci. Paris Ser. C* **275**, 309 (1972).
157. Guilbault, G. G., and Hrabankova, E., *Anal. Chim. Acta* **52**, 287 (1970).
158. Guilbault, G. G., and Nagy, G., *Anal. Chem.* **45**, 417 (1973).
159. Guilbault, G. G., Nagy, G., and Kuan, S. S., *Anal. Chim. Acta* **67**, 195 (1973).
160. Anfalt, T., Graneli, A., and Jagner, D., *Anal. Lett.* **6**, 969 (1973).
161. Guilbault, G. G., and Tarp, M., *Anal. Chim. Acta* **73**, 355 (1974).
162. Papastathopoulos, D. S., and Rechnitz, G. A., *Anal. Chim. Acta* **79**, 17 (1975).
163. Mascini, M., and Guilbault, G. G., *Anal. Chem.* **49**, 795 (1977).
164. Guilbault, G. G., and Shu, F. R., *Anal. Chem.* **44**, 2161 (1972).
165. Tran-Minh, C., and Broun, G., *C. R. Acad. Sci. Paris Ser. D* **276**, 2215 (1973).
166. Nilsson, H., Akerlund, A. C., and Mosbach, K., *Biochim. Biophys. Acta* **320**, 529 (1973).
167. Rechnitz, G. A., and Llenado, R. A., *Anal. Chem.* **43**, 283 (1971).
168. Llenado, R. A., and Rechnitz, G. A., *Anal. Chem.* **43**, 1457 (1971).
169. Mascini, M., *Anal. Chem.* **45**, 614 (1973).
170. Mascini, M., and Liberti, A., *Anal. Chim. Acta* **68**, 177 (1974).
171. Papariello, G. J., Mukherji, A. K., and Shearer, C. M., *Anal. Chem.* **45**, 790 (1973).
172. Cullen, L. F., Rusling, J. F., Schleifer, A., and Papariello, G. J., *Anal. Chem.* **46**, 1955 (1974).
173. Montalvo, J. G., and Guilbault, G. G., *Anal. Chem.* **41**, 1897 (1969).
174. Guilbault, G. G., and Hrabankova, E., *Anal. Lett.* **3**, 53 (1970).
175. Guilbault, G. G., and Hrabankova, E., *Anal. Chem.* **42**, 1779 (1970).
176. Guilbault, G. G., and Hrabankova, E., *Anal. Chim. Acta* **56**, 285 (1971).
177. Berjonneau, A. M., Thomas, D., and Broun, G., *Pathol. Biol.* **22**, 497 (1974).
178. Calvot, C., Berjonneau, A. M., Gellf, G., and Thomas, D., *FEBS Lett.* **59**, 258 (1975).
179. Guilbault, G. G. and Shu, F., *Anal. Chim. Acta* **56**, 333 (1971).
180. Guilbault, G. G., and Nagy, G., *Anal. Lett.* **6**, 301 (1973).
181. Hsiung, C. P., Kuan, S. S., and Guilbault, G. G., *Anal. Chim. Acta* **90**, 45 (1977).
182. Wawro, R., and Rechnitz, G. A., *J. Membr. Sci.* **1**, 143 (1976).
183. White, W. C., and Guilbault, G. G., *Anal. Chem.* **50**, 1481 (1978).
184. Davies, P., and Mosbach, K., *Biochim. Biophys. Acta* **370**, 329 (1974).
185. Ahn, B. K., Wolfson, S. K., and Yao, S. J., *Bioelectrochem. Bioenerg.* **2**, 142 (1975).
186. Papastathopoulos, D. S., and Rechnitz, G. A., *Anal. Chem.* **48**, 862 (1976).
187. Rechnitz, G. A., Papastathopoulos, D. S., and Saffran, M., *Fed. Proc. Fed. Am. Soc. Exp. Biol.* **36**, 687 (1977).
188. Kawashima, T., and Rechnitz, G. A., *Anal. Chim. Acta* **83**, 9 (1976).
189. Tran-Minh, C., Guyonnet, R., and Beaux, J., *C. R. Acad. Sci. Paris Ser. C* **286**, 115 (1978).
190. Durand, P., David, A., and Thomas, D., *Biochim. Biophys. Acta* **527**, 277 (1978).
191. Jensen, M. A., and Rechnitz, G. A., *J. Membr. Sci.* **5**, 117 (1979).
192. Shinbo, T., Sugiura, M., and Kamo, N., *Anal. Chem.* **51**, 100 (1979).
193. Tran-Minh, C., and Beaux, J., *C. R. Acad. Sci. Paris Ser. C* **287**, 191 (1978).
194. Tran-Minh, C., and Beaux, J., *Anal. Chem.* **51**, 91 (1979).
195. Montalvo, J. G., *Anal. Chem.* **42**, 2093 (1969).
196. Montalvo, J. G., *Anal. Biochem.* **38**, 357 (1970).
197. Crochet, K. L., and Montalvo, J. G., *Anal. Chim. Acta* **66**, 259 (1973).
198. Guilbault, G. G., *Handbook of Enzymatic Methods of Analysis*, p. 509, Marcel Dekker, New York (1976).

199. Nilsson, H., Mosbach, K., Enfors, S. O., and Molin, N., *Biotechnol. Bioeng.* **20**, 527 (1978).
200. Tran-Minh, C., and Broun, G., *Anal. Chem.* **47**, 1359 (1975).
201. Lubrano, G. J., and Guilbault, G. G., *Anal. Chim. Acta* **97**, 229 (1978).
202. Mosbach, K., and Mattiasson, B., in *Methods in Enzymology* (K. Mosbach, ed.), Vol. 44, p. 453, Academic Press, New York (1976).
203. Bouin, J. C., Atallah, M. T., and Hultin, H. O., in *Methods in Enzymology* (K. Mosbach, ed.), Vol. 44, p. 478, Academic Press, New York (1976).
204. Mosbach, K., Larsson, P. O., and Lowe, C., in *Methods in Enzymology* (K. Mosbach, ed.), Vol. 44, p. 859, Academic Press, New York (1976).
205. Blaedel, W. J., and Jenkins, R. A., *Anal. Chem.* **48**, 1240 (1976).
206. Mansson, M. O., Matiasson, B., Gestrelius, S., and Mosbach, K., *Biotechnol. Bioeng.* **18**, 1145 (1976).
207. Coughlin, R. W., Aizawa, M., Alexander, B. F., and Charles, M., *Biotechnol. Bioeng.* **17**, 515 (1975).
208. Jaegfeldt, H., Torstensson, A., and Johansson, G., *Anal. Chim. Acta* **97**, 221 (1978).
209. Malinauskas, A., and Kulys, J., *Anal. Chim. Acta* **98**, 31 (1978).
210. Berman, H. J., and Hebert, N. C., *Ion-Selective Micro Electrodes*, Plenum Press, New York (1974).
211. Kessler, M., Clark, L. C., Lubbers, D. W., Silvers, I. A., and Simon, W. (eds.), *Ion and Enzyme Electrodes in Biology and Medicine*, University Park Press, Baltimore (1976).
212. Czaban, J. D., and Rechnitz, G. A., *Anal. Chem.* **45**, 471 (1973).
213. Czaban, J. D., and Rechnitz, G. A., *Anal. Chem.* **47**, 1787 (1975).
214. Czaban, J. D., and Rechnitz, G. A., *Anal. Chem.* **48**, 277 (1976).
215. Fleet, B., Bound, G. P., and Sandbach, D. R., *Bioelectrochem. Bioenerg.* **3**, 158 (1976).
216. Caflisch, C. R., and Carter, N. W., *Anal. Biochem.* **60**, 252 (1974).
217. Pucacco, L. R., and Carter, N. W., *Anal. Biochem.* **90**, 427 (1978).
218. Himpler, H., M.S. thesis, The State University of New York at Buffalo (1976).
219. Sohtell, M., and Karlmark, B., *Pflugers Arch.* **363**, 1979 (1976).
220. Pui, C. P., Rechnitz, G. A., and Miller, R. F., *Anal. Chem.* **50**, 330 (1978).
221. Fritz, I., Nagy, G., Fodor, L., and Pungor, E., *Analyst* **101**, 439 (1976).
222. Enfors, S. O., *Proc. Analyt. Div. Chem. Soc.* **14**, 106 (1977).
223. Enfors, S. O., and Molin, N., *Process Biochem.* **13**, 9 (1978).
224. Enfors, S. O., Molin, N., Mosbach, K., and Nilsson, H., U. S. Patent 4024042 (1976).
225. Booker, H. E., and Haslam, J. L., *Anal. Chem.* **46**, 1054 (1974).
226. Gebauer, C. R., Meyerhoff, M. E., and Rechnitz, G. A., *Anal. Biochem.* **95**, 479 (1979).
227. Riechel, T. L., and Rechnitz, G. A., unpublished results, 1976.
228. Riechel, T. L., and Rechnitz, G. A., *Biochem. Biophys. Res. Commun.* **74**, 1377 (1977).
229. Blaedel, W. J., Kissel, T. R., and Boguslaski, R. C., *Anal. Chem.* **44**, 2030 (1972).
230. Racine, P., and Mindt, W., *Experientia Suppl.* **18**, 525 (1971).
231. Carr, P. W., *Anal. Chem.* **49**, 799 (1977).
232. Mosbach, K., and Mosbach, R., *Acta. Chem. Scand.* **20**, 2807 (1966).
233. Updike, S. J., Harris, D. R., and Shrago, E., *Nature* **224**, 1122 (1969).
234. Vieth, W. R., Wang, S. S., and Sani, R., *Biotechnol. Bioeng.* **15**, 565 (1973).
235. Kennedy, J. F., Barker, S. A., and Humphreys, J. D., *Nature* **261**, 242 (1976).
236. Larsson, P. O., Ohlson, S., and Mosbach, K., *Nature* **263**, 796 (1976).
237. Sato, T., Mori, T., Tosa, T., Chibata, I., Furui, M., Yamashita, K., and Sumi, A., *Biotechnol. Bioeng.* **17**, 1797 (1975).
238. Tanaka, A., Yasuhara, S., Osumi, M., and Fukui, S., *Eur. J. Biochem.* **80**, 193 (1977).
239. Keirstan, M., and Bucke, C., *Biotechnol. Bioeng.* **19**, 387 (1977).
240. Vandamme, E. J., *Chem. Ind.* (*London*) **18**, 1070 (1976).

241. Prave, P., *Angew. Chem. Int. Ed.* **16**(4), 205 (1977).
242. Chibata, I., and Tosa, T., in *Applied Biochemistry and Bioengineering* (L. B. Wingard, E. Katchalski-Katzir, and L. Goldstein, eds.), p. 342, Academic Press, New York (1976).
243. Divies, C., *Ann. Microbiol.* (*Paris*) **126A**, 175 (1975).
244. Karube, I., Matsunaga, T., Mitsuda, S., and Suzuki, S., *Biotechnol. Bioeng.* **19**, 1535 (1977).
245. Karube, I., Mitsuda, S., Matsunaga, T., and Suzuki, S., *Ferment. Technol.* **55**, 243 (1977).
246. Karube, I., Matsunaga, T., and Suzuki, S., *J. Solid Phase Biochem.* **2**, 97 (1977).
247. Matsunga, T., Karube, I., and Suzuki, S., *Anal. Chim. Acta* **98**, 25 (1978).
248. Mattiasson, B., Larsson, P. O., and Mosbach, K., *Nature* **268**, 519 (1977).
249. Rechnitz, G. A., Kobos, R. K., Riechel, S. J., and Gebauer, C. R., *Anal. Chim. Acta* **94**, 357 (1977).
250. Kobos, R. K., and Rechnitz, G. A., *Anal. Lett.* **10**, 751 (1978).
251. Rechnitz, G. A., Riechel, T. L., Kobos, R. K., and Meyerhoff, M. E., *Science* **199**, 440 (1978).
252. Jensen, M. A., and Rechnitz, G. A., *Anal. Chim. Acta* **101**, 125 (1978).
253. Riechel, T. L., and Rechnitz, G. A., *J. Membr. Sci.* **4**, 243 (1978).
254. Kobos, R. K., Rice, D. J., and Flournoy, D. S., *Anal. Chem.* **51**, 1122 (1979).
255. Matsunaga, T., Karube, I., and Suzuki, S., *Anal. Chim. Acta* **99**, 233 (1978).
256. D'Orazio, P., Meyerhoff, M. E., and Rechnitz, G. A., *Anal. Chem.* **50**, 1531 (1978).
257. Meyerhoff, M. E., and Rechnitz, G. A., *Science* **195**, 494 (1977).
258. Rechnitz, G. A., Arnold, M. A., and Meyerhoff, M. E., *Nature* **278**, 466 (1979).
259. Rechnitz, G. A., paper presented at the 30th Annual Summer Symposium on Analytical Chemistry, Amherst, Massachusetts, June 13, 1977.
260. D'Orazio, P., and Rechnitz, G. A., *Anal. Chem.* **49**, 2083 (1977).
261. Baum, G., and Ward, F. B., *Anal. Chem.* **43**, 947 (1971).

Chapter 2

Coated Wire Ion-Selective Electrodes

Henry Freiser

1. INTRODUCTION

Coated wire electrodes (CWE's) refer to a type of ion-selective electrode (ISE) in which an electroactive species is incorporated in a thin polymeric film coated directly to a metallic conductor. Although this definition is broad enough to include those instances when the electroactive species is heterogeneously dispersed in the polymeric matrix, both the earliest and the majority of CWE's involve a solution. The CWE devices are so simple and inexpensive that they can be made by students and used in an introductory laboratory course in analytical chemistry. They are capable of such extreme miniaturization that they should find application in biomedical and clinical research. In fact, the development of CHEMFET's (discussed in Chapter 3) may be considered as a logical extension of CWE's.

The inspiration for the discovery of CWE's came from the work of Hirata and Date,[1] who imbedded a copper wire in a polyvinyl chloride (PVC) disk in which a thick slurry of finely divided cuprous sulfide had been incorporated. The resulting device functioned very well as a sulfide ion ISE. What distinguished this electrode from earlier polymer-matrix-based ISE's, such as those of Pungor[2] or Shatkay,[3] is the bold elimination of an internal reference solution. Our research team, its curiosity piqued by the Hirata and Date work, attempted to make ISE's by the simple expedient of coating a platinum wire that was attached to the central conductor of a coaxial cable,

Henry Freiser • Department of Chemistry, University of Arizona, Tucson, Arizona 85721.

with a mixture of PVC solution in cyclohexanone with either a solution of a high-molecular-weight tetraalkylammonium salt in decanol[4] or a solution of calcium didecylphosphate in dioctylphenylphosphonate[5] and, when the resulting film was air dried, tightly wrapping the remainder of the exposed wire with a paraffin film. These "coated wire electrodes" were very well behaved and gave reproducible and reliable potential–log-a response curves over a wide concentration range ($0.1–10^{-5}\,M$). Although some fundamental studies designed to elucidate the mechanism of the CWE's have been conducted, their behavior cannot yet be considered to be fully understood. Nevertheless, CWE's have been prepared for a wide variety of ions and they have wide application in general, clinical, environmental, and industrial analysis.

In this chapter, the preparation, testing, fundamental characterization, and applications of CWE's will be described.

2. PREPARATION OF COATED WIRE ELECTRODES

A number of different polymers have been used for CWE's from PVC to epoxy resins. While the exact difference in electrode performance characteristics due to the specific polymer base has not been investigated, the effect is generally small, provided that the polymer is suitably plasticized and the film is not so thick that the electrical impedance of the CWE exceeds the capability of the meter ($<30\ \mathrm{M}\Omega$).

Such a wide variety of types of electroactive species have been found to be suitable for use in CWE's that it is fair to state that all those materials that found application in the earlier liquid membrane ISE's also can be used in CWE production.

Several variations in the preparation of CWE's have been successfully carried out. The major features include (a) obtaining the desired electroactive species, (b) introducing the electroactive species in the polymeric matrix, (c) use of plasticizers, and (d) construction of the CWE.

2.1. Preparation of Electroactive Species

Occasionally, the electroactive species is not commercially available, as might be the case when one is developing new electrode systems. A simple means of preparing the material used for anion-responsive electrodes is the extraction exchange using the readily available salt of the large cation. For example, the chloride of the large quaternary ion, tricaprylylmethylammonium ion, Aliquat 336S, sold by Henkel Corporation (Kankakee, Illinois) can be exchanged by the desired anion X^{-n} (where n is usually unity), by successive extractions of an n-decanol solution of the Aliquat 336S (10 vol % or more) with the portions of a concentrated aqueous solution of the sodium salt of X^{-n} until the final portion gives a negative test for Cl^-.[4] The number of extractions necessary for total exchange of Cl^- by the anion of interest will depend on the relative extractability of the corresponding ion pair. An alternative to using the sodium salt would be appropriate if the silver salt is reasonably soluble; then the formation of the insoluble AgCl facilitates the exchange considerably. In such a case, centrifugation is a convenient way to obtain a clear decanol solution.

A similar approach can be used to prepare the electroactive species for cation-responsive electrodes. Calcium ion can be extracted by di(2-ethylhexyl)phosphoric acid (DEHPA) in decanol at elevated pH values. Alternatively, the calcium chelate can be prepared by treating the DEHPA solution with an excess of CaO. It is not always necessary to obtain the electroactive species to be used prior to its incorporation into the polymeric matrix. For example, in cation-selective electrodes based on dinonylnaphthalene sulfonic acid (DNNS), the acid itself is dissolved in a PVC solution, the film electrode formed, and then converted to the desired cation form by immersing it in the appropriate cation salt solution.[6] This approach is, in effect, the method of activating "neutral-carrier" CWE's. The neutral carrier is dissolved in the polymeric matrix and the resulting film electrode is conditioned by immersing it in the appropriate salt solution.

2.2. Polymeric Matrix

A relatively wide variety of polymeric matrices have been found utilizable in CWE construction, including polyvinyl chloride (PVC), polymethyl methacrylate (PMM), and epoxy. In order to introduce the electroactive species into epoxy, a suitable weight of the salt was mixed with the curing agent and this in turn was combined with the epoxy resin. A metallic wire was dipped into the resulting mixture to form a small bead, which completed its polymerization in place. The addition of the electroactive species, e.g., Aliquat perchlorate, generally retarded the epoxy polymerization time, but 3–4 hr was adequate in all cases for completion of this step.

In the cases of polymers like PVC and PMM, for which suitable solvents were available, the polymer solution was mixed with the electroactive species, either neat or in solution, and the film was formed on the metallic substrate by evaporation. For example, PVC can be dissolved in solvents such as tetrahydrofuran (THF) or cyclohexanone, PMM is soluble in methyl acetate or trichloroethylene, and polystyrene is soluble (4 wt %) in chloroform.

2.3. Use of Plasticizers

In order for the polymer film to behave properly in those CWE's and similar ISE's which are based on liquid membrane principles, i.e., those in which the electroactive species is molecularly dispersed or dissolved rather than present in finely divided solid particles, the polymer film must have some attributes of a liquid phase. In particular, the temperatures at which the electrode is employed should be higher than the glass transition temperature of the polymer.

The glass transition temperature, t_g, or an amorphous polymer is the relatively well-defined temperature at which the rigid glassy solid material becomes a viscoelastic fluid. The most pronounced changes in the behavior of amorphous polymers occur at t_g, of which significant increases in conductivity and dielectric constant are most relevant.[7] The glass transition temperatures of unplasticized polymers increase with the average molecular weight of the polymer according to the Fox–Flory model[8] as well as with the specific nature of the polymer. For commercial grades of unplasticized polyvinyl chloride (PVC), t_g is about 80 °C. Below this temperature, the electrical conductivity of PVC is about 10^{-10} mho cm^{-1} and its dielectric constant is about 3; above this temperature, these values increase to about 10^{-9} mho cm^{-1} and 15, respectively.[7]

Plasticizers, as their name implies, are polymer additives whose purpose is to render the polymer more plastic, i.e., more viscoelastic at ambient temperatures. A wide variety of organic compounds, most frequently aromatic esters such as the phthalates, have been employed as plasticizers. If enough plasticizer is incorporated, it will lower the t_g to under room temperature. Properly plasticized films are necessary for proper CWE and other polymer-film ISE behavior. This was intuitively accepted early on, but Carmack first clearly demonstrated this correlation.[9]

Consideration of the role of plasticizers is most critical for neutral-carrier CWE's, where because of the very expensive nature of valinomycin and other carriers, only minimal amounts are used. As a result, plasticizers such as didecylphthalate are required. In contrast, with anion-selective electrodes based on the higher-molecular-weight quaternary ammonium salts such as Aliquat 336, the electroactive species comprises a major component in the film and thus provides sufficient plasticizer action so that no additional plasticizer is needed.

2.4. Construction of the CWE

A number of techniques have been used in the construction of coated wire electrodes. Originally, a platinum wire, soldered to the central conductor of coaxial cable wire, was used as the metallic substrate. Subsequently, it was found that the simpler procedure of applying the film to the copper central conductor directly was entirely equivalent, except for a significant shift in the E_0' of the electrode. Quite a number of metals have since been tested and, with the interesting exception of the value of E_0', which was a function of the metal, the electrode behavior was the same. The effect of the nature of the metal used as the conducting substrate on the E_0' of the CWE might be useful to shift response curves to more convenient potential ranges.

The current mode of construction of a CWE in our laboratory[10] follows:

(1) Cut an approximately 50–70 cm section of a single-conductor coaxial cable and attach a BNC connector (e.g., Amphenol 74868 UG-89A/U) to one end. BNC connectors are used because they are the least expensive shielded connector available. Since the pH/mV meter used to measure the potential will not, however, accept a BNC jack, a patch cord having an Amphenol 36775 connector on one end and a U.S. Standard ISE plug (Orion Research) on the other is needed. Solder the central conductor to the tip of the plug; solder the outer braid to the case.

(2) Remove approximately 5 cm of the outer insulation at the other end of the cable with a razor blade. Care must be exercised so as to avoid cutting the inner insulation. The exposed braided outer copper conductor is then cut away with a wire cutter.

(3) Carefully remove approximately 2 cm of the inner insulation of the coaxial cable. The tip of the copper wire thus exposed is sanded flat with a fine-grade emery cloth.

(4) Wash the exposed Cu wire with laboratory detergent and water, thoroughly rinse with deionized water, and dry with acetone. Then rinse the wire with chloroform and allow to dry.

(5) During the coating procedure, the Cu wire must be in a vertical position so that a uniform and symmetrical coat is obtained. This is conveniently accomplished by clamping the wire assembly in a ring stand during coating. Coat the wire by quickly dipping about 10 mm of the exposed Cu wire into the coating solution several times and then allow the film of PVC solution left on the wire to dry in air for about 1 min. This dipping and drying procedure is repeated until a plastic bead, approximately 2 mm in diameter, is obtained.

(6) Allow the plastic bead to dry in air overnight and wrap the remaining exposed Cu in Parafilm®. The Parafilm® seal should extend from the base of the bead to about 2 cm above the exposed portion of Cu wire. The finished CWISE is shown in Fig. 1.

(7) Test the suitability of the film by measuring the impedance of the electrode by dipping it either in a small pool of mercury or in an appropriate salt solution, using a metal wire in either case to complete the circuit. The impedance should be under 50 MΩ if the film is of the proper thickness. Unusually low impedances clearly signal incomplete coverages of the metal substrate by either the polymer film or the Parafilm®. If rewrapping is not effective, discard the electrode (i.e., cut the end of the cable) and start again.

3. PRECONDITIONING AND STORAGE PROCEDURES

The CWE should be preconditioned in a 0.1 M solution of the appropriate electrolyte (nitrate salts for cations, sodium salts for anions) for at

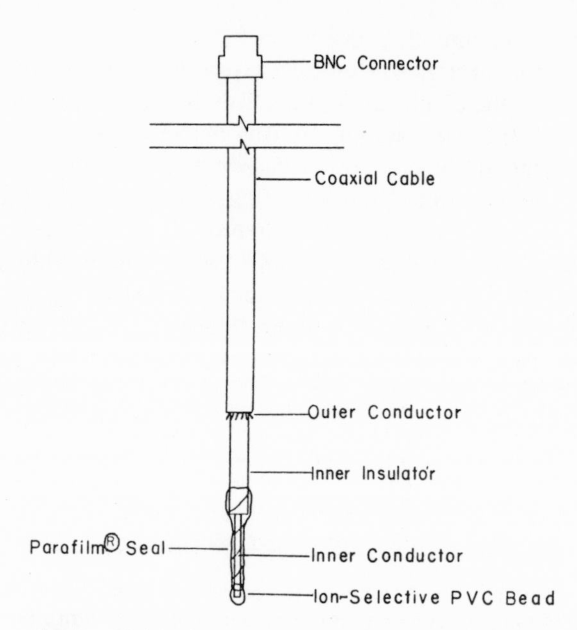

Fig. 1. Finished coated wire ion-selective electrode.

least 1 hr prior to use. It is often convenient to precondition the electrodes in the 0.1 *M* solution overnight. During periods of frequent use, the CWE should be stored in the 0.1 *M* solution which has been saturated with the electroactive species. This storage solution is prepared by shaking the 0.1 *M* solution with the decanol solution of the electroactive species, allowing the phases to separate, and centrifuging the aqueous phase. Prior to each use, the electrode should be soaked about 20 min in deionized water. The electrode may be stored dry when not needed, but should be preconditioned again prior to reuse.

4. CALIBRATION OF COATED WIRE ELECTRODES

The essential elements of CWE calibration are the same as those conventionally employed with any electrode. It is noteworthy, however, to observe the precaution of rinsing the electrode after removing it from the conditioning solution and soaking it for 10–20 min in deionized water prior to calibration. It is advisable, as with other types of electrodes, to progress from the most dilute to the most concentrated standard solutions in the calibration series.

There are very few restrictions concerning the choice of a reference electrode. Obviously, the electrolyte used in the salt bridge must not present any interference with respect to the ion under test. If such is the case, a flowing double junction may readily solve the problem.

In preparing the calibration curve, use is made of values of activity coefficients, δ_A, obtained from the extended Debye–Hückel equation

$$-\log \delta_A = \frac{0.5057 Z_A^2 I^{1/2}}{1 + 3.3 \times 10^7 a_A I^{1/2}} \tag{1}$$

where I is the ionic strength of the calibration solution and Z_A and a_A represent the ionic charge and size parameter, respectively.[11]

The linearity of the response-potential–log ion-activity curve over the range 0.1 to 10^{-4} or 10^{-5} M in a well-behaved CWE should be good to better than 0.5%. That is, the slope of the curve should be constant to within $0.5Z_A$ (mV)/log a_A. The $E^{0\prime}$, i.e., the intercept of the line, can vary from 1 to 3 mV from one day to the next. Since the slope of a particular electrode remains reasonably constant over the lifetime of the electrode, daily recalibration can be conducted by measuring a single standard solution. To obtain the highest reliability in the potentiometric determination of the concentration of an unknown solution, however, it is best to make measurements at the same time of two standards whose concentration span includes that of the unknown. Reliability of 1–2% is readily accessible by this means.[12]

5. DETERMINATION OF SELECTIVITY RATIOS

The selectivity ratio, K_{ij}, is defined by the Eisenman–Nicolsky equation

$$E = E_i^0 + \text{slope} \log(a_i + K_{ij} a_j^{1/Z_j}) \tag{2}$$

and may most readily and reliably be determined by a simplified form of the mixed-solutions method.[13] The method involves the comparison of the electrode response to a solution containing only the primary ion at activity a_i, with that containing a_j of the interferent as well as a_i of the primary ion. The observed difference in response, ΔE_{meas}, must be corrected for the change in activity coefficient corresponding to the higher ionic strength of the mixed solution, but this is simply done by use of the Debye–Hückel equation

$$\Delta E_{\text{cor}} = \Delta E_{\text{meas}} + \text{slope} \times \Delta \frac{I^{1/2}}{1 + 3.3 \times 10^7 a_i I^{1/2}} \tag{3}$$

Rearrangement of the Eisenman–Nicolsky equation gives the expression for K_{ij}:

$$K_{ij} = (10^{E_{\text{cor}}/\text{slope}} - 1) \frac{a_i}{a_j^{1/Z_j}} \tag{4}$$

TABLE 1. Selectivity Coefficients, K_i, for the Coated Wire Electrodes Compared to Those for the Liquid Membrane Electrodes

Electrode	Interfering anion: Chloride	Nitrate	Sulfate	Miscellaneous
Perchlorate	0.004 (0.18)[a]	0.028 (0.12)	<0.001 (<0.001)	ClO_3^-, 0.039 (0.20)
Chloride	—	2.0 (3.0)	0.12 (0.061)	Br^-, 1.2 (2.7)
Bromide	0.19 (0.40)	2.0 (1.1)	0.020 (<0.001)	I^-, 14.5 (3.6)
Iodide	0.0045 (0.064)	0.11 (0.23)	<0.001 (>0.001)	Br^-, 0.056 (0.22)
Thiocyanate	<0.001 (<0.001)	<0.046 (0.067)	0.001 (0.018)	I^-, 0.34 (0.42)
Benzoate	1.3 (0.15)	1.3 (0.48)	0.15 (<0.001)	Salicylate, 1.4 (0.29)
Salicylate	<0.001 (<0.001)	0.42 (0.030)	<0.001 (<0.001)	*m*-OH benzoate, 0.22 (0.13)
Oxalate	51 (≈80)	—	1.3	OAc^-, 11 (11)
Sulfate	16 (100)	30 (820)	—	
Phenylalanine	0.8 (1)	2 (1.6)	0.04 (0.1)	Gly, 0.025 (0.040) Leu, 0.13 (0.40)
Leucine	0.56 (0.50)	0.50 (0.30)	0.020 (0.025)	Gly, 0.032 (0.063) Val, 0.25 (0.25) Phe, 2.0 (1.6)

[a] Value of K_i for liquid membrane electrode is in parentheses.

An interesting comparison of the selectivity of CWE's and the corresponding liquid membrane ISE of the barrel type is shown in Table 1.[4]

It should be kept in mind that values of K_{ij} are not independent of total solution composition.

6. MECHANISM OF CWE BEHAVIOR

From their inception, CWE's represented something of an anomaly. To incorporate an electroactive species in a thin polymeric film that was coated directly on a metallic conductor without the conventional internal reference solution or rapid and reproducible reference electrode, did not appear to be a feasible approach. For CWE's not only to function properly but to demonstrate clear superiority in sensitivity and selectivity in many instances over their more conventional, bulky, barrel-type liquid membrane electrode counterparts was and, to a large extent, remains puzzling. One set of studies was carried out whose objective it was to ascertain the mode of electrical conduction through the film.

A series of sample films, either as beads or cast disks, were prepared containing Aliquat 336S salts in 1:6 weight ratio with a series of unplasticized polymers, and their dc electrical conductivities measured as a function of temperature from 15 to 85 °C.[14]

The most surprising feature of the conductivity behavior of the polymer-dispersed salts was the unusually high temperature coefficient. From the plots shown in Fig. 2, some of these materials compare favorably with commercially available semiconductor thermistors. Indeed the temperature dependence of the conductivity follows the operational semiconductor expression

$$\sigma = \sigma_0 e^{-E_a/kT} \tag{5}$$

As may be seen from Table 2, which lists the values of E_a calculated for a series of Aliquat salts in various polymeric matrices using this equation, the activation energies vary from 0.6 to almost 3 eV (or 15–65 kcal/mol). This is in sharp contrast to the behavior of cellulose acetate films impregnated with alkali metal salts, which was characterized not only by low conductivity ($\sim 10^{-14}\ \Omega\ \text{cm}^{-1}$) but also by low activation energies (<1.0 eV).[15] Polyaromatic and other organic semiconductors which have been observed to have E_a values similar to ours[16] show conductivities that are 10^{12}-fold or more lower than those observed with Aliquat-salt–polymer materials (see Fig. 3). The variation of behavior of the Aliquat-salt–polymer film on the nature of the anion might well have some bearing on the selectivities of the corresponding electrodes, although the nature of the relationship is not entirely clear at this time.

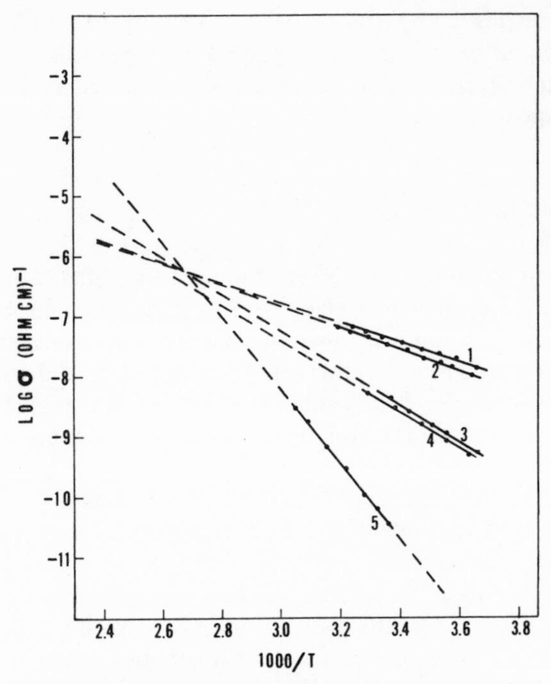

Fig. 2. Temperature dependence of conductivity of 0.3 *M* Aliquat$^+$-ClO_4^- in various polymeric matrices. (1) Polyethylene, (2) polystyrene, (3) polymethylmethacrylate, (4) nylon-66, (5) epoxy.

Another point of similarity in the behavior of these materials to that of organic semiconductors[16,17] is evident from Fig. 2 in which the extrapolated lines representing the application of the semiconductor equation above to Aliquat$^+$, ClO_4^- in various polymeric matrices are seen to intersect, not at

TABLE 2. Conductance Activation Energies for Aliquat-Salt–Polymer Films

Aliquat salt	Poly(styrene)		Poly(methylmethacrylate)		Poly(vinylchloride)		Epoxy	
	eV	kcal	eV	kcal	eV	kcal	eV	kcal
ClO_4^-	0.96	22.1	1.03	23.8	2.05	47.3	2.78	64.1
NO_3^-	0.88	20.3	0.66	15.2	1.53	35.3	—	—
SO_4^{2-}	0.60	13.8	—	—	—	—	2.56	59.0
OAc^-	0.70	16.1	0.67	15.5	1.33	30.7	—	—
CNS^-	0.50	11.5	0.67	15.5	—	—	2.13	49.1
Cl^-	0.74	17.1	0.64	14.8	1.10	25.4	2.08	48.0

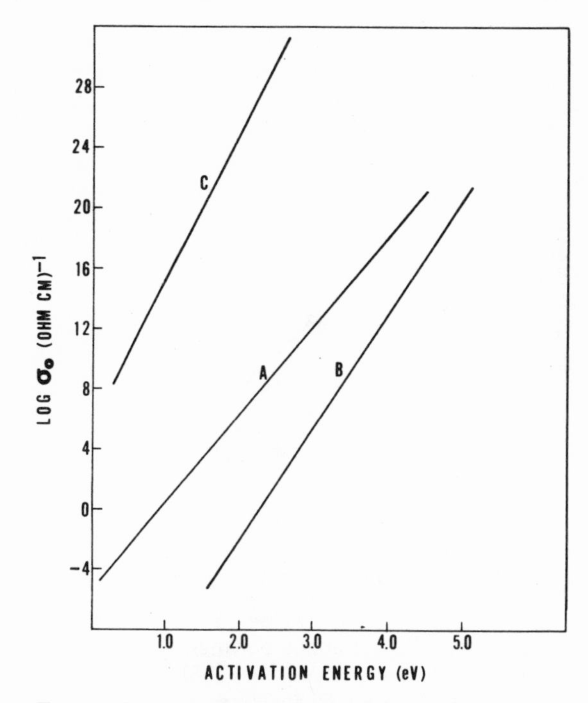

Fig. 3. Conductivity–action-energy relationships for various organic systems. (A) Aromatics, (B) proteins, (C) Aliquat-ClO_4.

$1/T = 0$ but rather at a "characteristic" temperature, T_0, here found equal to 375 °K. Such results, which conform to the relationship

$$\log \sigma_0 = \frac{E_A}{2kT_0} + \log \sigma_0' \tag{6}$$

developed by Rosenberg,[17] show a compensation between the energy and entropy of activation of the conduction process (Fig. 4).

A point of significant contrast in the behavior of these materials and those of organic semiconductors, however, is their unusually high conductivity in relation to their activation energies. Furthermore, it can be seen that as the concentration of the salt in the matrix increased, the increase in conductivity was accompanied by a decrease in E_a (Fig. 5). This certainly beclouds the adoption of a single mechanism of electrical conduction. Perhaps both electronic and ionic conduction mechanisms were operative.[18] In order to further exploit this equation, correlation of the conductivities of various membrane compositions with their dielectric constants were conducted (Tables 3 and 4).[19]

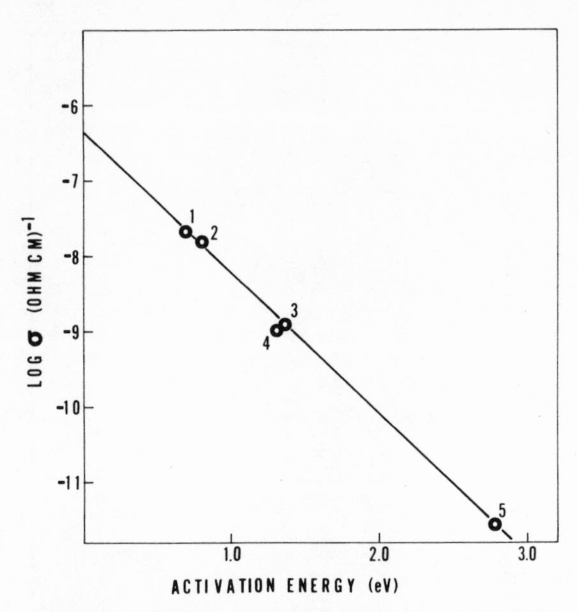

Fig. 4. Conductivity–activation-energy relationship for Aliquat$^+$-ClO_4^- in various polymeric matrices. (1) Polyethylene, (2) polystyrene, (3) polymethylmethacrylate, (4) nylon-66, (5) epoxy.

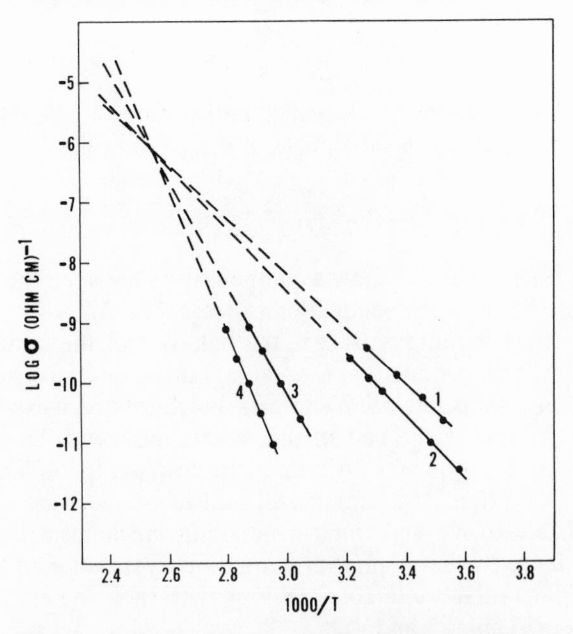

Fig. 5. Temperature dependence of conductivity of Aliquat$^+$-ClO_4^- in epoxy resin as a function of concentration. (1) 1.2 *M*, (2) 0.92 *M*, (3) 0.20 *M*, (4) 0.090 *M*.

TABLE 3. Conductivity of Aliquat-Cl Polymer Films as a Function of Concentration

Aliquat-Cl (wt %)	$(M)^a$	ε' (1 kHz)	$-\log \sigma$ 25 °C $(\Omega\ \text{cm})^{-1}$	E_a (eV)
Epoxy matrix				
11	(0.09)	3.5	12.4	2.27
25	(0.20)	4.6	10.2	1.79
50	(0.92)	7.0	7.6	1.44
67	(1.20)	8.0	7.0	1.00
PVC matrix				
1	—	3.2	13.4	2.34
5	—	5.5	9.3	1.70
11	—	6.7	7.2	1.08
50	—	8.1	6.9	0.58

[a] Approximate concentration in final mixture.

The difficulty in determining the mechanism of conduction in organic materials is compounded by the similar behavior of both ionic particles and electrons when they are thermally stimulated to promote electrical conduction. Regardless of the carrier types, a general description of the conductivity has the form

$$\sigma = \sum_i q_i n_i \mu_i \tag{7}$$

where q_i is the charge, n_i the number, and μ_i the mobility of the ith charge-carrying species. The incorporation of the high concentrations of weakly dissociated ionic salts into a polymer matrix will tend to "swamp out" the effects of residual ionic or electronic carriers found as impurities. The result of doping with such a high concentration of ionic material is that at least part, if not all, of the observed conductivity will be ionic. For an ionic

TABLE 4. Effect of Plasticizer on Conductivity of Aliquat-Cl/PVC Films

Dioctyl phthalate[a] (wt %)	ε' (1 kHz)	$-\log \sigma$ 25 °C $(\Omega\ \text{cm})^{-1}$	E_a (eV)
0.0	5.5	9.3	1.70
4.8	6.0	8.9	1.30
13.0	6.2	8.8	1.16
16.7	6.6	7.9	1.15

[a] All samples contain 0.01 g Aliquat-Cl + 0.19 g PVC.

process, the observed conductivity relationship [equation (7)] can be given by

$$\sigma = qn_0 \exp(-E_c/2kT)\mu n_0 \exp(-E_\mu/kT) \tag{8}$$

or combining terms

$$\sigma = \sigma_0 \exp[-(E_c + 2E_\mu)/2kT] \tag{9}$$

where E_c and E_μ are the activation energy of carrier generation and mobility, respectively, and σ_0 contains the terms μ and n_0. Models to describe the observed electrical conductivity must therefore include parameters for both carrier generation and mobility.

Turning to the aspect of carrier generation in our system, the primary source for the generation of ionic charge carriers is the dissociation of R_4N salts, e.g.,

$$\begin{array}{ccc} R_4N^+ \cdot Cl & \rightleftharpoons R_4N^+ & + \; Cl^- \\ n_0(1-\alpha) & n_0\alpha & n_0\alpha \end{array} \tag{10}$$

for which an equilibrium expression $K = \alpha^2 n_0/1 - \alpha$ can be written. For such cases, Barker and Sharbaugh[15] have developed a relationship of ionic conductivity and the equilibrium constant by using the equation

$$\sigma = [(Kn_0)^{1/2} \exp(-U'_0/2\varepsilon' kT)](\mu_+ + \mu_-)e \tag{11}$$

where ε' is the effective dielectric constant of the system, U'_0 is the energy needed to separate the ions in the vacuum, μ is the ion mobility, and e is the electronic charge.

As seen from equation (9), values of the activation energy can be described as comprising contributions from charge generation, i.e., dissociation of ion pairs, as well as from ion mobilities. From the Bjerrum theory one can expect, in a series of ion pairs having the same cation, that the dissociation energy would decrease with increasing anionic radius.[20] This situation is even more complicated because of the formation of micellar aggregates. Although ionic mobility depends on size, its temperature coefficient, and therefore E_μ, reflects the temperature variation of viscosity of the medium and is therefore substantially independent of ionic radius (which is independent of temperature). This relationship may be confused by solvation, as in the case of the alkali metals where Li^+, nominally the smallest ion, is large by virtue of hydration. Our results (Table 5) would indicate that solvation of ions may be significant even in these low-dielectric-constant polymeric matrices.

Pressure dependence of electronic conductivity, from which activation volumes for this process can be calculated, is a most useful tool of elucidating mechanisms of charge conduction.[19] The activation volumes for electronic

TABLE 5. Conductivity Parameters for R_4N^+,X/Polymer Films

	PVC[a]		Epoxy[b]	
Aliquat-336S salt	E_a (eV)	$-\log \sigma$ 25 °C (Ω^{-1} cm^{-1})	E_a (eV)	$-\log \sigma$ 25 °C (Ω^{-1} cm^{-1})
Cl^-	0.58	6.9	1.79	10.2
SCN^-	0.60	7.4	—	—
NO_3^-	0.92	8.2	—	—
ClO_4^-	1.25	8.8	—	—
I^-	1.35	9.0	—	—
Benzoate	0.68	7.6	2.28	9.4
Acetate	0.76	8.0	—	—
Laurate	0.82	7.9	2.04	8.8
1-Amino-2-naphthol-4-sulfonate	1.55	9.5	—	—
p-Toluene sulfonate	1.06	9.6	2.36	9.8

[a] 0.2 g PVC + 0.2 g Aliquat salt.
[b] 0.3 g epoxy + 0.1 g Aliquat salt.

conductivity can be calculated from the following equation:

$$V^* = \left(\frac{\partial \Delta G^*}{\partial P}\right)_T RT\left(\frac{\partial \ln \sigma}{\partial P}\right)_T \tag{12}$$

One might expect activation volumes involved in ionic (other than protonic) conduction processes to be reasonably large since they would be related to either a molar volume or an ion pair consisting of partners of about equal size, or at least to the volume of the smaller of two ions as a limiting case.[21] Activation volumes involving either protonic of electronic species, on the other hand, might be expected to be small. Proven cases involving electronic conduction have actually exhibited negative activation volumes.[22] Such values have been attributed to the increase in the degree of orbital overlap between adjacent molecules that can occur with even moderate pressure increases.

The conductivity of membranes containing 50 wt % Aliquat-Cl in both PVC and epoxy matrices, measured at 25 °C over a range of 1-2000 atm, was found to decrease progressively with increasing pressure (Fig. 6).[19]

The large negative pressure coefficients found for the Aliquat/polymer mixtures in this study are indicative of ionic carrier processes.[23,24] Activation volumes calculated for the materials used in this study gave a $\Delta V^* =$ 61 cm^3/mole for Aliquat/PVC and 87 cm^3/mole for Aliquat/epoxy. While it is difficult to assign an exact molecular correspondence to these values of ΔV^*, they are inconsistent in sign with electronic conduction, and in magnitude with protonic conduction.

As indicated by equation (11), the conductivity would be expected to vary with the square root of the concentration of the ion pair for 1:1 electrolyte-impregnated polymers. Experimentally such relationships are known.[25] On the other hand, as King and Medley[26] pointed out, addition of electrolytes to polymers in sufficient concentration increases the dielectric constant, which leads to greater dissociation. Moreover, ionic migration is facilitated by the plasticizing effect of the salt. Both phenomena thereby complicate the relationship between conductivity and the concentration of electroactive material.

We have used the large change in dielectric constant that occurs with changes in quaternary ammonium salt concentration to determine the influence of dielectric constant on conductivity. Of course, changing ε' by simply altering the salt concentration neglects the parameters n_0 and μ in equation (11). Nevertheless, as predicted by the dissociation hypothesis, a linear dependence of $\log \sigma$ of $1/\varepsilon'$ with a high degree of correlation was observed (Fig. 7). Interestingly, this relationship seems nearly independent of the polymer matrix. The slope of the line in Fig. 7, 0.35, can be compared to those observed in similar studies of alkali-metal salts in hydrated cellulose (slope = 0.6) or to the slope of such a plot for a large number of organic solvents (slope = 0.2).[27]

Hence, both the dielectric constant and pressure dependencies of the conductivity point unequivocally to an ionic conduction mechanism. This

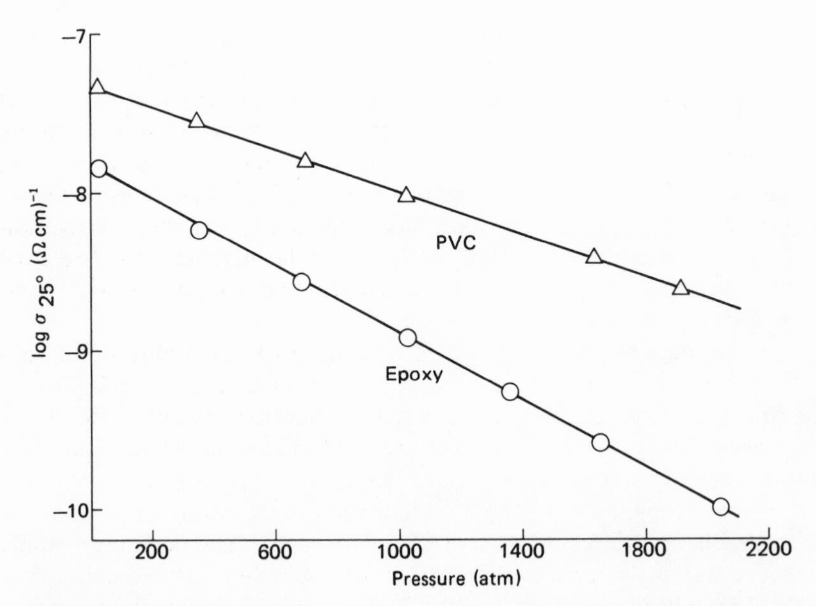

Fig. 6. Pressure dependence of conductivity of Aliquat-Cl/polymer (50 wt %) samples.

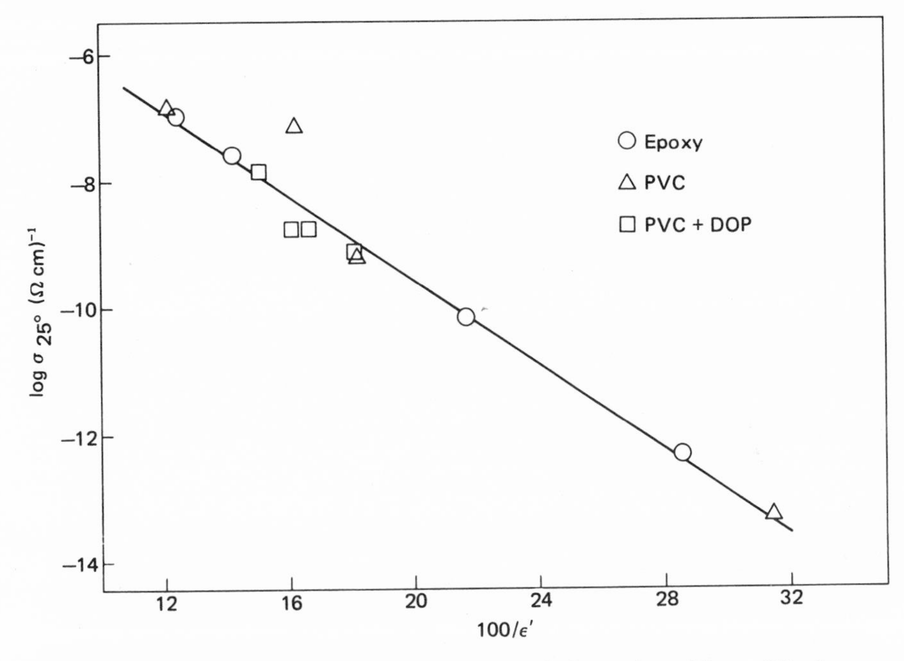

Fig. 7. Conductivity–dielectric-constant relationship for various Aliquat-Cl/polymer samples. Sample compositions are given in Tables 3 and 4.

still leaves unexplained the relatively high activation energies, which may have some precedence, however, in purportedly ionic systems.[28]

7. APPLICATIONS OF COATED WIRE ELECTRODES

It is probably reasonable to conclude that ion-selective electrodes of any sort can be constructed as a CWE, so that CWE's have application wherever ISE's have. Specific advantages in using CWE's derive from the following: (a) Their small size so that the volume of the test solution required can be minimized. In principle, with a sufficiently fine wire and film, intracellular ion activity measurements should be feasible. Miniaturization is being actively pursued in several ways, most notably in the CHEMFET field (see Chapter 3) in which the polymeric films are used to coat the gates of FET carried on part of an IC chip. (b) Their simple, rugged construction. Ease of CWE construction makes this device accessible to students as well as researchers at very low cost of either time or money.

As an indication of the applicability of CWE's, several illustrative examples may be cited.

7.1. Analytical Applications of Anionic CWE's

7.1.1. Determination of NO_x[12]

Even low levels of NO_x, such as are found in ambient air, can be determined by collecting a suitable volume of air in a gas-washing bottle containing 2% H_2O_2, treating the resulting solution with MnO_2 to destroy the excess peroxide, and determining the nitrate potentiometrically. The method is capable of good precision and compares well (1–3 relative %) with such accepted methods as the spectrophotometric xylenol procedure. The method can be used in the presence of at least a 40-fold excess of SO_2 or SO_3. With minor modifications, the method could be used for nitrates in atmospheric particulate matter.

7.1.2. Determination of Anionic Detergents[29]

Detergents of the alkyl sulfate and alkylsulfonate types can be reliably determined in the concentration range of about 10^{-3}–10^{-5} M by direct potentiometry. The CWE based on lauryl benzyl sulfonate responds almost as well to lauryl sulfonate (K_{ij} = 0.81), and somewhat better to lauryl sulfate (K_{ij} = 1.36), but at levels of 10^{-4}–10^{-5} M, the electrode will give a satisfactorily approximate estimation of total anionic detergent present.

7.1.3. Assay of Phenobarbital[30]

Since phenobarbital (5-ethyl-5-phenylbarbituric acid) is lipophilic, an ion pair complex of its anion and Aliquat 336S can be effectively used in a CWE for its estimation. A linear response to a PVC-based CWE was obtained in the concentration range 0.1–10^{-4} M phenobarbital with solutions buffered at pH 9.6 so that only the anion species is present. Interference from nitrate (K_{ij} = 0.2) or chloride (K_{ij} = 0.1) is manageable. Use of the electrode permits a rapid and convenient method for phenobarbital assay in tablet preparations. In contrast to the 4 hr required for assay by the USP method which which it compares favorably in precision and accuracy, the potentiometric assay can be accomplished in under 20 min, thus making it practical to perform the CWE procedure on single tablets, should tablet-to-tablet variation data be desired.

7.1.4. Amino Acids[4]

A variety of well-behaved CWE's responsive to anionic forms of amino acids have been prepared. Among these, the most interesting are those amino acids containing lipophilic substituents, notably phenylalanine. With the phenylalanine electrode, chloride presents substantial interference, $K_{ij} = 0.8$, but the smaller amino acids like glycine marginally interfere, $K_{ij} = 0.025$, and even leucine only moderately interferes, $K_{ij} = 0.13$.

7.2. Analytical Applications of Cationic Electrodes

7.2.1. Potassium[31]

A CWE version of the valinomycin-based electrode, prepared using a PVC matrix and di-*n*-decylphthalate as plasticizer, functioned well in the concentration ranges from 10^{-1} to 10^{-5} *M* potassium ion and could be used in the pH range from 2 to 10. Such an electrode can be readily used for the determination of potassium in a wide variety of materials ranging from whole blood to sea water. Aside from Rb^+ and Cs^+, whose K_{ij} values are 2.5 and 0.44, respectively, most cations do not appreciably interfere (except for Cu^{2+} and Zn^{2+}).

7.2.2. Iron(III)

Cattrall and Pui[32] have developed an ingenious principle that they have applied to the determination of polyvalent cations. By means of a suitable ligand, change a positively charged, highly hydrated cation to an anionic complex that is both singly charged and poorly hydrated. The production of such an anion results in the possibility of constructing an ion-selective electrode based on the complex of this metal-containing anion and a suitably lipophilic cation such as Aliquat 336S. The slope of the response curve will be close to 59 mV/decade rather than (59/3) mV/decade expected from the trivalent ion. Such an electrode in a PVC matrix gave a linear response to Fe(III) from 10^{-1} to 10^{-4} *M* in solutions whose total chloride concentration was maintained at 6 *M*. Of the common ions only Sn^{2+}, Hg^{2+}, and Zn^{2+} gave any interference. Use of the electrode permitted reliable determination of iron in pyrites and silicate rock ores.

7.2.3. Higher-Molecular-Weight Quaternary Amines[6]

Use of dinonylnaphthalenesulfonic acid in a PVC film results in CWE's for a number of cations. For inorganic ions generally, the electrodes are well behaved but possess relatively little selectivity. With respect to higher-molecular-weight quaternary ammonium cations, however, CWE's of great selectivity are seen. For example, the dodecyltrimethylammonium-DNNS electrode is capable of near-Nernstian linear response in the range 0.1–10^{-5} M cation and K_{ij} values of less than 10^{-4} with respect to simple inorganic cations. The high likelihood of preparing CWE's for a number of pharmaceutically and clinically interesting materials is currently being investigated. The sensitive (detection limit $\sim 10^{-5.1}$ M) and selective electrode for phencylidine (a drug of abuse known as "Angel dust") has been developed.[33]

7.3. Determination of pK_a and Other Equilibrium Constants

Because CWE's are responsive solely to ions and not neutral species, the pH profile of the potential response of a CWE for an anion that has proton affinity, e.g., acetate or leucine anion, can be interpreted as the change of the activity of the anion with pH. This would make it possible to calculate the corresponding pK_a. The procedure is exactly analogous to calculating pK_a values using a pH-responsive electrode. Similarly solubility products constants and complex formation constants are accessible.

Another equilibrium parameter, one applicable to colloid-forming species, that can be obtained using the CWE is the critical micelle concentration (CMC). In a study of anionic detergent solutions,[29] a sharp break was found in the otherwise linear calibration curve of, say, a lauryl sulfate CWE which occurred at precisely the CMS. The electrode responds to only the monomer because the micelle concentration in moles/liter is close to zero. This represents a much more rapid method for reliable CMC determination than the study of interfacial tension as a function of concentration.

8. CONCLUSIONS

Coated wire electrodes have proven to be useful devices for potentiometric sensing. Improvements in CWE design will undoubtedly be seen in the near future in (a) greater sensitivity, (b) greater selectivity, and (c) greater miniaturization.

One of the most likely reasons that the sensitivity of CWE's, as well as other liquid membrane ISE's, seem to be at 10^{-5}–10^{-6} M is the possibly inherent solubility of the electroactive species in the aqueous phase. This probably should then be significantly alleviated by chemically bonding the electroactive species to the polymeric matrix. With regard to improved selectivity, since this is related to solvent extraction parameters it would seem worthwhile to examine ways in which to improve selectivity in comparable extraction systems. In addition, electrodes in which chemical interactions that are stronger than ion pair might be incorporated because of the significant improvement in selectivity that such reactions, e.g., acid–base, complexation, etc., entail.

Two avenues of investigation for improved miniaturization of CWE's are already in sight. Use of finer wires with thinner coats or, much more importantly, the coating of appropriate polymer films on FET's to provide so-called CHEMFET's, are being actively pursued.

ACKNOWLEDGMENT

I am happy to acknowledge the support of the Office of Naval Research.

REFERENCES

1. Hirata, H., and Date, K., *Talanta* **17**, 883 (1970).
2. Pungor, E., and Toth, K., in *Ion-Selective Electrodes in Analytical Chemistry* (H. Freiser, ed.), Vol. 1, Ch. 2, Plenum Press, New York (1978).
3. Shatkay, A., *Anal. Chem.* **39**, 1056 (1967).
4. James, H., Carmack, G., and Freiser, H., *Anal. Chem.* **44**, 856 (1972).
5. Cattrall, R. W., and Freiser, H., *Anal. Chem.* **43**, 1905 (1971).
6. Martin, C. R., and Freiser, H., *Anal. Chem.* **52**, 562 (1980).
7. Hedvig, P., *Dielectric Spectroscopy of Polymers*, Wiley and Sons, New York (1977).
8. Fox, T. G., and Flory, P. J., *J. Polym. Sci.* **14**, 315 (1954).
9. Carmack, G., Ph.D. thesis, University of Arizona (1977).
10. Martin, C. R., and Freiser, H., *J. Chem. Educ.*, in press.
11. Kielland, J., *J. Am. Chem. Soc.* **59**, 1675 (1937).
12. Kneebone, B. M., and Freiser, H., *Anal. Chem.* **45**, 449 (1973).
13. Srinivasan, K., and Rechnitz, G. A., *Anal. Chem.* **41**, 1203 (1969).
14. Carmack, G., and Freiser, H., *Anal. Chem.* **45**, 1975 (1973).
15. Barker, R. E., and Sharbaugh, A. H., *J. Polym. Sci. Part C* **10**, 139 (1965).
16. Eley, D. D., *J. Polym. Sci. Part C* **17**, 73 (1967).
17. Rosenberg, B., Bhowmik, B. B., Harder, H. C., and Postow, E., *J. Chem. Phys.* **49**, 4108 (1968).
18. Heyne, L., in *Fast Ion Transport in Solids* (W. van Gool, ed.), p. 123, North-Holland, Amsterdam (1973).
19. Carmack, G., and Freiser, H., *Anal. Chem.* **47**, 2249 (1975).
20. Monk, C. B., *Electrolytic Dissociation*, Academic Press, New York (1961).
21. Hamann, S. D., *Aust. J. Chem.* **18**, 1 (1965).
22. Datta, P. K., *J. Sci. Ind. Res.* **30**, 222 (1971).
23. Sasabe, H., Sawamura, K., Saito, S., and Yoda, K., *Polym. J.* **2**, 518 (1970).
24. Saito, S., Sasabe, H., Nakajima, T., and Yada, K., *J. Polym. Sci. Part A* **6**, 1297 (1968).
25. Sandrolini, F., Pietra, S., and Manarisi, D., *Chim. Ind. (Milan)* **53**, 755 (1971).
26. King, G., and Medley, J. A., *J. Colloid Sci.* **8**, 148 (1958).
27. Barker, R. E., and Sharbaugh, A. H., *J. Polym. Sci. Part C* **10**, 139 (1965).
28. Amborski, L. E., *J. Polym. Sci.* **62**, 331 (1962).
29. Fujinaga, T., Okazaki, S., and Freiser, H., *Anal. Chem.* **46**, 1842 (1974).
30. Carmack, G., and Freiser, H., *Anal. Chem.* **49**, 1577 (1977).
31. Cattrall, R. W., Tribuzio, S., and Freiser, H., *Anal. Chem.* **46**, 2223 (1974).
32. Cattrall, R. W., and Pui, C. P., *Anal. Chem.* **47**, 93 (1975).
33. Martin, C. R., and Freiser, H., *Anal. Chem.* **52** (1980).

Chapter 3

Chemically Sensitive Field Effect Transistors

Jiří Janata and Robert J. Huber

1. INTRODUCTION

Sometimes unjustified, but nonetheless ever present need for more information continues to stimulate the development of new sensors and detectors. One of the more recent additions to the armory of these devices is the chemically sensitive field effect transistor (CHEMFET). It was born in the early seventies out of two very successful technologies: solid state integrated circuits and ion-selective electrodes (ISE). It is still in its infancy but already out of the teething stage and with a very bright prospect ahead.

The chemically sensitive field effect transistor is a member of the family of chemically sensitive solid state devices for which the acronym CSSD was coined by Zemel.[1] Another acronym which will be used throughout this review is IGFET, standing for insulated gate field effect transistor. In order to avoid confusion, it should be mentioned here that the *metal* oxide field effect transistor (MOSFET) and the metal nitride oxide field effect transistor (MNOSFET) are just special cases of IGFET

A considerable amount of work has been done on a MOSFET sensitive to electrically neutral molecules, namely, hydrogen. This transistor is both a MOSFET and CHEMFET because the metal gate is made of palladium, which can be looked at also as a specific hydrogen-sensitive layer.

Jiří Janata • Department of Bioengineering, University of Utah, Salt Lake City, Utah 84112. *Robert J. Huber* • Department of Electrical Engineering, University of Utah, Salt Lake City, Utah 84112.

By far the most widely studied CHEMFET is the ion-sensitive field effect transistor (ISFET), which is often regarded as a second-generation ion-selective electrode (ISE). The similarity between these two types of sensors is a good starting point for an explanation of the operation of the ISFET. The evolutionary chain is depicted in Fig. 1, which shows (a) a schematic representation of a conventional measurement with an ISE (with the electroactive membrane M) and a reference electrode R. The ISE is connected by

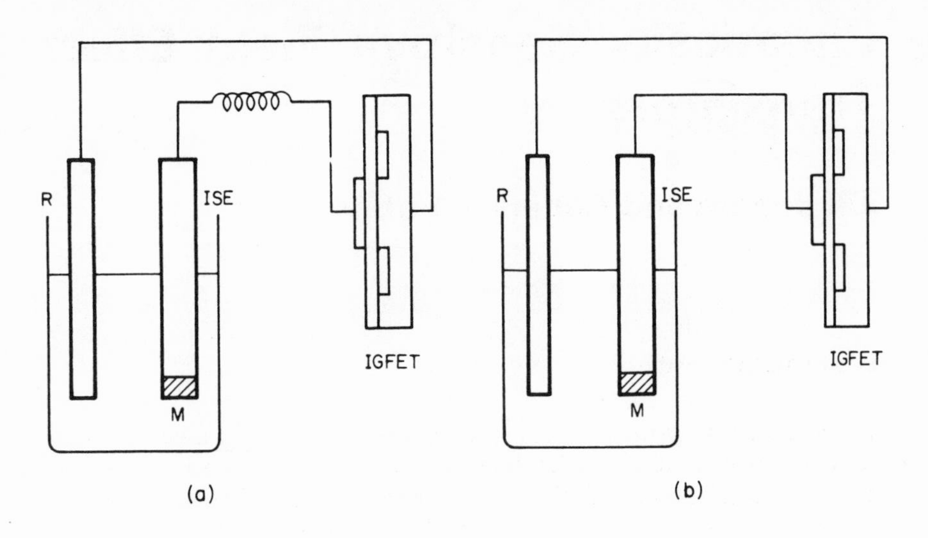

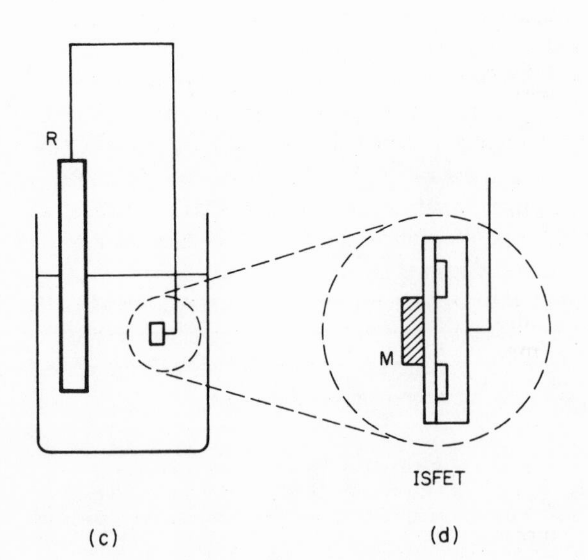

Fig. 1. Evolution of ISFET from ISE and IGFET.

a wire to an FET input of a pH meter. The readout part of the meter is not shown for convenience. In Fig. 1b the lead connecting the ISE is made shorter until it is completely gone; the membrane M is placed directly on the insulator of the FET. Next, the input transistor is placed in the measured solution (Fig. 1c). In this state the integration of the ion-sensitive membrane and the solid state amplifier is accomplished. Figure 1d shows the details of the ISFET, which is very, very small. Thus, in this first crude approximation, the main difference between the ISE and the ISFET is the length of the conductor connecting the membrane to the input of the meter, and the size of the whole assembly. It is interesting to note that all three types corresponding to this diagram exist. There are, of course, many ISE's with solid internal contact (e.g., coated-wire electrodes). A glass membrane connected with a short length of metallic conductor to the FET amplifier input has been described.[2] Similarly, Nagy *et al.*[3] made a fluoride sensor by bonding a small LaF_3 chip to an FET through a short silver lead. Finally, there is now a rapidly growing family of ISFET's which will be discussed in this review.

It must be realized that the thermodynamics of the cell composed of a reference-electrode–solution–ion-sensitive device is not affected by the geometry or size of any of its components (i.e., the length of the conductor between the membrane and the amplifier). In order to fully understand the mechanism of operation of ISFET's it is necessary to perform a complete analysis of the whole measuring circuit, i.e., to move a hypothetical test charge from the semiconductor of the FET through the external circuit, reference electrode, solution, and to the membrane and to describe separately the processes in each phase and at each interface. This will be done in Section 2.3.

Historical Survey

The first report on the chemically sensitive field effect transistor is a short communication by Bergveld.[4] Shortly afterwards Matsuo *et al.*[5] reported a similar device. The first full description of the ISFET sensitive to hydrogen and sodium ions was presented by Bergveld two years later.[6] He used the device for measurement of pH and sodium ion activity as well as for recording physiological transient action potentials. Another pH-sensitive ISFET was described by Matsuo and Wise.[7] There are two major differences between Bergveld's and Matsuo's work that proved to be important for the future development of these devices. Firstly, Bergveld did not use an external reference electrode and indeed has claimed that one is not needed for operation of ISFET.[6] Secondly, Matsuo used silicon nitride as the top layer of the gate insulator, while Bergveld used silicon dioxide. Although Bergveld later moderated his stand on the reference electrode (*vide infra*) his claim attracted wide attention and stimulated a very useful discussion.

The use of silicon nitride proved to be essential for development of ISFET's with ion-sensitive layers deposited over the gate insulator. The first device of this kind was the potassium-ion-sensitive ISFET[8] in which the ion-sensitive layer was the PVC membrane containing valinomycin, which is normally used in conventional potassium-sensitive ISE's. Several other reports describing pH-, K^+-, and Ca^{2+}-sensitive ISFET's have appeared; they will be discussed in the appropriate sections of this review.

Almost every paper on ISFET's published so far gives at least a rudimentary theory of operation of a field effect transistor in its introduction. There are, however, several papers surveying the theory and the mechanism of the operation of ISFET's in greater depth. The first review of the area of chemically sensitive semiconductor devices was written by Zemel.[1] A narrower review focusing on chemically sensitive field effect transistors (CHEMFET's) was written by Janata and Moss,[10] and the mechanism of operation of ISFET was addressed by Kelly.[11] Another article by Zemel[12] more or less reiterates the main points of his previous review, and a paper by Revesz[13] deals particularly with the role of surface states in the mechanism of operation of ISFET. Theoretical analysis of the structure: solution/membrane/insulator/semiconductor supplemented by a thorough experimental study of the system

$$Ag^+ \text{ (or Br-) solution/AgBr/SiO}_2\text{/Si}$$

was published by Buck and Hackleman.[14]

An event of major significance for the development of CSSD's was the International Workshop on Theory, Design, and Biomedical Applications of Solid State Chemical Sensors held at Case Western Reserve University, Cleveland, Ohio, in March 1977. Many important papers addressing theory, practical aspects, and applications of CSSD's are included in the symposium proceedings.[15]

2. THEORY OF IGFET'S

There are several kinds of CSSD discussed in the literature, all of which depend in some way on the semiconductor *field effect* for their operation. The ISFET is no exception. Structurally it is very similar to the IGFET (insulated gate field effect transistor), a device which has been very highly developed as the basis for much of the integrated circuit revolution presently taking place in the field of electronics. The theory of IGFET's is very well developed, although it is generally presented in the language of the electronics engineer or solid state physicist, not the chemist. The purpose of this section is to outline those elements of the theory which are of most use to the reader in understanding and using ISFET's. We begin with a short discussion of the metal–insulator–semiconductor structure.

2.1. MIS Structure

The IGFET theory is based on the analysis of the metal–insulator–semiconductor (MIS) structure, a most useful tool in the study of semiconductor surfaces. The MIS structure is shown in Fig. 2. In its simplest form it consists of a "parallel-plate capacitor," where the metal forms one plate and the semiconductor forms the other. The insulator, of thickness d, is assumed to be a "perfect" insulator, i.e., no charged species can be transported through it. This structure is considered to be a single system for the purpose of a thermodynamic analysis. Equilibrium then refers to the condition in which the free energy of the electron in the metal and in the semiconductor are equal. The free energy of electrons in a solid is usually described in terms of the Fermi energy level, E_F. We adopt the same terminology here. Therefore, the MIS structure is in equilibrium when the Fermi level in the metal and the Fermi level in the semiconductor are equal.

The Ideal MIS Structure

The starting point for the analysis of the MIS structure is the "ideal" MIS, which by definition has the following characteristics: (1) The electron work function of the metal, Φ_m, and that of the semiconductor, Φ_s, are the same, i.e., the metal–semiconductor work function difference, Φ_{ms}, is zero; (2) the charge density in the insulator is everywhere zero; (3) there is no transport of any charged species through the insulator; and (4) the ideal band structure of the bulk semiconductor extends to the surface, i.e., there are no surface states.

The analysis of the ideal MIS structure consists of calculating the charge distribution, electric field distribution, and potential distribution throughout this structure as a function of an externally applied voltage difference between the metal and semiconductor. Once that is done, nonideal effects which always occur in any real system can be easily taken into account and the same three distributions, charge, field, and potential, recalculated. These nonideal effects are nonzero work function difference, nonzero charge density in the insulator, interface charge layers, and deviations from the ideal semiconductor band structure at the surface (surface states).

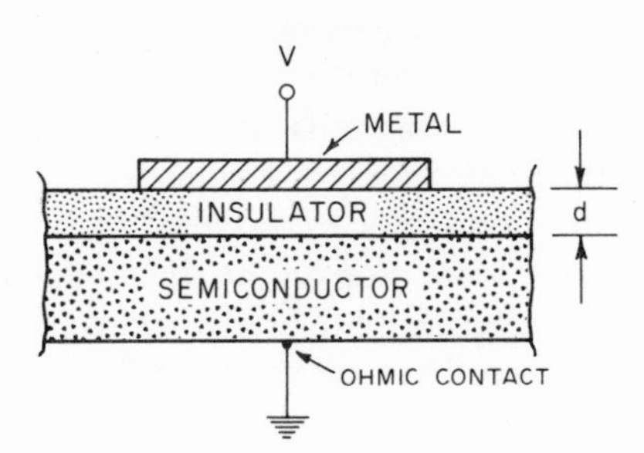

Fig. 2. Metal–insulator–semiconductor structure.

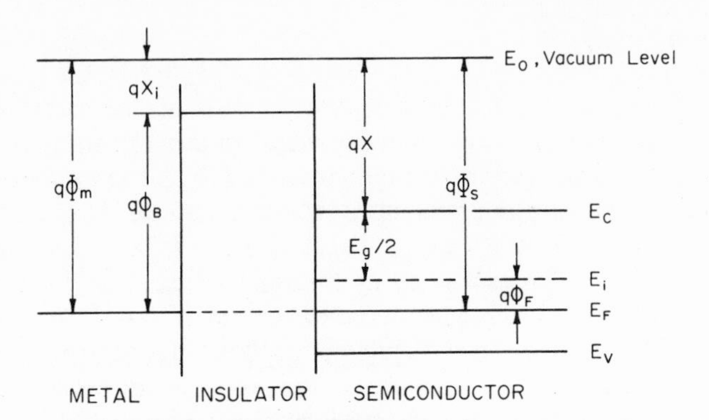

Fig. 3. Energy band diagram of MIS structure for a *p*-type semiconductor. q, electronic charge; Φ_m, metal work function; Φ_s, semiconductor work function; ϕ_B, barrier potential; ϕ_F, Fermi potential; E_g, band-gap energy; χ, electron affinity of the semiconductor; χ_i, insulator electron affinity.

The energy band diagram for the ideal MIS structure is shown in Fig. 3 for a *p*-type semiconductor.

The metal–semiconductor work function difference, Φ_{ms}, is given by

$$\Phi_{ms} = \Phi_m - \left(\chi + \frac{E_g}{2q} + \phi_F\right) \tag{1}$$

It is important to note here that χ is equivalent to the standard chemical potential of the electron, $\mu^{e,0}$, which will be used later.

Again for the ideal MIS, $\Phi_{ms} \equiv 0$. The electron density, hole density, and the net charge density in the semiconductor can be calculated in terms of the energy difference between the Fermi level ϕ_F and any of the other defined levels, E_c, E_v, or E_i. The reader is referred to one of the standard texts on semiconductor physics for a detailed treatment, for example, the text by A. S. Grove.[16]

When a voltage, V_G, is applied between the metal and semiconductor, the "flat-band" condition of Fig. 3 is changed. The Fermi levels in the metal and in the semiconductor become separated by the applied voltage. Since in this treatment it is customary to show increasing electron energy upwards on the band diagrams, applications of a positive voltage to the metal relative to the semiconductor displaces the metal Fermi level downward relative to the semiconductor Fermi level!

Depending on the sign and magnitude of the applied voltage, there are three basically different states of charge distribution in the semiconductor surface. They are shown in Fig. 4. In this diagram the (+) symbols represent holes, mobile positive charges. The (−) symbols just above E_v represent the negatively charged, immobile, acceptor atoms, while the (−) symbols just above E_c represent mobile electrons in the conduction band. Under "accumulation" conditions, the negative voltage applied to the metal attracts additional holes to the semiconductor–insulator interface, i.e., holes accumulate at the interface. Because a very high density of holes is allowed near the top of the valence band, the accumulated layer of the holes need extend only a very short distance into the semiconductor. It behaves much like a metal.

Under "depletion" conditions, the applied voltage V_G drives the mobile holes away from the surface, leaving behind a charge density composed of only the negatively charged acceptor atoms. For semiconductors of use in the fabrication of IGFET, this charge density is very much lower than the corresponding density of holes under accumulation; therefore, the depletion space-charge region extends a considerable distance into the semiconductor. Simple electrostatic theory shows that under these conditions, the energy bands shift an appreciable amount over the extent of the space-charge region, as shown in Fig. 4.

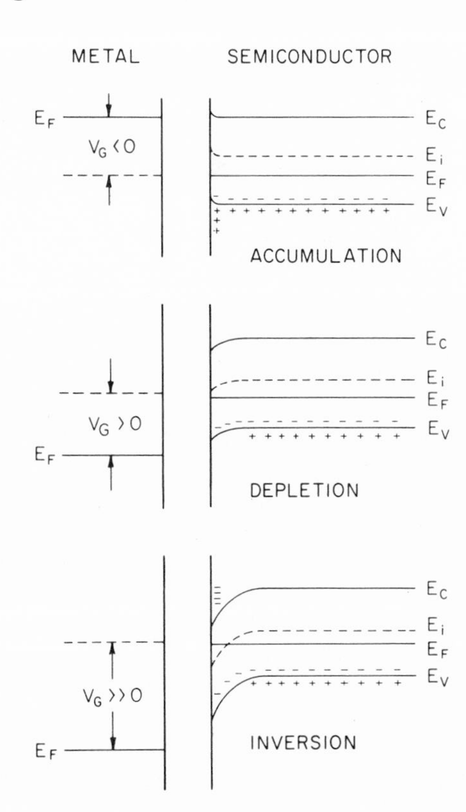

Fig. 4. Energy band diagrams of MIS structure with a *p*-type semiconductor for different biasing conditions.

A sufficiently large positive voltage applied to the metal relative to the semiconductor will bend the bands far enough that the intrinsic level, E_i, crosses the Fermi level, E_F. In general, the hole density depends exponentially on the quantity $(E_i - E_F)$, while the electron density has the same exponential dependence on the quantity $(E_F - E_i)$. Once E_i crosses E_F the electron density exceeds the hole density even though the dopant atoms normally make it *p*-type. The semiconductor is said to be inverted. As the band displacement is greatest at the surface, this inversion occurs at the surface. The band displacement corresponding to inversion can be seen in Fig. 3 to be $q\phi_F$. For the analysis of the IGFET here, a more useful definition is that of "strong inversion," which occurs when the band displacement at the surface is $2\phi_F$. The exponential dependence of electron density on $(E_i - E_F)$ leads to the fact that the surface electron density increases very rapidly beyond the point of strong inversion. In other words, voltages applied to the metal beyond that required for strong inversion increase the electron density in the surface inversion layer but do not increase the band displacement at the surface appreciably. The applied voltage just sufficient to bring about strong inversion is known as the threshold voltage, V_T, and for the ideal MIS structure given by[16]

$$V_T = -\frac{Q_B}{C_0} + 2\phi_F \qquad (2)$$

where Q_B is the charge per unit surface area contained within the surface space-charge region, and C_0 is the capacitance per unit area of the insulator.

The number of mobile electrons per cm^2 in the surface inversion layer becomes a linear function of $(V_G - V_T)$ for values of V_G greater than V_T. The principle of operation of the IGFET is the modulation of electrical conductivity of the surface channel by the applied voltage V_G.

The nonideal effects that exist in a real MIS structure generally result in band bending at the semiconductor surface even with no externally applied voltage V_G. A nonzero work function difference, $\Phi_{ms} \neq 0$, means that a charge flow must take place between the metal and semiconductor to bring about thermal equilibrium. This results in a net charge separation in the system. The excess charge in the metal appears at the metal–insulator interface, while the excess semiconductor charge appears near the insulator–semiconductor interface. The band displacement in the semiconductor is the same as though a voltage equal to Φ_{ms} were applied to the ideal MIS structure. In the equation (2) above V_T is the applied voltage necessary for strong surface inversion starting from the flat-band condition of Fig. 3. If $\Phi_{ms} \neq 0$, an applied voltage equal to Φ_{ms} must first be applied to bring about the flat-band condition, then an additional voltage equal to that given in equation (2) must be applied to bring about strong surface inversions. In this case

$$V_T = \Phi_{ms} - \frac{Q_B}{C_0} + 2\phi_F \quad (3)$$

A nonzero charge density in the insulator can be handled in a similar manner. A charge in the insulator will induce an image charge in both the metal and semiconductor. The induced charge in the surface of the semiconductor will again result in band bending at the surface even for no externally applied voltage. However, it is easy to show[16] that by applying the proper voltage, the image charge in the semiconductor can be canceled out, again bringing about the flat band condition. When both Φ_{ms} and insulator charge distribution are considered,

$$V_T = \Phi_{ms} - \frac{1}{C_0}\int_0^d \frac{x}{d}\rho'(x)\,dx - \frac{Q_B}{C_0} + 2\phi_F \quad (4)$$

where d is the insulator thickness, x the distance measured from the metal–insulator interface, and $\rho'(x)$ is the charge density in the insulator from all sources.

By far the most common MIS structures include thermally grown SiO_2 on silicon. There is a charge distribution associated with this structure that consists of a dense layer of positive charge Q_{ss} in the SiO_2 next to the silicon (effectively at $x = d$). The usual practice in developing this theory defines a new insulator charge density $\rho(x)$ such that

$$\rho'(x) = \rho(x) + \delta(d)Q_{ss} \quad (5)$$

where $\delta(d)$ is the usual "delta" function.

Q_{ss} is the charge per unit area next to the Si/SiO_2 interface. Considering this nonzero charge density,

$$V_T = \Phi_{ms} - \frac{1}{C_0}\int_0^d \frac{x}{d}\rho(x)\,dx - \frac{Q_{ss}}{C_0} - \frac{Q_B}{C_0} + 2\phi_F \quad (6)$$

There is one other nonideal effect that must be considered. It is the deviation from the ideal band structure at the surface in the form of a near-continuous distribution of allowed energy levels across the "forbidden" band at the surface. These are known as "surface states." The influence of surface states on the voltage V_T is complex and is treated elsewhere.[17] Fortunately, modern semiconductor technology can reduce the density of surface states to a negligible level, at least for thermally grown SiO_2 on silicon. The CHEMFET which is the subject of this review is built with this SiO_2/Si system so we will not further consider surface states.

The three terms in the equation for V_T which arise from the nonideal properties of the real MIS are customarily grouped together and collectively called the "flat-band voltage," V_{FB}. This is the voltage which must be applied to a real MIS structure to bring about the flat-band condition:

$$V_T = V_{FB} + 2\phi_F - \frac{Q_B}{C_0} \tag{7}$$

where

$$V_{FB} = \Phi_{ms} - \frac{Q_{ss}}{C_0} - \frac{1}{C_0}\int_0^d \frac{x}{d}\rho(x)\,dx \tag{8}$$

The IGFET consists of a MIS structure with provisions for making electrical contacts to the surface inversions layer. The current voltage relations for the devices are described in Section 2.2. below.

Before proceeding further it may be instructive to comment on the term "field effect transistor." The ability to deplete the density of mobile charge carriers and subsequently invert the surface is known as the surface field effect. The above equations which describe the phenomena could equally well have been derived in terms of electric fields at the surface instead of voltages across a MIS structure. In fact, the density of mobile charges in a surface inversion layer can be considered to be a measure of the electric field normal to the surface of the semiconductor. The surface field can be produced in alternative ways. The semiconductor can be built up into a MIS capacitor and an external potential applied as in the IGFET, or the field can arise from the electrochemical effects between different materials, as in the ISFET.

In either case, changes in the surface electric field change the density of mobile charge carriers in the surface inversion layer. The physical effect that is measured is the change in electric current carried by the surface inversions layer, the drain current. Obviously, these devices must be operated under conditions that cause the surface inversion layer to form.

2.2. Current–Voltage Relationships for the IGFET

Figure 5 shows the typical construction and biasing arrangement for the n-channel IGFET. In effect the semiconductor substrate (3) forms one side of a parallel-plate capacitor and the metal gate field plate (4) forms the other. In the normal operating mode the gate voltage, V_G, is applied between the semiconductor substrate (3) and the gate (4). The polarity and magnitude of V_G are such that the semiconductor field effect gives rise to an inversion layer on the surface of the substrate under the gate. In fact, the surface of the p-type substrate (3) becomes n-type. This n-type inversion layer forms a conducting channel between the source (2) and drain (1) regions. If this inversion layer is not present, application to the drain (1) of a positive

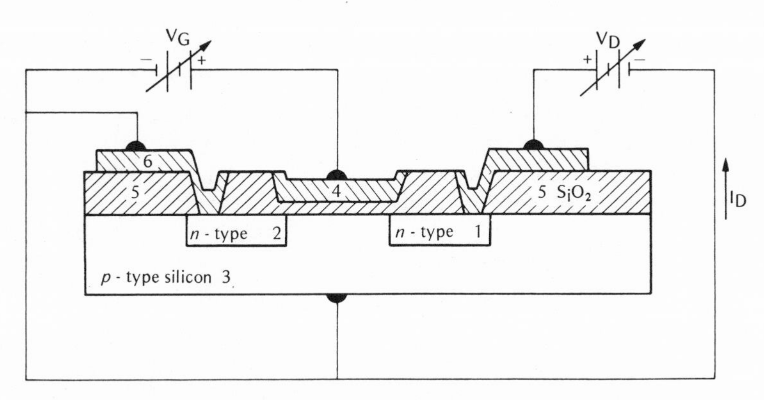

Fig. 5. Schematic diagram of IGFET. (1) Drain, (2) source, (3) substrate, (4) gate, (5) insulator, (6) metal contacts.

voltage V_D with respect to the source (2) and substrate (3) results in no appreciable current flow into drain (1), because the drain-to-substrate p–n junction diode is reverse biased. However, if the n-type inversion layer exists along the surface of the semiconductor between the source (2) and drain (1), a continuous path for the flow of current I_D exists. The current I_D shown in Fig. 5 flows through the channel connecting the drain (1) and source (2). The magnitude of the drain current I_D will be determined by the electrical resistance of the surface inversion layer and the voltage difference, V_D, between the source and drain.

The basis for the derivation of the current–voltage relationships is the calculation of the density of the mobile electrons in the surface inversion layer as a function of the applied voltages V_G and V_D, and position along the channel.

Figure 6 is a schematic representation of the channel region of the IGFET. At point y along the channel there is a charge density $Q_n(y)$ per unit

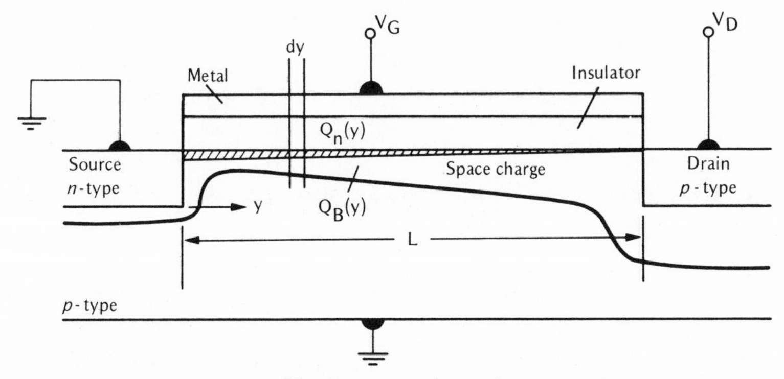

Fig. 6. IGFET channel.

area of mobile electrons in the surface inversion layer, and a charge density $Q_B(y)$ of ionized dopant atoms in the space-charge region. There, charge densities are both a function of y because of the voltage change along the channel due to the drain current. $Q_s(y)$, the sum of $Q_n(y)$ and $Q_B(y)$, is equal and opposite to the charge per unit area of the metal plate which forms the other side of the capacitor (the gate):

$$Q_s(y) = Q_n(y) + Q_B(y) \tag{9}$$

Q_s may be related to the applied gate voltage and the capacitance C_0 of the gate insulator by

$$V_G - V_{FB} = -\frac{Q_s}{C_0} + \phi_s \tag{10}$$

where ϕ_s is the surface potential of the semiconductor.

The mobile electron charge in the surface inversion layer is then seen to be given by

$$Q_n(y) = -[V_G - V_{FB} - \phi_s(y)]C_0 - Q_B(y) \tag{11}$$

In writing these equations we assume that a surface inversion layer does in fact exist. At this point in the derivations the assumption is made that the surface potential is given by the conditions of strong inversion

$$\phi_s(y) = V(y) + 2\phi_F \tag{12}$$

$V(y)$ is the "reverse bias" applied to the "field-effect-induced" p–n junction composed of the n-type surface inversions layer and the p-type substrate.

$Q_B(y)$ is the charge in the space-charge region. Under the "depletion approximation" which assumes that there are no mobile charges at all in the space-charge regions

$$Q_B(y) = -\{2K_s\varepsilon_0 qN_A[V(y) + 2\phi_F]\}^{1/2} \tag{13}$$

Collecting terms,

$$Q_n(y) = -[V_G - V_{FB} - V(y) - 2\phi_F]C_0 + (2K_s\varepsilon_0 qN_A[V(y) + 2\phi_F)]^{1/2} \tag{14}$$

Now referring to Fig. 6, the voltage drop along the element of channel, dy, can be written as

$$dV = I_D\ dR = \frac{I_D\,dy}{W\mu_n Q_n(y)} \tag{15}$$

W is the width of the channel and μ_n is the electron mobility in the channel.

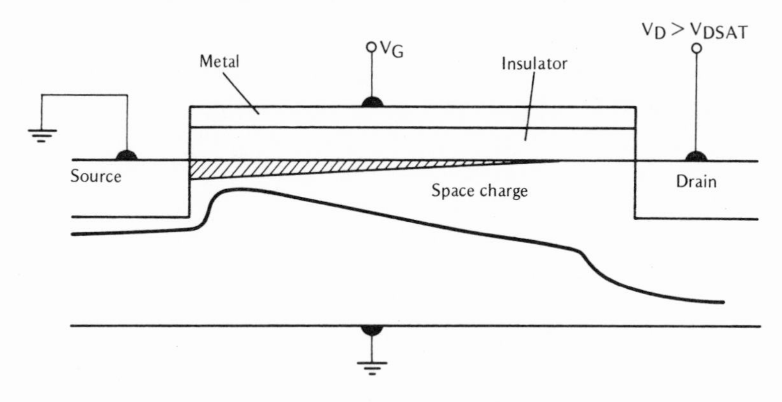

Fig. 7. IGFET in saturation.

Substitution of the above expression for $Q_n(y)$ and integrating over the channel length yields the basic current–voltage relationship:

$$I_D = \frac{W}{L}\mu_n C_0 \left\{ \left[V_G - V_{FB} - 2\phi_F - \frac{V_D}{2} \right] V_D - \frac{2}{3}\left(\frac{2K_s\varepsilon_0 q N_A}{C_0}\right)^{1/2} \left[(V_D + 2\phi_F)^{3/2} - (2\phi_F)^{3/2} \right] \right\} \tag{16}$$

In the derivation of this equation it was assumed that the surface inversion layer actually existed at all points along the channel. In equation (13) it is seen that $Q_B(y)$ increases with $V(y)$, the voltage drop along the channel. This increase is at the expense of $Q_n(y)$, the mobile channel charge. $V(y)$ must change from 0 at the source to V_D at the drain so if V_D is large enough, the conducting channel will disappear near the drain end of the channel. When this happens equation (16) is no longer valid and the IGFET is said to become "saturated."

Figure 7 is a schematic diagram showing the saturated IGFET. V_{DSAT} is the drain voltage at which the channel just disappears at the drain end. Any increase in V_D beyond V_{DSAT} results in a short space-charge region between the drain end of the channel and the drain. Current continues to flow under these conditions because electrons in the channel see no potential barrier restricting flow from the channel across the space-charge region to the drain. However, the number of electrons arriving at the drain end of the channel is determined by the voltage between the source and the end of the surface channel. This voltage is V_{DSAT}. It can be shown to be[16]

$$V_{DSAT} = V_G - V_{FB} - 2\phi_F + \frac{K_s\varepsilon_0 q N_A}{C_0^2}\left[1 - \left(1 + \frac{2C_0^2(V_G - V_{FB})}{K_s\varepsilon_0 q N_A}\right)^{1/2}\right] \tag{17}$$

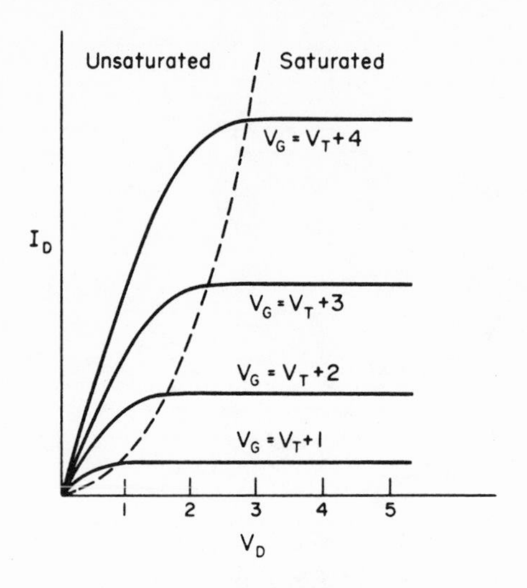

Fig. 8. Change of drain current I_D with drain voltage V_D.

For drain voltages greater than V_{DSAT}, the drain current I_D is given by equation (16) with V_D replaced by V_{DSAT}, provided the channel length L is not appreciably shortened by the space-charge region between the end of the channel and the drain.

The curves in Fig. 8 are divided into the saturated and unsaturated regions by the dotted line. Note that V_{DSAT} is a function of the applied gate voltage.

Even though these equations (16) and (17) are derived using several simplifying assumptions, they are quite combersome because of their complexity. A much simpler set of equations may be derived if the dependence of $Q_B(y)$ [equation (13)] on $V(y)$ is neglected. In this case

$$Q_B = (2K_s\varepsilon_0 qN_A 2\phi_F)^{1/2} \tag{18}$$

and the integrations required to derive the current–voltage relationships can be simplified to yield

$$I_D = \mu_n C_0 \frac{W}{L}\left[(V_G - V_T)V_D - \frac{V_D^2}{2}\right], \qquad V_D < V_{DSAT} \tag{19}$$

where

$$V_T = V_{FB} + 2\phi_F - \frac{Q_B}{C_0} \tag{20}$$

and

$$V_{DSAT} = V_G - V_T$$

If there are no distributed charges in the insulator and no silicon surface states, the flat-band voltage V_{FB} reduces to just two terms:

$$V_{FB} = \Phi_{ms} - \frac{Q_{ss}}{C_0}$$

Equation (19) is the current–voltage relationship for the unsaturated case. For drain voltages greater than V_{DSAT}, the current is given by equation (19) with V_D replaced by V_{DSAT}:

$$I_D = \mu_n C_0 \frac{W}{L} \frac{(V_G - V_T)^2}{2}, \qquad V_D > V_{\text{DSAT}} \tag{21}$$

Equations (19) and (21) are the IGFET current–voltage relations most often quoted in the literature. Their use can lead to significant quantitative errors in calculated currents. They do, however, contain the correct qualitative features of the device, i.e., the saturated and unsaturated regions, the near-constant drain current beyond saturation, and the dependence of the saturation voltage on the gate and threshold voltages. Equations (19) and (21) have the distinct advantage of simple form.

Figure 9 is a plot of I_D vs. V_G for an n-channel depletion mode device. Depletion mode refers to the fact that the device has a conducting channel for zero applied gate voltage. Depletion mode n-channel devices are the result of a sufficiently negative flat-band voltage V_{FB}. In the saturated regions, i.e., $V_D > V_G - V_T$, or alternatively $V_G < V_D + V_T$, equation (21) predicts a quadratic dependence of I_D on V_G as shown. In the unsaturated region, i.e., $V_D < V_G - V_T$ or alternatively $V_G > V_D + V_T$, equation (19)

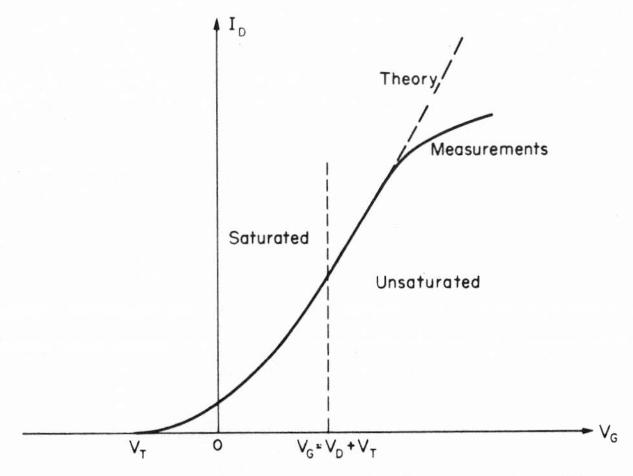

Fig. 9. Change of drain current I_D with gate voltage V_G.

predicts a linear dependence of I_D on V_G as shown. Actual data, however, show a significant departure from linearity in the unsaturated region as indicated in Fig. 9. One cause of this departure from theory is the variation of μ_n, the charge carrier's channel mobility, with the electric field normal to the surface of the semiconductor. Surface channel mobility is primarily determined by surface scattering. Strong electric fields that increase the probability of the carrier interacting with the surface reduce the surface mobility. Therefore, as the gate voltage increases, the electric field normal to the surface increases and μ_n decreases.

A second cause of the departure of the I_D–V_G curve from linearity is series resistance between the end of the channel and the point at which the drain voltage is actually measured. Schematically this is represented as shown in Fig. 10.

The resistive voltage drop actually reduces the effective drain voltage below that applied to the device. The magnitude of this resistance depends on both the individual device geometry and processing parameters, but it may be quite significant. In the case of the ISFET's reported by Janata and Moss,[10] the effective R of Fig. 9 can be 200 Ω. Their experimental data begin to depart from linearity for drain currents of about 2 mA, a value at which the 200 Ω has an appreciable effect.

As a result of the effects of nonideal structures, second-order effects in parameters, and the numerous approximations made in the derivation of the current–voltage equations, equations (19) and (21) can only serve as a qualitative description of the actual device. Each individual design must be experimentally characterized.

For these reasons, and others to be discussed in Section 2.3.4, it is advantageous to operate the ISFET in the constant drain current mode.

Summary of Assumptions

The derivation of equations (19) and (20) contains the most simplifying assumptions of any of the widely used sets. The most important are discussed below.

(a) It is assumed that there exists a well-defined threshold voltage, and that the formation of the surface inversion layer begins suddenly as the gate voltage is increased. It is equivalent to stating that there is a sharply defined semiconductor surface potential, dividing surface depletion and surface inversion. In fact, this transition is continuous. In conventional structures this is a good approximation if the gate voltage exceeds the "threshold voltage" by about 0.5 V.

(b) The voltage dropped due to a surface channel current (drain current) flowing has no effect on the thickness of the surface space-charge region. This approximation can lead to relatively large errors in the

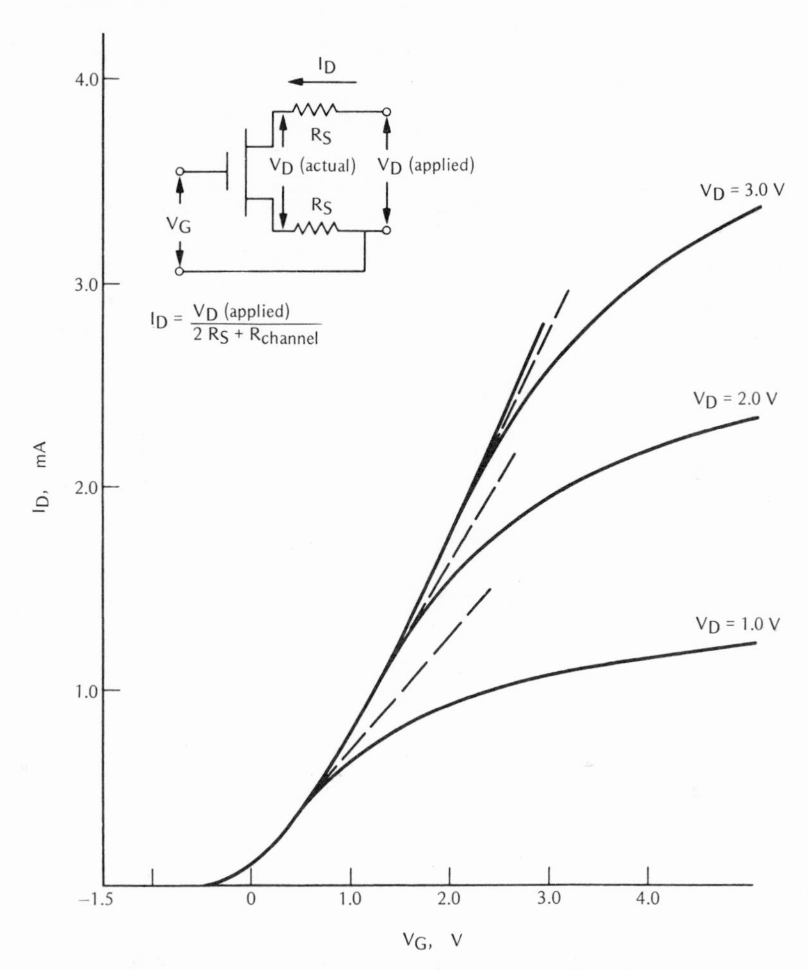

Fig. 10. Effect of series drain resistance. Theoretical response for $R_S = 0$ (– – –).

magnitude of the predicted drain current in the saturation region, but the general shape of the drain current vs. drain voltage curves is satisfactory, i.e., the qualitative features of the device are not affected by this approximation.

(c) The doping of the semiconductor is constant near the surface where the channel is formed. This is not a good approximation for many ion-implanted structures.

(d) The channel length is large compared to the thickness of the depletion region surrounding the *p–n* junctions.

(e) Both the source and substrate of the device are connected to the same point of the external circuit.

Under assumption (a) above an equation relating the threshold voltage, V_T, to the device structure is given by equation (20).

2.3. CHEMFET

In the case of the CHEMFET the gate metal is replaced by a more complex structure consisting of chemically sensitive layer (which may be part of the uppermost layer of the gate insulator), the solution containing the ions of interest, and a reference electrode. Those terms in equation (11) which are unique to the metal gate must be replaced with quantities which describe the more complex CHEMFET system.

According to the nature of the interaction between the species to be detected and the chemically sensitive layer, CHEMFET's can be divided into three groups: ones that will measure electrically neutral molecules, ISFET's that are analogs of ISE's, and CHEMFET's with polarized solution/membrane interfaces that do not have a conventional electroanalytical counterpart.

2.3.1. CHEMFET Sensitive to Neutral Species

One CHEMFET structure sensitive to neutral species is the IGFET with a palladium gate.[9] Neutral hydrogen molecules under suitable conditions are reversibly absorbed by the Pd which[18] changes the flat-band voltage, V_{FB}, of the Pd–SiO_2–Si structure, equation (2). As can be seen by inspection of equation (8), there is a corresponding change in the IGFET drain current, all other parameters remaining fixed.

It is important to note that even though the hydrogen enters the Pd as a neutral species and changes V_{FB}, the Pd gate H_2-sensitive IGFET must be operated with a charge return path in order that the gate voltage be well defined in equation (8) above.

The flat-band voltage V_{FB} is the sum of several terms. They are the metal–semiconductor work function difference, electrostatic effect of distributed charges in the insulator layer, and interface states at the insulator–semiconductor boundary. By far the best-developed IGFET structure, and that currently used for all CHEMFET's of practical significance, is the SiO_2–Si system. The microelectronics industry has achieved a detailed empirical, if not theoretical, understanding of this system.

It has been suggested[19] that "the change in energy barrier at the Pd–SiO_2 interface is attributed to an interfacial polarization layer and not to a change of the bulk metal Fermi level."

This cannot be the case because the flat-band voltage, V_{FB}, is not a function of this barrier energy. This is because charge equilibrated in the IGFET system takes place via the charge return path which consists of an electrical connection from the metal gate to the silicon. Charge equilibration does not take place by charge transport through the insulator; therefore, the barrier height for electrons between the Pd and SiO_2 cannot effect V_T.

If the hydrogen induces a charge separation across the whole SiO_2 layer the situation is different. This would introduce a net charge on both the metal and semiconductor, thereby changing the equilibrium charge distribution on the system. A simple electrostatic analysis[16] shows that the effect on V_{FB} of a net charge in the oxide is proportional to the distance of the charge from the metal–oxide interface. If the charge in the oxide is very close to the metal, the change in V_{FB} is negligible. Thus a dipole located only at the metal/oxide interface has no effect on V_{FB}.

It has also been suggested[20] that neutral hydrogen atoms from the Pd–H_2 interaction enter the SiO_2 where they migrate to the SiO_2–Si interface and alter the interface state conditions and that this is the mechanism that results in the change in V_{FB}. However, this is contrary to the general experience of the silicon MOS integrated circuits industry. Hydrogen is widely used to "quench" the interface states in the Si–SiO_2 system. Proper annealing of the structure in H_2 at elevated temperatures (above 400 °C) will reduce the electrical effects of interface states to negligible levels and the long-term stability (years) of the devices at 150 °C or below attests to the stability of the hydrogen-quenched interface states. It seems unlikely that H atoms generated by the H_2–Pd interactions would have greater mobility in SiO_2 and be less stable at the SiO_2–Si interface than H from the more conventional annealing steps. If the H enters the oxide as a charged species with sufficient mobility to account for the rapid response of the H_2-sensitive IGFET, it would also drift rapidly under the applied gate voltage, making V_{FB} dependent on the magnitude of V_G. The effect of H_2 on the flat-band voltage is probably not a result of charged hydrogen species in the oxide.

It appears that the only acceptable mechanism involves the change of the work function difference between palladium and silicon.[9,19–22] This change can be affected only through either the change of the bulk chemical

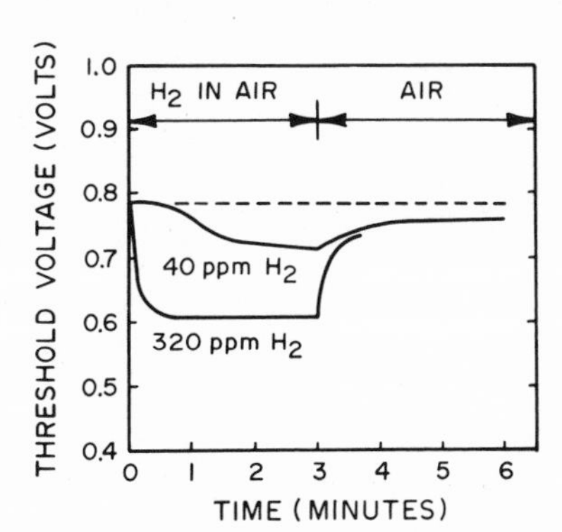

Fig. 11. Typical transient response of Pd gate MOSFET to hydrogen and 150 °C. (Reprinted with permission from reference 9.)

potential of electron in palladium or, more likely, through the change of the contact potential between palladium and the contact metal.

The response of the Pd gate transistor to hydrogen is shown in Fig. 11. The application of this sensor to smoke detection has been described.[23] Another secondary response is that to gaseous H_2S.[24] Hydrogen sulfide readily dissociates at the palladium surface to sulfur and atomic hydrogen, which then diffuses into Pd. The sensitivity of this device in air at 150 °C is 1 mV (ΔV_T) per 2 ppm H_2S. The use of these devices for detection of other dissociable molecules such as NH_3 has been also suggested.[25]

2.3.2. CHEMFET with Nonpolarized Interface—ISFET

In the ISFET (Fig. 12) the metal gate is replaced with a reference electrode (7), solution (8), and an ion-sensitive membrane (9). The rest of the device is protected by a suitable encapsulant (10). The heart of the ISFET is the gate. It has been shown in the preceding section how the gate voltage

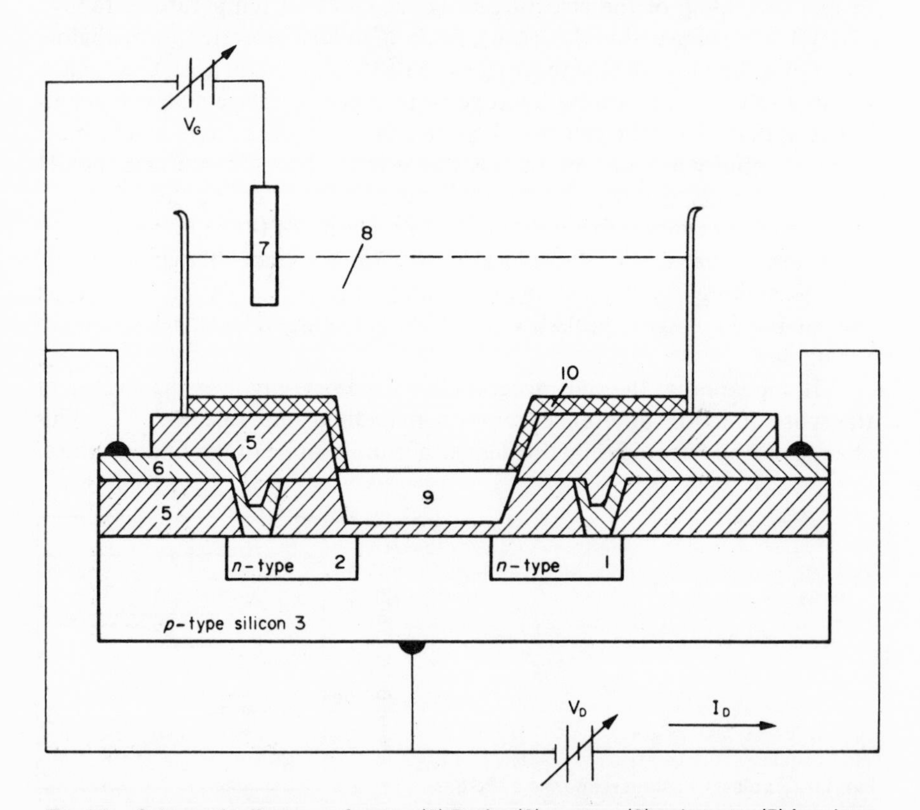

Fig. 12. Schematic diagram of ISFET. (1) Drain, (2) source, (3) substrate, (5) insulator, (6) metal lead, (7) reference electrode, (8) solution, (9) membrane, (10) encapsulant. Note that 7, 8, and 9 replace the metal gate 4 of the IGFET (Fig. 5).

V_G controls the drain current I_D in the transistor equations (19) and (21). In order to describe quantitatively the mechanism of operation of the ISFET it is necessary to perform the thermodynamical analysis of the structure shown in Fig. 13, which represents the cross section through the *whole* measuring circuit including the reference electrode and the connecting leads. It is the simplest case, one where the reference electrode (1) is of the first kind, described by the equilibrium

$$M \rightleftharpoons M^+ + e^-$$

A typical example of such an electrode would be a silver wire immersed in solution of silver ions. Let us further assume that the solution (2) also contains a small amount of ions which can permeate reversibly into and out of the membrane (3) which, therefore, forms a nonpolarized interface. A possible example would be a solution (2) containing 0.1 *M* $AgNO_3$ and 1×10^{-4} *M* KNO_3 and the membrane (3) being potassium ion sensitive such as used in the equivalent ISE. The insulator (4) is assumed to be ideal, i.e., no charge can cross it and it is thicker than electron tunneling distance ($d \geq$ 100 Å).† Layer (5) is the transistor semiconductor (such as silicon). The metal (6) will be identical with metal (1). A switch Sw represents operation with (Sw closed), and without (Sw open) reference electrode. The charge, field, and potential profiles across this structure are also shown in Fig. 13. Note that this is a very simplified case. A liquid junction, dual-layer insulator, trapped charges in the insulator, surface states at the insulator/semiconductor interface, channel doping, and a multitude of connecting metals have been omitted for the sake of simplicity. Similar charge, field, and potential profiles taking into account some of these elements have been published.[10,11,14,26]

From the thermodynamics point of view, this is a multiphase system for which the *Gibbs equation* must apply at equilibrium:

$$\sum_i dn_i \, \tilde{\mu}_i = 0 \tag{22}$$

where dn_i is the number of species i transported across individual interfaces and $\tilde{\mu}_i$ is the electrochemical potential of species i. The equation (22) can be expanded to yield the relationship

$$\tilde{\mu}_3 \equiv \tilde{\mu}_2 \equiv \tilde{\mu}_1 \equiv \tilde{\mu}_6 \equiv \tilde{\mu}_5 \tag{23}$$

† In their recent paper, Bergveld *et al.*[20] proposed a schematic diagram representing all CSSD's including ISFET (Fig. 2 in reference 20). In their structure the measured species are shown to move within the *whole* structure of the device. We suggest that if these species carry an electric charge Bergveld's structure cannot represent CHEMFET in any form because charge cannot move within the gate insulator nor can it cross its boundaries with the adjacent layers. This point has implications on the equilibrium condition of CHEMFET (*vide infra*).

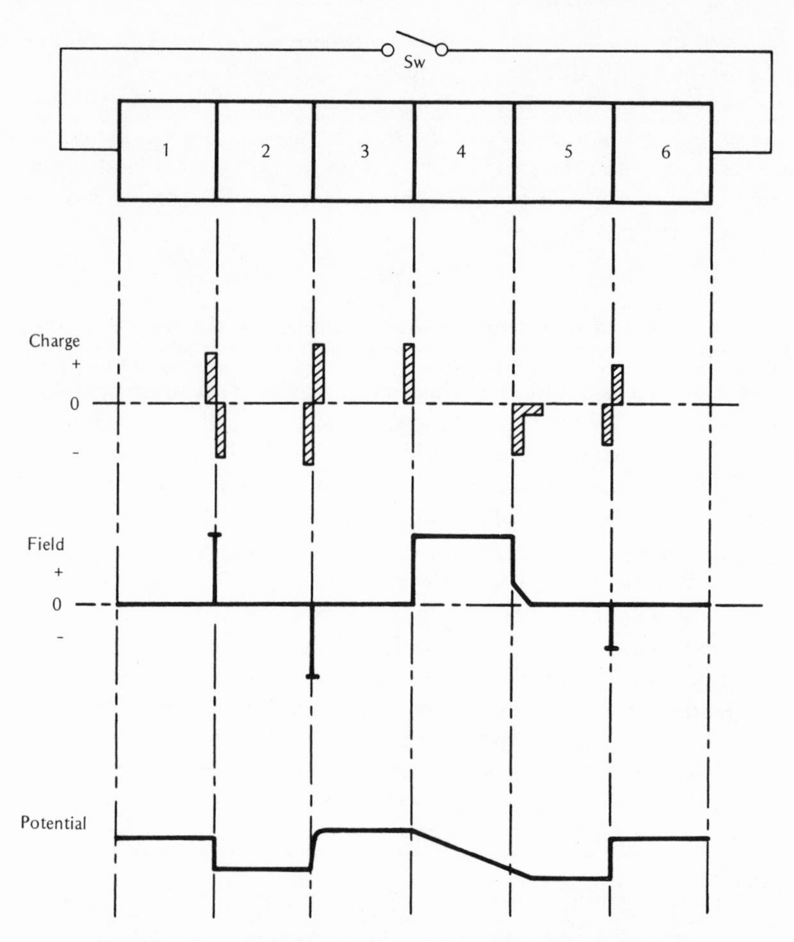

Fig. 13. Charge, field, and potential profiles of the ISFET gate.

which simply expresses the fact that in the case of ideal insulator (4) the only way this system can reach equilibrium is through the external parthway, (1), (6). If the switch Sw is open, a situation equivalent to operation without a reference electrode (or a signal return), then

$$\tilde{\mu}_1 \neq \tilde{\mu}_6 \tag{24}$$

The inequality of Fermi levels in metals (1) and (6) results in inequality of $\tilde{\mu}_3$ and $\tilde{\mu}_5$. Thus the basic condition for stable operation of ISFET is not satisfied.

As was mentioned in the Introduction, it was claimed originally[1,4,6,12,13] that ISFET's can be operated without a reference electrode as a single-ended probe. This point of view has been contested[8,10,11,14] and it is now agreed that a signal return (reference electrode) is needed for a stable, dc operation of ISFET's. However, controversy still exists as to whether ISFET's sensitive to electrically *neutral species* can or cannot be operated without a reference electrode. This point has been addressed in Section 2.3.1.

Let us now analyze the circuit in Fig. 13. The inner potential in the semiconductor (5) can be expressed as

$$\phi_5 = \frac{1}{F}(\mu_5^e - \tilde{\mu}_5^e) \tag{25}$$

where μ_5^e is the chemical potential of an electron in the semiconductor (the electron–lattice interaction energy) and $\tilde{\mu}_5^e$ is the electrochemical potential of an electron in phase 5, normally known as the Fermi level. Similarly, for membrane (3) the inner potential ϕ_3 is

$$\phi_3 = \frac{1}{z^i F}(\tilde{\mu}_3^i - \mu_3^i) \tag{26}$$

where z^i, the number of elementary charges, is positive for cations and negative for anions, and $\tilde{\mu}_3^i$ and μ_3^i are the electrochemical and chemical potentials of species i in phase 3, respectively. The potential difference across the insulator and the semiconductor space-charge region is then

$$\phi_5 - \phi_3 = \frac{1}{F}(\mu_5^e - \tilde{\mu}_5^e) - \frac{1}{z^i F}(\tilde{\mu}_3^i - \mu_3^i) \tag{27}$$

It is now essential to identify the relationship between species i in the membrane (3) and electron in the semiconductor (5).

We know that ion i can transfer from the solution (2) into the membrane (3); thus, according to equation (23), its electrochemical potentials in the two phases must be equal:

$$\tilde{\mu}_3^i = \tilde{\mu}_2^i = \mu_2^i + z^i F \phi_2 \tag{28}$$

where ϕ_2 is the inner potential in the solution. Similarly, Fermi levels in the semiconductor and the metal (6) are equal. Because we defined the metal (6) to be the same as the metal (1) (the reference electrode) we can write

$$\tilde{\mu}_5^e = \tilde{\mu}_1^e = \mu_1^e - F\phi_1 \tag{29}$$

There is an equilibrium in the metal between the cations and electrons

$$M_1 \rightleftharpoons M_1^+ + e_1^-$$

for which we can formally write

$$\mu_1^M = \mu_1^{M+} + \mu_1^e \tag{30}$$

Substituting for μ_1^e in equation (29) we have

$$\tilde{\mu}_5^e = \mu_1^M - \mu_1^{M+} - F\phi_1 \tag{31}$$

Combining equations (27), (29), (31), and rearranging yields

$$\phi_5 - \phi_3 = \frac{1}{F}(\mu_5^e - \mu_1^M + \mu_1^{M+}) - \frac{1}{z^i F}(\mu_2^i - \mu_3^i) + (\phi_1 - \phi_2) \tag{32}$$

The first term on the right-hand side of equation (32) represents the *contact potential* between semiconductor and metal:

$$\phi_5 - \phi_1 = \frac{1}{F}(\mu_5^e - \mu_1^e) = \Delta\phi_{\text{cont}} \tag{33}$$

The second term can be related to the solution activity of the ion (Nernst equation):

$$\frac{1}{z^i F}(\mu_2^i - \mu_3^i) = E_0^i + \frac{RT}{z^i F}\ln a_2^i \tag{34}$$

where a_3^i is assumed to be constant and is included in term E_0^i. Finally, $\phi_1 - \phi_2$ is the reference electrode potential E_{ref}. Equation (32) can now be written as

$$\Delta\phi_{3/5} = \phi_3 - \phi_5 = \Delta\phi_{\text{cont}} + E_0^i + \frac{RT}{z^i F}\ln a_2^i - E_{\text{ref}} \tag{35}$$

This equation is essentially identical to that derived by Buck and Hackleman.[14] The voltage across the gate insulator $\Delta\phi_{3/5}$ can be superimposed on the externally applied voltage V_G, which has the same meaning and function as defined in the theory of operation of the IGFET (Section 2).

Let us now combine equations (19) and (35):

$$I_{DS} = \frac{\mu_n C_0 W V_D}{L}\left(V_G + \Delta\phi_{\text{cont}} + E_0^i + \frac{RT}{z^i F}\ln a_2^i - E_{\text{ref}} + \frac{Q_{ss}}{C_0} - 2\phi_F + \frac{Q_B}{C_0} - \frac{V_D}{2}\right) \tag{36}$$

Define V_T^* for the ISFET,

$$V_T^* = -\Delta\phi_{\text{cont}} - E_0^i - \frac{Q_{ss}}{C_0} + 2\phi_F - \frac{Q_B}{C_0} \tag{37}$$

The inclusion of the term E_0^i (but not E_{ref}) in the threshold voltage is rather arbitrary. The reason is that the membrane is *physically* part of the ISFET and thus its standard potential should be included in the constant V_T^*.

On the other hand, the reference electrode is a completely separate structure. The final equation for the drain current of the ISFET sensitive to the activity of ions i is

$$I_D = \frac{\mu_n C_0 W V_D}{L}\left(V_G - V_T^* \pm \frac{RT}{z^i F}\ln a_2^i - E_{\text{ref}} - \frac{V_D}{2}\right) \tag{38}$$

for operation in the nonsaturation region and

$$I_D = \frac{\mu_n C_0 W}{2L}\left(V_G - V_T^* \pm \frac{RT}{z^i F}\ln a_2^i - E_{\text{ref}}\right)^2 \tag{39}$$

for operation in the saturation mode.

In the previously derived drain current equations the reference electrode potential was missing[6] and the poorly defined "membrane-semiconductor"[10,27] or "liquid-semiconductor"[28] *work function* difference was used instead of the *contact potential.*

We have frequently encountered the opinion that the thermodynamics of ISFET's is undefined because the "reversible potential cannot be defined at the membrane/insulator interface." Electroanalytical chemists often refer to *nonpolarized* electrodes as "reversible." A *polarized* electrode is also thermodynamically reversible, but has one more variable (usually charge) than a *nonpolarized* electrode, which means that the interfacial potential can be *measured* directly *only if* the charge density can be kept constant. If it is not, the potential is observed to "drift." This point has been discussed in greater detail elsewhere.[26] What is meant by "nonreversible membrane/insulator interface" is in fact that it is a polarized interface, i.e., that a charged species cannot cross it. It is indeed true because the insulator is considered to be *ideal* and for the voltages involved in these measurements it *behaves like an ideal* insulator. In short, we do not need a *nonpolarized* interface because the charge transport stops at this interface.

2.3.3. CHEMFET with Polarized Solution/Membrane Interface

If charge cannot cross from the bulk of the solution to the bulk of the solid (or immiscible liquid) the interface between these two phases is said to be polarized and is electrically equivalent to a capacitor. In the ensuing discussion we shall consider such an interface to be between liquid and solid phases. The polarizability of such an interface depends on the electronic state of the solid phase, on the composition of the liquid phase and on the difference between the bulk potentials of the two phases. Thus at constant temperature and pressure and for constant composition of both solid and liquid phases, the interface can be polarized, nonpolarized, or partially polarized depending on the potential difference. There is, however, a significant difference between the CHEMFET and a conventional polarized metal or semiconductor electrode. Although a net charge transfer does not take place across the solid/solution interface in either case, a conventional electrode is a *conductor*. On the other hand, the interface between the CHEMFET and the solution is of the *insulator*/solution type. It is, of course, possible to deposit a metal or a semiconductor layer on top of the gate insulator and to study it as a truly *blocked* electrode. This subtle but important difference between the conventional and CHEMFET electrodes introduces some changes into the standard treatment of the polarized electrode. The overall approach, however, is the same.

The simplest case of a transistor with a polarized interface is one with the gate insulator in contact with solution. It must be stressed, however, that a material which is an insulator in the dry state can hydrate when exposed to water. If that happens, the hydrated layer is invariably so thick that it represents a new phase which has a nonpolarized solution interface. Silicon dioxide and silicon nitride, which are discussed in Section 3.2.1, are prime examples of this situation. It is therefore assumed in the ensuing discussion that the insulator *remains* insulating, which means that the insulator/solution interface is polarized.

The polarized interface is described by the Gibbs–Lippmann equation, which at constant pressure and temperature is

$$-d\gamma = Q_i\,dE + \sum_i \Gamma_i\,d\mu_i \tag{40}$$

where γ is the surface energy, and Q_i is the charge density at the polarized solid/solution interface. Surface excess, Γ_i, is defined with respect to a component of the interface which is in large excess, typically the solvent

$$\Gamma_i = \Gamma_i^* - \frac{n_i}{n_o}\Gamma_o^* \tag{41}$$

where $\Gamma_{i,o}^*$ are surface concentrations of *i*th and *o*th species. It should be noted that compared to the Nernst equation, the Gibbs–Lippmann equation has an additional variable, charge, which corresponds to an additional degree of freedom of the polarized interface.

By definition an interface represents a nonhomogeneity which results in a different distribution of charged species from that in the bulk of the phase. The exact mode of interaction depends on specific long-range forces of attraction. As an illustration, we shall consider two extreme cases, nonelectrostatic and electrostatic interactions.

If the free energy of adsorption is dominated by a nonelectrostatic contribution, if there are no lateral interactions between the adsorbing species, *h*, and if all adsorbing sites, *S*, are equal, then the adsorption equilibrium

$$h + S \overset{K}{\rightleftharpoons} A$$

can be expressed by the Langmuir adsorption isotherm. In a crude approximation the total excess charge density Q_i at the surface is the sum of charge contributed by the sites and the charge contributed by the adsorbed species:

$$Q_i = Q_A + Q_S \tag{42}$$

According to our assumptions the surface activity of the binding sites, a_S, is

$$a_S = a_S^T - a_A \tag{43}$$

where a_S^T is the initial (total) surface activity of binding sites and a_A is the surface activity of adsorbed species. At equilibrium

$$K = \frac{a_A}{a_h a_S} \tag{44}$$

Equation (44) combined with equations (42) and (43) gives

$$Q_i = (z_A - z_S)F\frac{Ka_h a_S^T}{1 + Ka_h} + z_S F a_S^T \tag{45}$$

If the surface coverage is low, $Ka_h \ll 1$, then

$$Q_i = (z_A - z_S)FKa_h a_S^T + z_S F a_S^T \tag{46}$$

If the interaction of the adsorbing species with the solid/solution interface is governed mainly by long-range electrostatic forces, the change of the interfacial potential difference with the change of bulk activity of adsorbing species is expressed by the Esin–Markov coefficient,

$$\left(\frac{\delta E}{\delta \ln a_i}\right)_{q_i} = -RT\left(\frac{\delta \Gamma_i}{\delta q_i}\right)_{a_i} + \frac{RT}{z^i F} \tag{47}$$

where E is the potential of the polarized electrode measured against a constant potential reference electrode (e.g., calomel electrode) and q_i is the total charge at the interface, typically on metal. The exact form of the partial differential $(\delta\Gamma_i/\delta q_i)_{a_i}$ depends on the type of adsorption isotherm. The Esin–Markov coefficient is therefore an important tool in the study of the structure of the double layer. The full discussion of this topic can be found in reference 29. It is important to note that the Esin–Markov coefficient is defined for constant interfacial charge [equation (47)]. With conventional metal electrodes it can be evaluated only indirectly, by measuring the differential capacitance at different applied potentials and different concentrations of adsorbing species. It will be shown below that the CHEMFET can be used for a direct determination of the Esin–Markov coefficient.

2.3.4. Measurement of Charge with CHEMFET

The insulated gate field effect transistor is basically a charge-measuring device. The relationship between charge, applied voltage, and capacitance of the gate is expressed by equation (10).

In the case of a strong specific adsorption which follows a Langmuir adsorption isotherm, the change of drain current with change of bulk concentration of adsorbing species can be obtained by combining equations (36) and (46):

$$I_D = \frac{W\mu_n C_0 V_D}{L}\left[V_G - V_T^* - E_{\text{ref}} + \frac{(z_A - z_S)FKa_h a_S^T}{C_0} - \frac{V_D}{2} + \frac{z_a F a_S^T}{C_0}\right] \quad (48)$$

where

$$V_T^* = \phi_{\text{cont}} - \frac{Q_{ss}}{C_0} + 2\phi_F = \frac{Q_B}{C_0} \quad (49)$$

Similarly, for saturation

$$I_D = \frac{W\mu_n C_0}{2L}\left[V_G - V_T^* - E_{\text{ref}} + \frac{(z_A + z_S)FKa_h a_S^T}{C_0} + \frac{z_S F a_S^T}{C_0}\right]^2 \quad (50)$$

In the more general case where long-range electrostatic forces affect the structure and composition of the double layer, the requirement of constant interfacial charge, as derived for the Esin–Markov coefficient, must be observed.

The charge and potential profiles across the CHEMFET gate with a polarized interface are shown in Fig. 14. The applied gate voltage V_G is the sum of the reference electrode potential ϕ_R, double-layer potential ϕ_{dl}, potential drop across insulator ϕ_i, and the surface potential ϕ_s. Thus

$$V_G = \phi_R + \phi_{dl} + \phi_i + \phi_s \quad (51)$$

The double-layer potential ϕ_{dl} is identical to the potential of the polarized electrode E in equation (40).

$$\phi_{dl} = E = V_G - \phi_0 - \phi_R - \phi_s \quad (52)$$

The potential ϕ_i is related to the gate capacitance and the total charge q_i at the semiconductor side of the insulator:

$$\phi_i = \frac{q_i d}{\varepsilon_i} \quad (53)$$

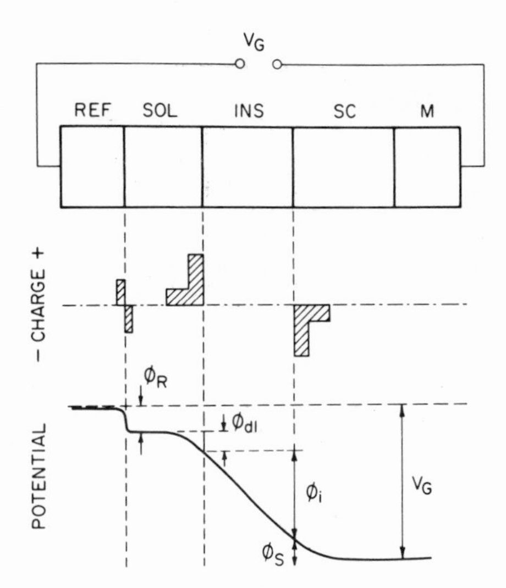

Fig. 14. Charge and potential profiles across CHEMFET gate with polarized interface. REF, reference electrode; SOL, solution; INS, insulator; SC, semiconductor; M, metal.

where d is the insulator thickness and ε_i is its permittivity. It is important to note that when charge q_i is held constant (therefore ϕ_0 is constant), the change of ϕ_{dl} with bulk activity of adsorbing species is the Esin–Markov coefficient, because ϕ_R and ϕ_s are constant. Thus

$$\left(\frac{\delta E}{\delta \ln a_i}\right)_{P,T,q_i} \equiv \left(\frac{\delta \phi_{dl}}{\delta \ln a_i}\right)_{P,T,q_i} = \left(\frac{\delta V_G}{\delta \ln a_i}\right)_{P,T,q_i}$$

$$= -RT\left(\frac{\delta \Gamma_i}{\delta q_i}\right)_{\mu} + \frac{RT}{z^i F} \tag{54}$$

The principle of its *direct* measurement is shown in Fig. 15. For a given composition of the solution and a given applied gate voltage V_G, the value of the drain current is set at I_D = const (curve 1). When more cations adsorb at the interface the potential ϕ_{dl} increases, which would result in the corresponding increase of the drain current (curve 2). In order to maintain a constant interfacial charge the drain current is held constant at its original value by adjusting V_G (curve 3). It can be seen that the drain current thus serves as a third experimental variable, the first two being bulk activity and applied gate voltage, respectively. Therefore the three degrees of freedom in the Gibbs–Lippmann equation, charge, potential, and activity (at constant P and T), are matched by three experimental variables. In this way the transistor is used in a constant-current mode, which is a preferred mode of operation even for ISFET's. A simple measuring circuit that allows this mode of operation is discussed in Section 3.8.

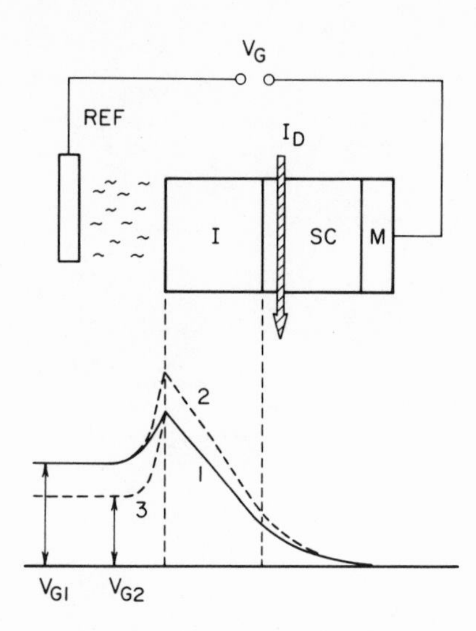

Fig. 15. Measurement of the interfacial charge with CHEMFET.

2.3.5. Temperature Dependence of the ISFET

This discussion is based on the strong similarity between the IGFET and the ISFET.

As both drain current equations (19) and (21) depend only on the quantities $\beta = \mu_n C_0(W/L)$, V_T, and applied voltages, any temperature dependence must be through the quantities β and V_T.

Since C_0 is the gate insulator capacitance/unit area,

$$C_0 = \frac{\varepsilon_i}{d_{\text{ins}}}$$

where ε_i is the permittivity of the insulator and d_{ins} is the thickness of the insulator.

These parameters are relatively constant with temperature in the range of interest so any temperature variation of β must be through μ_n, the charge carrier mobility in the surface inversion layer. This quantity has been experimentally studied as a function of temperature.[78,79] In the temperature region of practical interest, -55 to $+120\,°\text{C}$, the variation of μ_n with temperature is satisfactorily approximated by a T^{-1} dependence. Therefore, the variation of mobility with absolute temperature, T, will introduce a T^{-1} dependence into I_D.

Of the terms in the equation for V_T, the temperature dependence of ϕ_F and ϕ_B must be considered. The fixed oxide charge Q_{ss} is observed to be relatively independent of temperature over the temperature range of interest.[80] When dealing with the IGFET, Φ_{ms} can be considered independent of temperature. However, when the expression for V_T is modified for the

ISFET, the terms which replace Φ_{ms} in the threshold equation are temperature dependent [see equation (37), Section 2.3.2].

The ISFET drain current equation (38) may now be differentiated with respect to temperature to show the temperature dependence[10]

$$\frac{dI_D}{dT} = I_D\left\{\frac{1}{\mu_n}\frac{d\mu_n}{dT} - \frac{V_D[dV_T/T - d(\text{EMF})/dT]}{V_G - V_T^* \pm (RT/z^iF)(\ln a_2^i)_{,} - E_{\text{ref}} - V_D/2}\right\} \quad (55)$$

The term $d(\text{EMF})/dT$ is the temperature dependence of the electrochemical cell consisting of the reference electrode, the electrolyte, and the ion-sensitive membrane.

The term dV_T/dT in the above equation accounts for the temperature dependence of the semiconductor band gap and bulk Fermi level. It is the same as the temperature variation of the IGFET threshold equation. It is always negative for n-channel structures.

Inspection of equation (55) shows that given $d\mu_n/dT$ always negative and the term $[dV_T/dT - d(\text{EMF})/dT]$ always negative[10] it is possible to choose an operating point around which I_D is, to first order, independent of temperature. However, it is known that the term $d(\text{EMF})/dT$ varies with the activity of the measured ion and is equal to zero only for one value of activity. The effect of temperature on the drain current can therefore be minimized by the proper selection of V_G, but not completely eliminated.

3. PRACTICE

3.1. CHEMFET Geometry and Encapsulation

There are many different chip designs. Much of the work at the University of Utah has been done using a single silicon chip containing two CHEMFET devices and two conventional metal gate IGFET devices. The two metal gate devices provide a method of monitoring the basic device parameters independent of the chemically sensitive gate. The two CHEMFET devices allow use of either a multisensor device or a reference device measuring standard conditions.

A microphotograph of the dual-device chip is shown in Fig. 16a. The overall dimensions of this chip are 1.28 × 2.16 mm. There are four field effect transistor (FET) structures on this chip identified as Q_1, Q_2, Q_3, and Q_4. Two of them, Q_2 and Q_4, are conventional metal gate IGFET's and the other two, Q_1 and Q_3, are devices fabricated without the metal gate. Numbered regions, 1–9, are metal contact pads to which wires are bonded for connections to the external circuitry. The chip is so designed that the CHEMFET gates are well removed from the wire bonding pads to facilitate

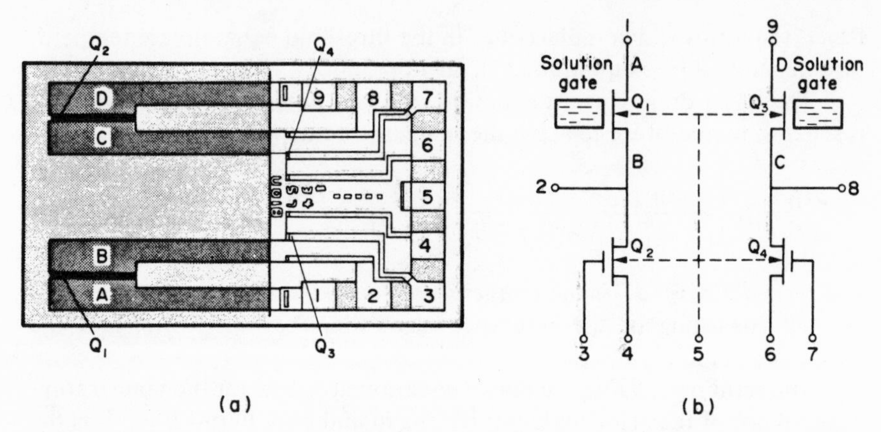

Fig. 16. (a) Photomicrograph and (b) circuit schematic of a multidevice ISFET chip. (Reprinted with permission from reference 27.)

deposition of the ion-sensitive membranes, and from the area of the chip containing the electrical connections to be coated with an insulating material.

The *n*-type regions which form the source and drain of the two CHEMFET gates are elongated and form the current-carrying paths to the bonding pads for the two CHEMFET devices. These elongated *n*-type areas, designated A, B, C, and D in Fig. 16a, contain appreciable resistance and form the parasitic resistors shown in Fig. 10.

Figure 16b shows the circuit diagram of the devices on the CHEMFET chip. The identified points correspond to the areas of the photograph having the same number. Point 5 is the connection point to the substrate silicon. Other chip designs have been described,[5,28,30–34] and the original papers should be consulted for details.

Encapsulation

Since the beginning of the use of integrated solid state devices, the electronics industry has been trying to provide better protection against moisture and ionic contaminants. The very nature of CHEMFET's require that they be used in solutions of electrolytes, which, from the electronic point of view, is a hostile environment. The electrical integrity of the whole CHEMFET assembly is critical and prerequisite for a successful operation of these devices.

There is no general recipe for encapsulation. The choice of materials is governed by the general device package (e.g., catheter, transistor header, etc.), by the geometry and layout of the chip, and by the intended use.

From the point of view of the operation of the device, there are two distinct encapsulation problems:

(1) The encapsulation and electrical integrity of the gate and its immediate surrounding. The electrical leakage in this area will usually have an adverse effect on the electrochemical behavior of the transistor and must be prevented.

(2) The encapsulation and electrical integrity of the rest of the package. This affects the lifetime of the probe and can be an important consideration in electrical safety in medical applications.

The package that has been used by our group is shown in Fig. 17. The top lumen of a 6 French dual lumen PVC catheter (1) is cut off about 5–7 mm from the end. This provides a platform on which the chip (2) is attached with a cyanoacrylate contact adhesive. The connecting laquered copper leads (3) are threaded through the lower lumen (4), anchored with a layer of epoxy (5), and flattened to provide bonding pads. The end of the lower lumen is closed with epoxy. The upper lumen (6) is usually used as a reference electrode compartment, housing an Ag/AgCl wire (7), and a glass capillary (8) which provides a liquid junction. After wire bonding, the chip and the bonding wires are covered with the final encapsulant (9).

The choice of materials in a package like this has to be based on the following criteria: (a) low bulk permeability for water and electrolytes, (b) good adhesion between materials used, and (c) the mechanical and rheological properties of the noncured encapsulant, suitable for the fabrication.

The most satisfactory combination of materials used in our package has been found by trial and error over several years. Silicone rubbers and epoxies are generally recognized as good encapsulants for electronic components. Although the uptake of water by silicone rubber is low, its permeability is considerably higher than that of epoxies. Another reason for preference of epoxy as the encapsulant instead of silicone is that the latter does not adhere well to PVC (membrane or catheter). The adhesion of epoxy to the silicon chip coated with silicon nitride can be improved by silanization. It also improves the adhesion of the PVC membrane (*vide infra*) to the gate area of the chip.

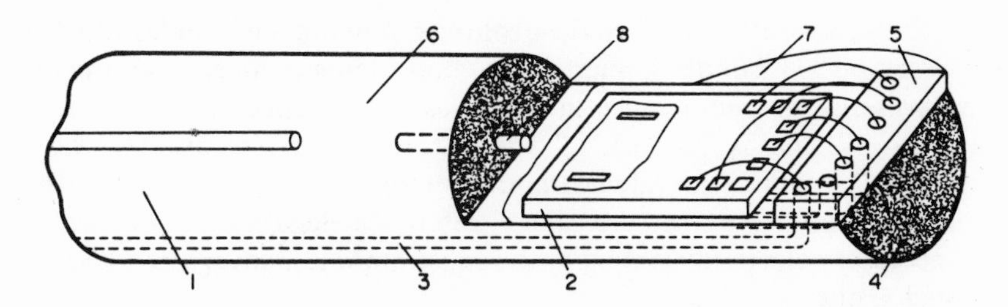

Fig. 17. ISFET chip mounted on dual-lumen 6F catheter (see text for description).

A radically different approach to encapsulation has been suggested by Wise and Weissman.[35] It involves covering of the prebonded chip with a thin layer of Si_3N_4. Similarly, vapor deposited polymers such as Parylenes can be used for encapsulation provided that a suitable technique for opening the gate area can be found. Esashi and Matsuo[30] have insulated the sides of the chip by thermally growing SiO_2 after separation of the chips by anisotropic etching. Bergveld *et al.*[20] suggest the fabrication of devices with leads on the opposite side from the ion-sensitive gates. To our knowledge no report on the device with this geometry has yet been published.

The electrical integrity of the gate is critically important for predictable operation of the ISFET's with membranes (i.e., other than bare SiO_2 or Si_3N_4 gate). Any passage of current through the gate "insulator" and the overlaying membrane causes polarization of the latter. This leads to instability of the potential and after prolonged polarization sometimes to the degradation of the membrane. This situation is equivalent to using a low-input impedance amplifier to measure voltage across a high-resistance ISE membrane.

We have undertaken a detailed study[36] of the electrical integrity of SiO_2 and Si_3N_4 prepared by chemical vapor deposition and RF sputtering and found that Si_3N_4 prepared by either technique offers an excellent insulator. These materials can withstand exposure to 0.15 M NaCl for 90 days. On the other hand thermally grown SiO_2 lost its insulating property within hours of immersion in the solution. The mechanism of this breakdown appears to be consistent with the formation of microfissures in the top layer of the SiO_2 which are filled with the solution, and not with the total hydration of the SiO_2 layer as was originally thought.[6,8,10] When a sufficiently high voltage is applied, the electric field at the tip of the fissure is much greater than on the flat surface of the insulator. This leads to a catastrophic breakdown of the remaining insulator beneath the fissure. Depending on the polarity of the applied voltage one or the other of the following reactions then takes place:

$$\text{silicon negative:} \quad H_2O + e^- \rightarrow \tfrac{1}{2}H_2 + OH^- \qquad \text{(I)}$$

$$\text{silicon positive:} \quad Si + 2H_2O \rightarrow SiO_2 + 4H^+ + 4e^- \qquad \text{(II)}$$

The first reaction leads to the evolution of hydrogen bubbles which can be observed[6,8,10] under a microscope, thus aiding in location of the leak. The presence of microcracks in SiO_2 has been suspected by Esashi and Matsuo[30,31] and by Schenck.[32] Their presence can also explain why the first ISFET[6] could be operated without a reference electrode, that is, without a physically separate structure called "reference electrode." It is likely that the necessary electrical contact to the substrate was inadvertently provided by such a microcrack(s).

The ultimate test of the electrical integrity of the whole package is to connect all wires of the ISFET (i.e., drain, source, and substrate) together, and to polarize the device from +3 to −3 V against a low-resistance ($R < 100\ k\Omega$) reference electrode. The shape of the current–voltage curve (Fig. 18) as well as the absolute value of the leakage current reveals the possible leak. The increase of leakage current on the cathode side corresponds to evolution of hydrogen (reaction I). The value of the decomposition potential depends on what material this reaction takes place on (a wire or Si) and also on the previous polarization history. The anodic branch of the leakage current is due to reaction II, if the leak is through Si. On the metal, the corresponding reaction can be the formation of oxide, insoluble salt, or eventually evolution of oxygen. Since all these reactions are pH dependent, it follows that the leak does not make a good reference electrode if the pH of the solution is changing. This can account for inconsistent data obtained with the SiO_2 gate pH ISFET's. It is also apparent from Fig. 18 that testing between +1 and −1 V cannot tell anything about the encapsulation, the point which is not appreciated by many researchers in this field.

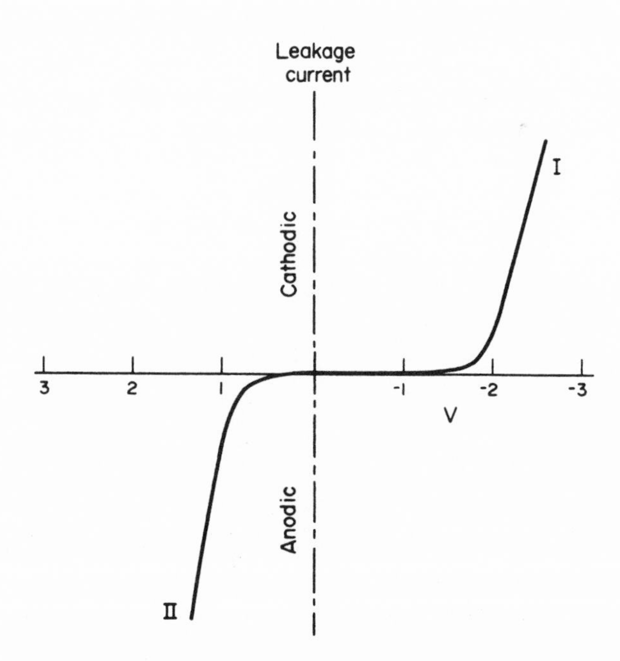

Fig. 18. Leakage current of poorly encapsulated ISFET package. The polarity of the applied voltage is that of the ISFET.

The requirements on the longevity of encapsulation vary from one application to another. With our encapsulation package, described above, the electrical integrity of the encapsulation is maintained for at least two months.

3.2. Hydrogen-Ion-Sensitive Field Effect Transistor

The hydrogen-ion-sensitive FET with a bare insulator gate is the most widely reported ISFET. Because of this and because of its different mechanism of operation it will be treated separately.

There are two types of bare insulator gate ISFET's: one with only silicon dioxide and the other silicon dioxide/silicon nitride gate structure.

3.2.1. Silicon Dioxide Gate Insulator

The original mechanism of operation of the pH ISFET with a SiO_2 gate was based on the assumption that the thermally grown silicon dioxide exposed by etching off the Al metal gate has ion exchange properties similar to a pH glass electrode.[6] The observed sensitivity of these devices to changes in concentration of NaCl was based on the same assumption. Schenck[32,33] reported sensitivity values of 47 mV/pH at pH 7.5 and 0.01 N NaCl and 37 mV/pH at pH 7.5 for 0.1 N NaCl. At the same time it was observed that hydration of the SiO_2 layer caused a degradation of its insulating properties, so much so that evaluation of hydrogen was observed with the ISFET, which was biased sufficiently negative with respect to another electrode. The electrical integrity of SiO_2 upon immersion in aqueous electrolyte has been questioned by Esashi and Matsuo,[30,31] who suggested the existence of pinholes rather than bulk hydration as the cause of the electrical breakdown. We have confirmed[36] the existence of pinholes in thermally grown SiO_2, as was discussed in Section 3.2. It is very likely that the devices with very thin (~25 Å) SiO_2 suffered from the same problem but no data on gate leakage current have been given.[32]

It is known that quartz[37] as opposed to amorphous SiO_2 does not hydrate extensively. However, the surface silanol groups dissociate:

$$\equiv SiOH \leftrightarrows SiO^- + H^+$$

and because of this equilibrium the excess charge density on the surface of quartz is pH dependent.[38] Porous silica and silica gel were found to have a surface charge density in excess of the surface density of silanol groups,[39] which indicates the formation of a thin hydrated surface layer. This was also suggested by Perram *et al.*,[40] who proved that most oxides hydrate to the depth of 40–50 Å. Our capacitance study[36] of the effect of prolonged exposure of 400-Å-thick SiO_2 film failed to detect extensive hydration, although the insulation properties of this layer were lost within hours of immersion in solution. The sensitivity of our bridge was, however, insufficient to detect changes below 50 Å.

Based on these facts, and in the absence of detailed experimental information on SiO_2 gate ISFET's, it is difficult to formulate a coherent model of operation for these devices. The observed pH response can be partly due to the presence of the reversible double layer, i.e., the dissociation of the surface silanol groups as supported by the different pH sensitivity for high and low concentrations of inert electrolyte.[33] The possible presence of an alumino silicate layer in Bergveld's devices[6] could explain the weak Na^+ sensitivity, although the increased conductivity of the electrolyte could also have been responsible.

Because of the proven poor insulation quality of the thermally grown SiO_2 and because of the uncertain mechanism of operation, the practical value of pH ISFET with SiO_2 gate is doubtful.

3.2.2. Silicon Nitride Gate Insulator

The mechanism of operation of pH ISFET's with silicon nitride gates seems to be clearer. Properly prepared silicon nitride does not form pinholes and does not hydrate, which removes one uncertainty from the interpretation of the data. It has been shown,[28] however, that if a "leaky" nitride is prepared the leakage current affects the pH sensitivity. It was found that if the device is biased within the potential range where the Si/electrolyte interface is *polarized* (+1.5–1.0 vs. SCE) the pH sensitivity remains at 50–51 mV/pH. An attempt has been made to explain quantitatively the effect of the leakage current on the pH sensitivity by incorporating the overvoltage term in the drain current equation.[28] However, it is neither necessary nor desirable to complicate the operation of ISFET's with a gate leakage current which introduces a nonequilibrium condition, particularly when silicon nitride gate devices with good insulation can be made.[7,8,27,30,31,34]

The important contribution to the formulation of the mechanism of operation of Si_3N_4 gate ISFET is the excellent paper by Esashi and Matsuo.[31] They have used Auger spectroscopy combined with argon ion sputter etching to determine the concentration profiles of Si_3N_4 exposed to water for several hours. Their results are in complete agreement with an earlier Auger–Ar sputtering study of silicon nitride[41,42] shown in Fig. 19. They have attributed the relatively high level of oxygen in the bulk of the nitride layer to an oxygen leak in their nitride deposition apparatus. We have found exactly the same oxygen profile in our ESCA–Ar sputtering study of dry Si_3N_4.[43] The etching of one of the samples was interrupted after removing the top 200 Å silicon nitride (out of 400 Å) and the sample was

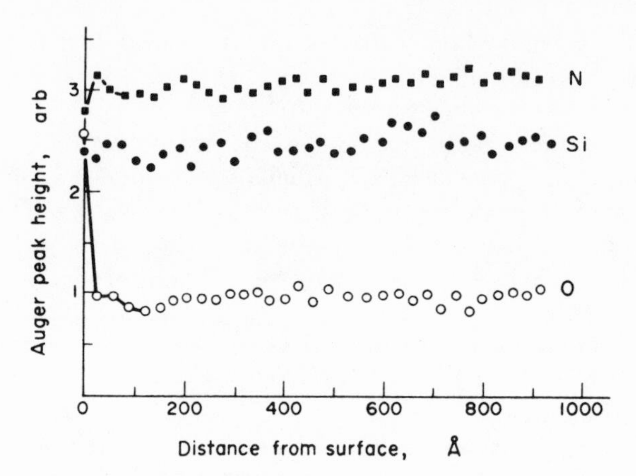

Fig. 19. Compositional profile of the silicon nitride surface (after aging in water for 14 hr). (Reprinted with permission from reference 31.)

then exposed to air for several days. The subsequent ESCA analysis and further etching show the O, N, and Si profile to be identical to the one obtained originally, i.e., high oxygen content gradually decreasing over 25–30 Å of the new top layer. These results indicate that the identical concentration profiles obtained in the three independent laboratories may represent the equilibrium state of "silicon nitride," which in fact is silicon oxynitride.[42] This agreement could also explain the relatively uniform pH response (50–60 mV) of Si_3N_4 gate ISFET's prepared in different laboratories and at different times (Table 1).

If in fact the hydrated layer is represented by a 25–30-Å *interphase* it can be responsible for the pH dependence of the surface charge of $Si_3N_4(O)$ in the same way as it is responsible for pH dependence of SiO_2.[33] It is also interesting to note that the pH response of a Si_3N_4 gate ISFET which was oxidized at 1050 °C with wet oxygen was reduced to 20–35 mV/pH. This result can be explained by the formation of a surface layer of SiO_2, which is

TABLE 1. Some pH Responses of Si_3N_4 Gate ISFET's Prepared in Different Laboratories

Reference	Sensitivity (mV/pH)	Measured range pH
31	54	1–14
28	50	1–13
30	60	1–13
34	59	3–10

known to have lower pH response.[34] It should be mentioned here that the pH dependence of surface charge and its effect on the flat-band potential of the semiconductor–electrolyte interface has been reported for other oxides. Thus, for example, the flat-band potential of the Ge/electrolyte interface was found to shift at the rate of 60 mV/pH.[44,45] The similar shift of the flat-band potential was reported for the TiO_2/electrolyte interface, while *no* pH dependence was found for CdS/electrolyte over the range of 2–12 pH units, which is in agreement with the absence of H^+ bonding sites on surface of CdS.[46]

The pH equilibrium at the nitride surface can be written as

$$SH \overset{K}{\leftrightarrows} S^- + H^+$$

where S^- is an unspecified negatively charged site. It follows from the equality of the electrochemical potentials for H^+ in solution and on the surface (SH), that

$$\mu_{sur}^{H+} + F\phi_{sur} - \mu_{sol}^{H+} + F\phi_{sol} \tag{56}$$

Assuming that the surface is stoichiometric and that the dissociation constant K is very small,

$$E = \phi_{sol} - \phi_{sur} = \text{const} + \frac{RT}{F} \ln a_{H^+} \tag{57}$$

Equation (57) substituted into equation (38) or (39) gives the response of the drain current to a change of hydrogen ion activity in solution. Although this equation simply describes the observed behavior, we realize that the above assumptions may not be justified and that the actual mechanism is probably more complicated.

The wide range of pH sensitivity of Si_3N_4 may be due to the silanol groups in different strata of the hydrated layer being subjected to different interactions. This would affect their dissociation constant over a broad range. On the other hand, if they were all on the surface in a very similar energy state, then the expected pH response would be centered ±2 pH units around the pK value of the dissociation of the surface silanol group. There are three pieces of experimental evidence to support this model:

1. The surface silanization of a hydrated Si_3N_4 gate ISFET with amino propyl trimethoxyethyl silane has no measureable effect on the pH response. If only surface silanol groups were responsible for the pH response, then the sensitivity of the drain current to pH should be reduced.

2. If only one kind of weakly acidic group on the surface of the pH ISFET were present, the pH response of this device should be ±2 pH units with respect to the pK of that group. This is demonstrated in Fig. 20, which shows the response of an ISFET with immobilized fatty acid (curve a) on the

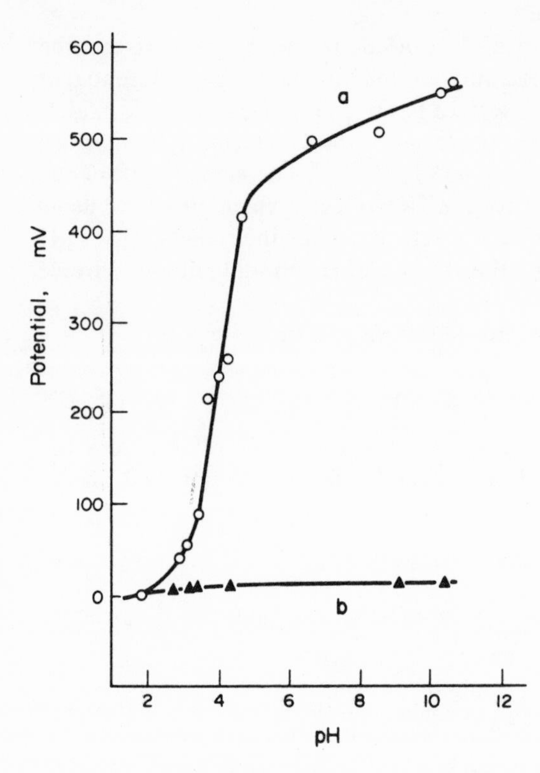

Fig. 20. pH response of hydrophobic polymer (a) with, and (b) without surface carboxyl groups.

surface of a hydrophobic polymer and the pH response of the same polymer without the carboxylic acid (curve b).[47] Curve a is a typical titration curve corresponding to the titration of a weak acid with a strong base. The lack of significant pH response of an ISFET coated with Parylene film was also reported by Matsuo *et al.*,[48] who suggested the use of this device as a reference electrode. Parylene/solution is a blocked interface which could be subject to adsorption of various ions. The suggestion that such an interface can be used as a reference is based on the erroneous assumption that no response to hydrogen ions is synonymous with a stable reference electrode potential.

3. We have observed that the drain current of the Si_3N_4 gate ISFET is virtually motion insensitive, as would be expected from a nonpolarized interface. On the other hand, the transistor described in point 2 above is strongly motion sensitive, which suggests the presence of a diffused charge on the solution side of the polymer/solution interface, which is disturbed by the movement of the liquid in the shear plane.

The above results amply confirm the earlier suggestions[26,27] that chemically sensitive transistors can be used as a tool for a study of solid/liquid interfaces. This topic is, however, beyond the scope of the present review. The reason for a remarkable hydrogen ion selectively in the presence of K^+, Na^+, and other ions is not clear and would merit further investigation.

3.3. ISFET's with Solid State Membranes

The main difference between bare gate insulator pH ISFET's and those which will be discussed in this and the following two sections is the thickness of the ion-sensitive layer. In this case it is much greater than the Debye length. The thermodynamics analysis presented in Section 2.3 applies directly to these devices. Naturally, the mechanism by which the inner potential difference between the solution and the bulk of the membrane is formed is different for each membrane. A detailed discussion of the various mechanisms of operation of ISE membranes can be found, for example, in reviews by Koryta[49] and Buck.[50]

The use of solid state membranes with ISFET's is particularly attractive because the deposition techniques, e.g., high vacuum evaporation, dc or rf sputtering and chemical vapor deposition are compatible with integrated circuit fabrication. Although this is a very promising approach, relatively little has been accomplished as yet.

Buck and Hackleman[14] have not made an ISFET; however, the results of their detailed study of the structure $AgBr/SiO_2/Si$ are directly applicable to the general theory of ISFET's. Esashi and Matsuo[31] describe fabrication, testing, and use of pH and Na^+ ISFET's. Their sodium-ion-sensitive layer is either aluminosilicate[51] or borosilicate[52] glass deposited by chemical vapor deposition techniques. The response of the Na^+ ISFET with an aluminosilicate gate is shown in Fig. 21. The device exhibits an equivalent potential change of 55 mV/pNa in the range pNa 0–3 and measureable response down to pNa 5. Devices with a borosilicate membrane show a similar response.

Auger electron spectroscopy combined with argon sputter etching of fully hydrated alumino silicate has shown that the hydrated layer extends approximately 200 Å from the surface (Fig. 22). This finding is in agreement with the Eisenmann's theory for sodium ion selectivity of alumino- and borosilicate glasses.[37] The particularly attractive feature of this sensor is its small size (200-μm tip, 1.5-mm shank) and the fact that it has been fabricated entirely by processes used in integrated circuits production. The only other ISFET with a solid state membrane is the F^- ISFET which was reported in a preliminary form by Moss.[53]

There seem to be many problems which are encountered in preparation of other ISFET's with solid state membranes. Many materials which form membranes in conventional ISEs cannot be deposited by high-vacuum techniques. An example of such a material is Ag_2S, which is used in silver-

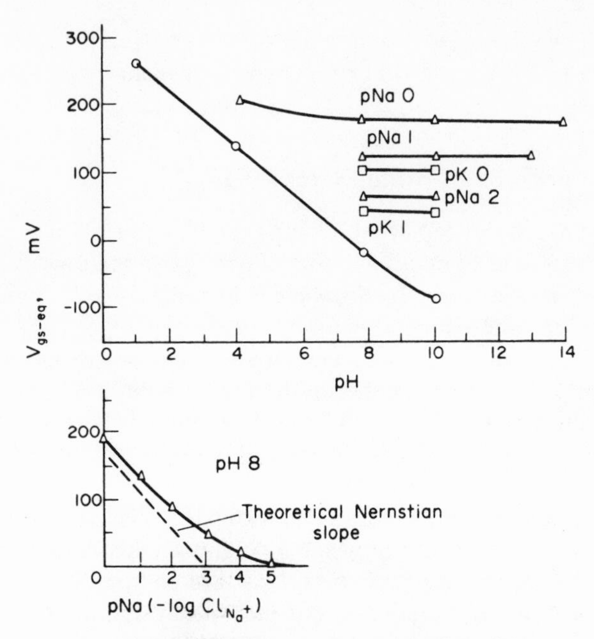

Fig. 21. Measured ion selectivity of the pNa sensor. (Reprinted with permission from reference 31.)

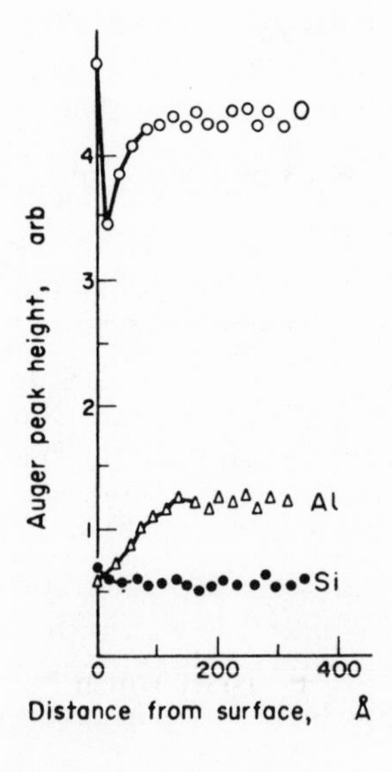

Fig. 22. Compositional profile of the alumino silicate surface (after aging in water for 24 hr). (Reprinted with permission from reference 31.)

and sulfide-sensitive electrodes, and which is a basic matrix for several other solid state electrodes. Many potentially useful ion-selective materials undergo dissociation and other chemical changes when evaporated.

3.4. ISFET's with Homogeneous Polymeric Membranes

The common denominator of the three types of sensors based on ion-selective field effect transistors (ISFET) which are described in this section is the mode of application of the ion-selective membrane: solvent casting. It has been pointed out before that the electrical integrity of ISFET's with membranes is a necessary prerequisite for stable operation of these devices. Our recently concluded study has shown that properly prepared silicon nitride provides a stable gate insulator. The surface silanization process combined with an epoxy encapsulant then provides sufficient protention for the transistor chip against the aqueous electrolyte.

In the first device of this type[(8)] a PVC-valinomicyn-dioctyladipate membrane[(54)] cast out of a tetrahydrofurane solution has been used. Although the principle of operation has been demonstrated with this device, it showed a poor long-term stability and relatively short lifetime. These problems have been attributed to the gradual leaching out of the ionophore and the plasticizer from the membrane. It was also estimated that a minimum membrane thickness of approximately 100 μm is required for a reliable performance. Membranes thinner than 50 μm suffer from the presence of pinholes which fill up with the electrolyte and eventually short out the membrane. Similar difficulties have been encountered in the construction[(27)] of Ca^{2+}-sensitive ISFET's utilizing a conventional PVC-based membrane.[(55)] Further development and the use of better membranes produced K^+, Ca^{2+}, and H^+ ISFET's with considerably improved performance characteristics.[(56)]

ISFET *Fabrication.* The general procedure for fabrication of ISFET's with polymeric membranes has been published.[(56)] The outline of this procedure is given below.

The semiconductor chip[(8)] is eutectically attached to a 1.9×3.2-mm Kovar substrate coated with an approximately 0.80-μm layer of vacuum-deposited gold to provide an electrical connection of low resistance between the chip and substrate. This chip–substrate package is silanized with 3-aminopropyltriethyoxysilane in a toluene reflux system, ultrasonically cleaned in toluene and dried in a desiccator. The chip–substrate package is then attached to a specially prepared dual-lumen 6-French PVC catheter with Eastman 910 adhesive, and electrical connections are made with an ultrasonic wire bonder by means of Al–1%-Si wires. The devices are encapsulated with epoxy (EPON 825 [Shell] with a J230 cross-linking agent [General Mills]) to provide electrical insulation for the bonding wires and exposed silicon regions of the chip as well as to provide a suitable configuration around the chemically sensitive gates for casting membranes. The entire sensor package is finally cured at 60 °C for 12–24 hr. Potassium and calcium ion-sensitive membranes are cast from a 1 : 1 (v/v) mixture of tetrahydrofurane and cyclohexanone solution. The valinomycin-based K^+ membrane is prepared according to the formulation developed by Band *et al.*[57] The calcium (II) membrane utilizes *t*-HDOPP [*p*-(1,1,3,3-tetramethylbutylphenyl)phosphoric acid] as the electroactive material.[58] The pH membrane developed by LeBlanc *et al.*[59] is cast from a chlorobenzene–dichloromethylene solution. The membranes are applied to the CHEMFET gates by a sequence of 2–3 solution castings. All three types of membranes are allowed to cure for 8–12 hr between membrane applications, and for approximately 24 hr before solution testing begins.

pH ISFET. The pH response of the pH ISFET was tested by titrating a solution containing 0.1 *M* sodium acetate and 0.05 *M* sodium tetraborate

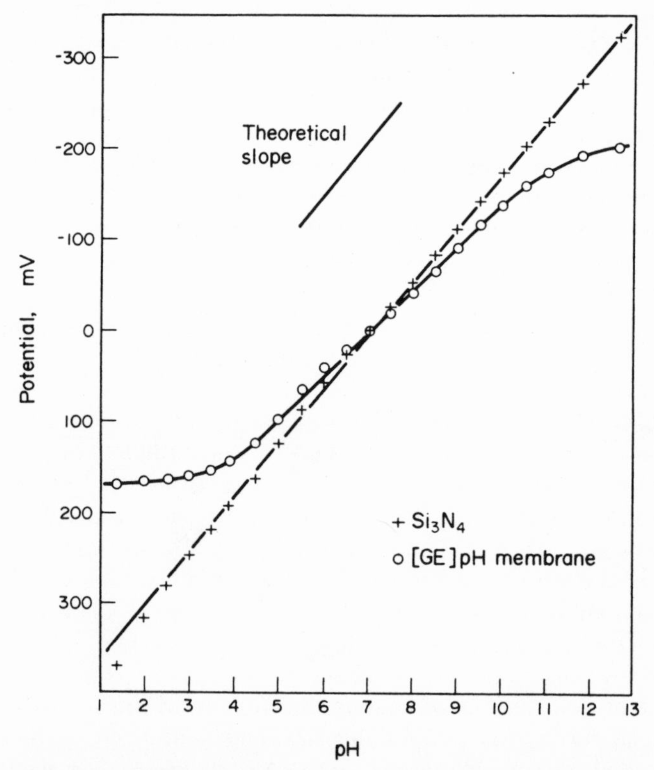

Fig. 23. pH response of pH ISFET with (+) Si_3N_4 surface and (○) Ge polymeric membrane. (Reprinted with permission from reference 56.)

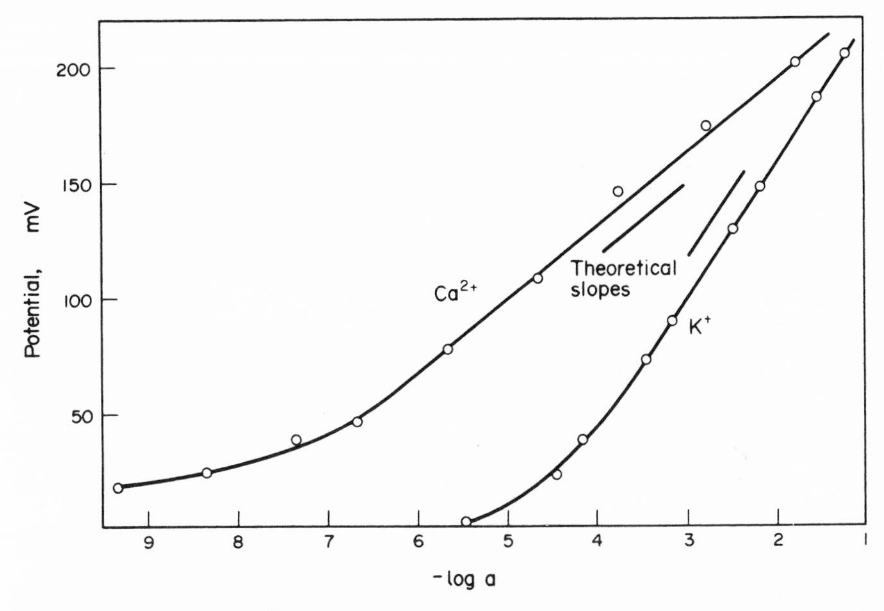

Fig. 24. Response of Ca^{2+} and K^+ ISFET. (Reprinted with permission from reference 56.)

with 0.1 *M* HCl. Because of the several dissociation equilibria, the pH changes gradually from approximately pH 12 to pH 2. The results of this measurement are shown in Fig. 23, from which it can be seen that the usable linear range is between pH 4 and pH 10 with a slope of 47 mV/pH. An average calibration slope between buffer solutions (pH 7 and pH 10) yields a value closer to 55 mV/pH. Bare gate pH sensors, utilizing the solid state Si_3N_4 gate insulator as a pH membrane, have also shown good response from pH 4 to pH 10; often with a greater slope than those obtained with the General Electric membrane. In agreement with LeBlanc *et al.*[59] the interference observed from the phthalate ion prevents the use of commercial pH 4 buffer for calibration. This problem can be easily circumvented, however, by using a sodium acetate buffer which does not interfere.

K^+ and Ca^{2+} ISFET's. Two types of test procedures were used for these sensors: either the sensor with an internal microreference electrode was dipped into solutions of known concentrations of K^+ or Ca^{2+}, or a 1.0 *M* solution of KCl or $CaCl_2$ was added from a micrometer syringe to 20.0 ml of 0.15 *M* NaCl while the response of the sensor was continuously recorded. For values of Ca^{2+} below 10^{-6} *M*, calcium buffers were used.[58] The results of these tests are shown in Fig. 24, and the performance characteristics of the three ISFET's are summarized in Table 2. Potassium ISFET's were used continuously for more than 30 days without any detectable degradation of their performance.

TABLE 2. Performance Characteristics of Three ISFET's

Membrane	Sensitivity (mV/decade)	Range	Stability (mV/hr)	Avg. lifetime (days)
pH	56 @ 37 °C	3–12 pH	1.0	10–20
K^+	57 @ 25 °C 59 @ 37 °C	$10^{-1} \rightarrow 10^{-5}$ *M*	0.1	>20
Ca^{2+}	27 @ 25 °C	$10^{-1} \rightarrow 10^{-9}$ *M*	1.0	≤10

In the case of Ca^{2+} ISFET's, the membranes become opaque after approximately ten days of continuous exposure to the solution with a concurrent loss of response. Recovery of the sensors was observed after the catheter had been dried in a desiccator and a drop of cyclohexanone placed on the membrane itself. Following this recovery, the devices were usable for approximately five days.

On calcium and potassium sensors, the thickness of the membrane has an effect on the response of the device. Membranes thinner than about 40 μm sometimes produced a substantially lower slope (40 mV/pK^+ or less). The same pattern was exhibited for calcium. This behavior is attributed to the existence of "pinholes" in the membrane which provide a solution path through the membrane shorting out the potential. With membranes 80–150 μm thick, this phenomena was not observed, and there was no deterioration of response over periods up to one month for K^+ sensors.

As was discussed previously, a well-behaved pH ISFET can be obtained by exposing the partially oxidized silicon nitride (Si_3N_4) surface of the gate insulator to solution. The main reason for the study of the membrane-based pH ISFET is the relatively high thrombogenicity of the silicon nitride surface. The linear range of both these types of devices, between pH 4 and 10, makes either type of device suitable for most biomedical applications.

The vastly improved stability of the K^+ and Ca^{2+} ISFET's reported here over those described previously[8,55] is due to better membranes and encapsulation. While the progressive deterioration of the calcium membrane appears to be the limiting factor for the lifetime of this type of sensor, the ultimate lifetime of the pH and K^+ ISFET's is determined by the adhesion of the membrane to the surface of the chip and the ability of the encapsulants to protect the chip during long exposures to aqueous media. With the present encapsulation techniques, the average usable lifetime of the sensors (pH and K^+) is approximately 20 days. Similarly, the stability and temperature sensitivity of the ISFET's are apparently determined by the transistor itself rather than by the ion-selective membrane.

3.5. Heterogeneous Membranes

The use of heterogeneous membranes in ISE's has been pioneered by Pungor and co-workers.[60] In general, the membrane consists of a semiconducting electrode material, usually an inorganic salt of a low solubility, which is dispersed in a suitable elastomer (e.g., silicone rubber).

Heterogeneous membranes offer a simple solution to the problem of manufacturing ISFET's with silver halides, mixed silver halides–silver sulfides, etc. In order to demonstrate this possibility, we have prepared[61] ISFET's with these materials using silicon rubber (SR) or polyfluorinated phosphazene (PNF) as the elastomer material. Because of the high intrinsic viscosity of SR it was necessary to dilute it with chloroform. Devices prepared by this method had very sub-Nernstian responses (5–27 mV/decade) to either Ag^+ or Cl^- ions.

A considerable improvement has been obtained with PNF as a matrix. This polymer[62] dissolves readily in ketones and other common solvents.

Silver-Chloride–PNF Membrane. The optimum composition of this membrane was found to be 75% AgCl, 25% PNF with methyl isobutylketone (MIBK) as the casting solvent. The response to chloride ion was 52 mV/decade in 0.01 *M* KNO_3 with selectivity constants K_{Cl^-/Br^-} (at $7.5 \times 10^{-5} a_{Br^-}$) = 1.87, $K_{SO_4^{2-}/Cl^-}$ (at 0.01 *M* SO_4^{2-}) = 0.01, and $K_{NO_3^-/Cl^-}$ (at 0.1 *M* NO_3^-) = 1.3×10^{-4} and range 10^{-1}–10^{-6} *M*. It was irreversibly poisoned by exposure to I^- and by the exposure to serum.

Silver-Chloride–Silver-Sulfide Membrane. The inorganic salt for this membrane was prepared by coprecipitation of mixtures of chloride and sulfide with silver ions. The optimum molar ratio of $Cl^- : S^{2-}$ was found to be 4 : 1 with 75% of this salt present in the PNF membrane. The casing solvent was again MIBK. The response was 53–58 mV/decade at 37 °C, range 10^{-1}–10^{-6} *M*. The selectivity constants determined by the fixed interference method are given in Table 3 and the response is shown in Fig. 25. After

TABLE 3. Selectivity Constants Determined by the Fixed Interference Method

Ion	Concentration	Selectivity constant $K_{Cl/i}$
Bromide	7.5×10^{-5}	5.4
Iodide	7.5×10^{-5}	0.42
Sulfate	0.01	4.5×10^{-3}
Nitrate	0.1	5.1×10^{-4}
pH	5–9	No sensitivity

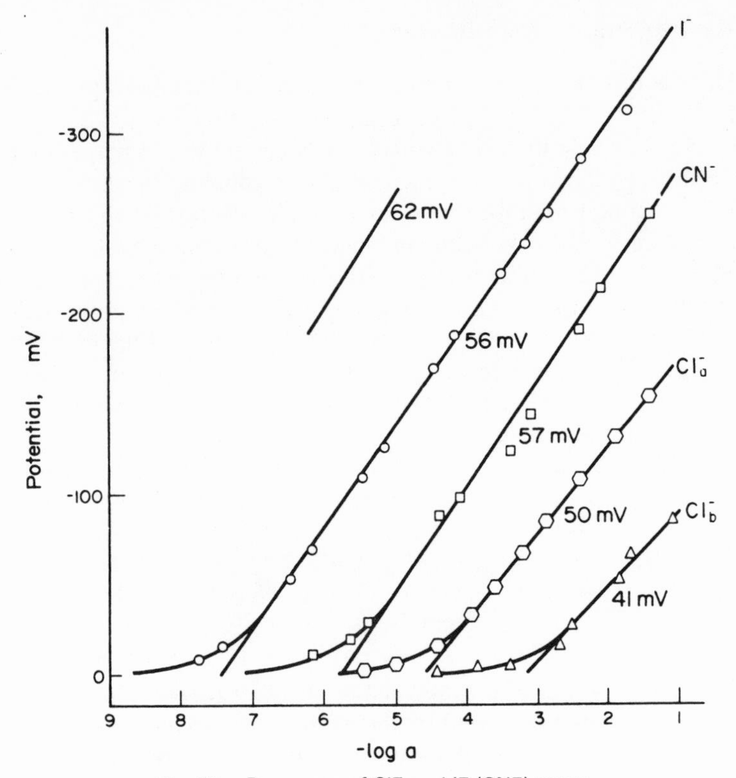

Fig. 25. Response of Cl^- and I^- (CN^-) ISFET.

exposure to iodide ions the sensitivity to chloride ion was reduced but not lost, as was observed with the AgCl–PNF membrane, it was, however, poisoned by serum proteins.

One of the attractive biomedical applications for the Cl^- ISFET is the chloride analysis in sweat for the detection of cystic fibrosis.[63] We have found that the response of our Cl^- ISFET is not affected by sweat.

Iodine and Cyanide PNF Membrane ISFET. The membrane for this ISFET was prepared from coprecipitated $AgI–Ag_2S$ (6:1) 75% and PNF 25%. The casting solvent was again MIBK. The sensitivity for I^- was from 58 to 62 mV/decade at 37 °C in 0.1 *M* KNO_3 solution and 62 mV/decade for CN^- in 0.01 *M* NaOH solution with range 10^{-1}–10^{-7} for both ions. The presence of 0.15 *M* NaCl in the NaOH solution somewhat reduces the response to CN^- ion (51 mV/decade). The selectivity constants were $K_{I^-/Cl} = 1.14 \times 10^{-5}$ and $K_{CN^-/Cl} = 2.00 \times 10^{-4}$. The CN^- response was not affected by 2 hr exposure to human serum although the response became slightly sluggish.

Time Response

It has been postulated[6,10] that the *in situ* impedance transformation should shorten the response time of an ISFET compared to an equivalent ion-selective electrode. With a typical ISFET gate capacitance of 3 pF and the combined resistance of the reference electrode and membrane of 500 kΩ, the expected time constant would be 1.5 μsec compared to the 25 μsec observed here for the Si_3N_4 gate. The considerably longer time responses of the present membrane sensors—40 msec to full response, as shown in Fig. 26—indicates that the response-limiting factor is the formation of a diffusion layer at the membrane surface. The diffusion has been shown to be the limiting factor in the response time for ion-selective electrodes.[64,65] Thus it is unlikely that an ISFET can be a faster probe than the corresponding ion-selective electrodes. However, because of their small size, localized, rapid changes of activity can be registered much more easily.

The dynamic response of ISFET's with and without a membrane was tested. The response to a step change in the externally applied gate voltage is

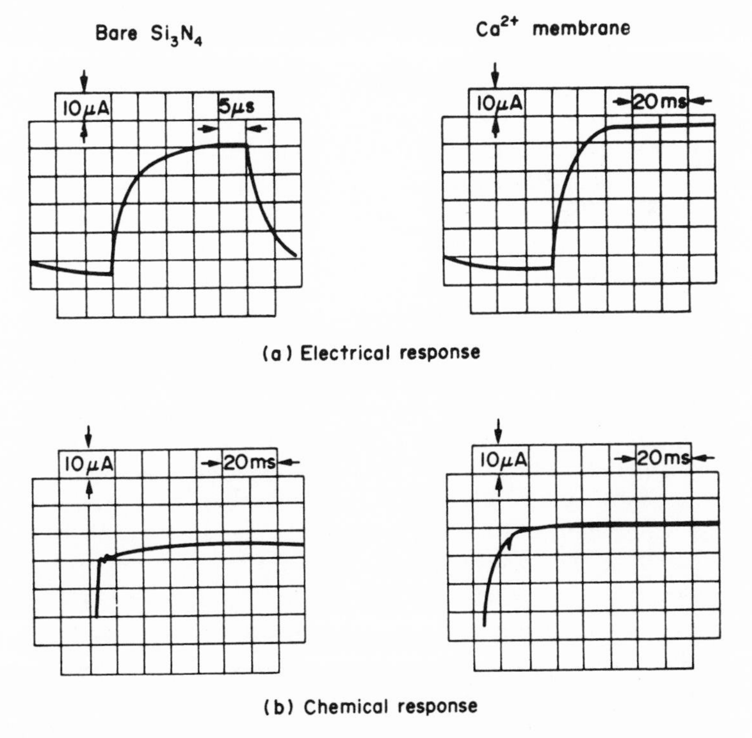

Fig. 26. Time response of Si_3N_4 and Ca^{2+} membrane ISFET. (Reprinted with permission from reference 56.)

shown in Fig. 21a, and the time response to a change in solution activity of the relevant ion is shown in Fig. 21b. This experiment was performed by pointing the effluent jet from a syringe containing a higher or lower concentration solution against the active gate of the ISFET. The signal from the sensor was fed into a storage oscilloscope and photographed with a Polaroid oscilloscope camera.

It was interesting to observe that when a jet of solution of the same composition as the background electrolyte was applied against the Ca^{2+} or K^+ ISFET, very little change in drain current was observed. This indicates that these two types of ISFET's appear to be relatively insensitive to motion.

3.6. Immunochemically Sensitive Field Effect Transistor

The unique capability of a CHEMFET with a polarized interface to measure directly the interfacial charge density can be utilized in the direct-reading, immunochemically sensitive probe. The biggest incentive for developing such a probe lies in the fact that the immunochemical reactions are second to none in terms of specificity. While ISE's respond mainly to small inorganic ions (with the exception of enzyme substrate electrodes which are based on ISE) the main domain of various immunochemical techniques is organic chemistry and biochemistry with a very large number of species to be measured. Although the high specificity and flexibility which is inherent in immunochemical systems satisfies this need it must be realized that all the available techniques are distontinuous (batch) as opposed to "direct reading" like ISE's. Our theoretical analysis,[26] supported by our preliminary experiments which are discussed below, indicates that such probes are feasible.

There has been much speculation on the desirability of coupling an immunochemical reaction to an electrochemical measurement.[66,67] A conventional immunoassay followed by a total destruction of the protein and measurement of the sulfide ion, produced from mercapto amino acids, with an ISE has been described.[68] There have been several attempts to develop a direct-reading immunochemical probe. We have used[69] a model immunochemical system, concanavalin A immobilized at the surface of a platinum wire coated with rigid PVC, to follow the change of surface charge as the result of the change of solution concentration of various polysaccharides. Measurable changes in surface charge have been observed although we now recognize that this arrangement is not suitable for practical use. Changes of potential measured across a membrane containing dissolved lipophilic antigen as the result of a specific binding of the antibody have been described.[70] It is presumed that the Donnan effect is responsible for the observed phenomenon. Similarly, a sharp increase of the conductance of a lipid bilayer membrane in the presence of an antibody, antigen, and a complement in the neighboring compartments has been described.[71,72] The appearance of pinholes in the membrane, consistent with the general mechanism of the immunochemically triggered lysis, is used to explain these observations. The idea of employing the immunochemically produced change of the interfacial charge for detection purposes has been described in two patents. In the first one[73] the charge is determined by measuring the change of the differential capacitance of the immunochemically reactive membrane using an ac bridge technique. In the second[74] it is proposed that the charge is measured directly, using a field effect transistor with an immunochemically reactive membrane.[75]

A direct-reading immunochemical probe must be based on a surface reaction. Large molecules such as antibodies (or antigens) cannot selectively permeate into a membrane and create a potential gradient by the mechanism which is typical for many ion-selective electrodes. In order to make such a probe, three conditions must be satisfied:

(a) There must be a technique for direct measurement of interfacial charge density.
(b) All (or most) of the surface charge must orginate from the immunochemical reaction and not from the reversible double layer at the solid surface.
(c) A technique for a covalent attachment of antibodies (or antigens) to the nonionic, inert surface must be available.

The measurement of interfacial charge with the CHEMFET has been discussed in Section 2.3.4. If an antibody (or antigen) is covalently attached to the surface of a nonionic substrate it becomes part of the double layer at the substrate/solution interface (Fig. 27) forming specific binding sites inside the double layer. Immunoglobulins being polyelectrolytes carry charge Z_1, the polarity and magnitude of which depend on the composition of the solution, mainly on pH. When they react with the corresponding

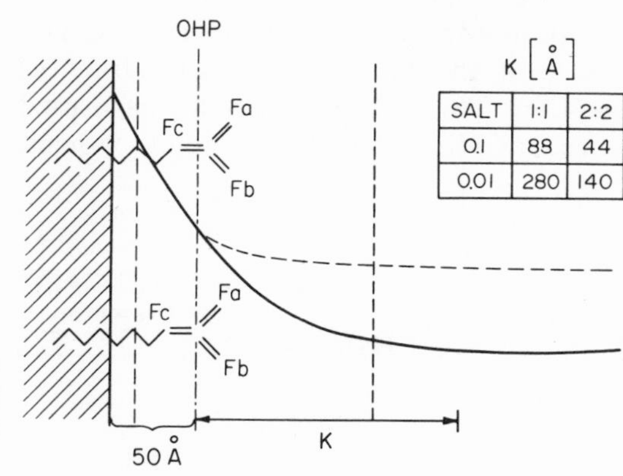

Fig. 27. Model of the double layer at the insulator/solution interface with immobilized antibody.

antigen (charged or neutral) the following reaction applies:

$$Ag^{Z_1} + Ab^{Z_2} \overset{K}{\leftrightarrows} [AbAg]^{Z_3}$$

The charge, Z_3, of the resulting complex is different from Z_1 and Z_2. Because there are other counterions associated with the protein, Z_3 does not necessarily equal Z_1 and Z_2. By making an assumption that there are no lateral interactions between immobilized antibodies (binding sites) and that the binding equilibrium is sufficiently mobile, the Langmuir adsorption isotherm can be applied. Equation (45), which was derived, for the change of interfacial charge density due to the adsorption of the ions, and the drain current equation (48), can then be applied. In that case a_S^T is identical to the surface density of the immobilized antibodies and a_h corresponds to the bulk concentration of antigen.

The estimate of the theoretical sensitivity can be based on the following parameters. Let us assume that the surface density of the immobilized antibody is 10^{12} molecules cm^{-2} (i.e., each molecule occupying 100 $Å^2$ of surface). Let us further assume that the immunochemical reaction will result in a unit charge change per molecule. The theoretical limit of detection of a field effect transistor is 10^9 cm^{-2}, which is three orders of magnitude below the maximum estimated charge density. In order to translate the surface charge change to the bulk concentration of binding species one must then assume a value for the binding constant K. Thus for $K = 10^8$ the detection limit would be 10^{-12} mol/liter.

3.6.1. Immobilization

The second prerequisite of an IMFET is the suitable immobilization technique for the antibody on the *surface* of an inert hydrophobic substrate. Since a hydrophobic substrate would generally not have polar groups which could be used for covalent attachment such groups must be introduced to the surface of the substrate without substantially altering its hydrophobic character. The technique that we have used[75] is a modification of the procedure of quasicovalent binding of heparin to various polymer surfaces[69]:

$$\text{R, R} \sim\sim CH(CH_2)_n—X \xrightarrow[\text{(toluene + ligroin + ethanol)}]{\text{swelling solvents}} \text{R, R} \sim\sim CH(CH_2)_n—X$$

step 1

$$\text{CH(CH}_2)_n\text{—X} + \underset{\text{protein}}{\text{Ⓟ}} \xrightarrow[\text{buffer}]{4^\circ\text{C}} \text{CH(CH}_2)_n\text{—Ⓟ}$$

step 2

In step 1, the inert (hydrophobic) membrane is exposed to a mixture of solvents that swell it. This mixture also contains a compound with one or more straight aliphatic chains of at least ten carbon atoms long and a polar ionic group X that can react with the protein. A typical example of such a compound can be bisdodecanoic acid. The dipolar character of this compound causes its orientation at the surface of the swollen polymer in such a way that the aliphatic chain interacts with the lipophilic membrane and can get trapped in the surface when the solvent is removed and the membrane shrinks. The aliphatic chains thus act as one or more points of attachment for the protein-reactive group X. In the case of carboxylic acid the final step involves reaction with carbodiimide.[76]

3.6.2. Results

We have first tested the basic premise that the CHEMFET with polarized interface can be used for measurement of interfacial charge. The pH response of a hydrophobic polymer with and without carboxylic groups immobilized at the surface (Fig. 20) has been already mentioned in Section 3.2. In the second experiment sodium dodecyl sulfate was adsorbed on rigid poly(vinyl chloride) film from 0.1 *M* NaCl. The part of the adsorption isotherm is shown in Fig. 28. The highest concentration tested was just

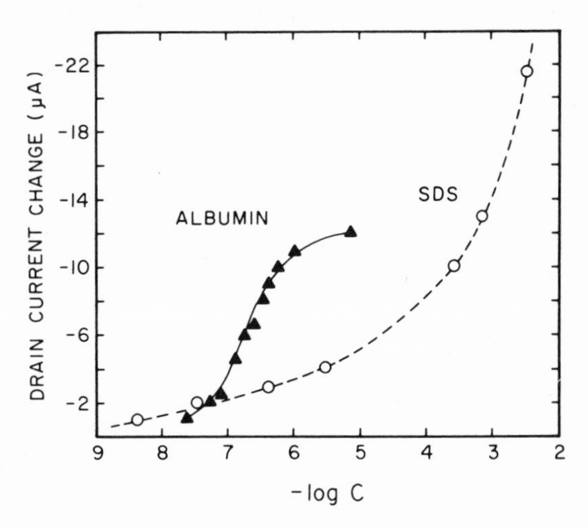

Fig. 28. Change of drain current with adsorption of albumin and SDS from 0.15 *M* NaCl at 37 °C.

below the critical micelle concentration. In the same diagram is also shown adsorption of fatty-acid-free Cohn fraction V of human serum albumin from 0.15 *M* NaCl. All these results unequivocally indicate that the CHEMFET can be used for a direct measurement of interfacial charge.

A simple immunochemically reactive surface can be prepared by incorporating a lipophilic antigen into the bulk of a hydrophobic polymer and applying it in the form of a thin film to the surface of the transistor. Cardiolipin, cholesterol, and lecithin are known to bind to syphilis antibodies. This reaction is the basis of the well-known Wassermann test and has been used by Aizawa *et al.*[70] in their qualitative potentiometric probe. We have incorporated this antibody into a PVC membrane which was cast on the gate area of our transistor. The response of this IMFET to the addition of VDRL positive and VDRL negative serum to 0.15 *M* NaCl is shown in Fig. 29.

It is known that albumin adsorbs strongly on hydrophobic surfaces. We found that only a fraction of adsorbed albumin can be desorbed from the surface by washing it with various buffers. We have used this fact to test a very crude IMFET. We reasoned that an IMFET with adsorbed human albumin should be antigenic to rabbit–(human serum) antiserum. To prove this idea we have added 25-μl increments of rabbit control serum and rabbit antiserum to 2.0 ml of 0.15 *M* NaCl. The result is shown in Fig. 30. Initially (point A), rabbit antiserum was added and again the same amount at point B. The different magnitude of the response is a result of the hypobolic shape of the adsorption isotherm (similar to albumin, Fig. 6). The third addition was 25 μl of normal serum (C) but here the response was *higher* than the preceding addition and not *lower*, as would be expected from the saturation curve, the fourth addition (D) (antiserum) produced a very small response, followed by a large response (E) when control serum was added and a small

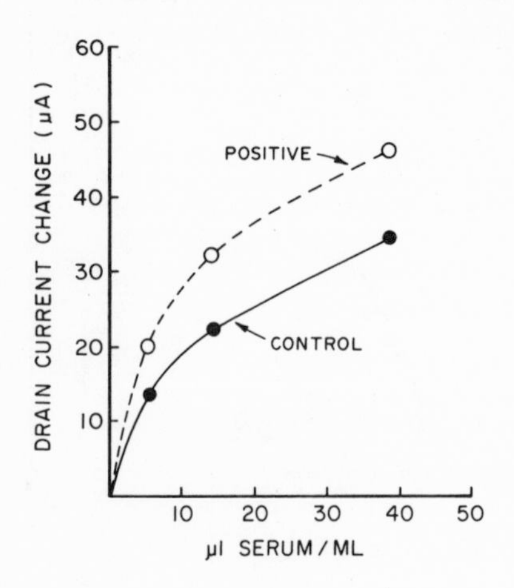

Fig. 29. Response of the IMFET with Wassermann antigen to VDRL control and VDRL positive serum. Reference electrode: Ag/AgCl, 37°C.

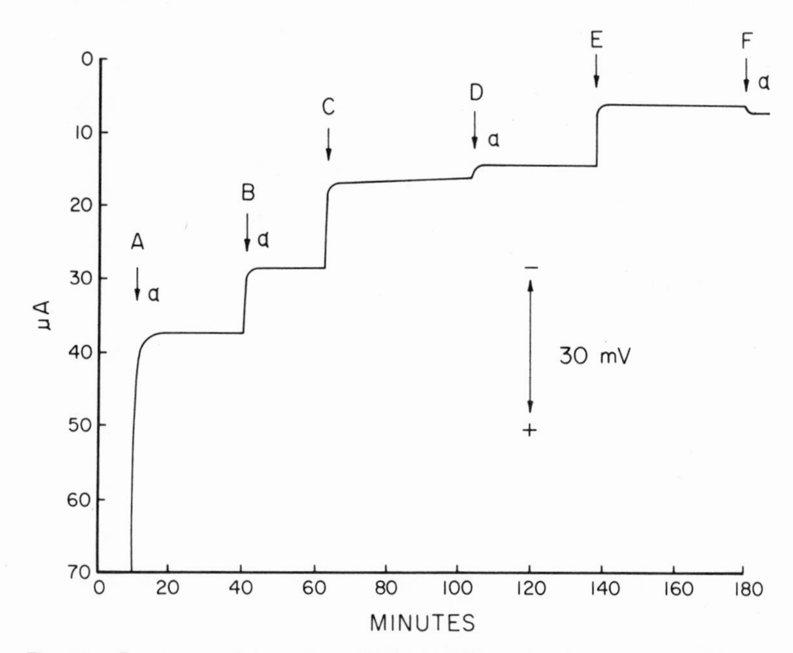

Fig. 30. Response of the IMFET with immobilized human serum albumin to addition of rabbit anti-human serum (points α) and normal rabbit serum. Measured against Ag/AgCl reference electrode at 37 °C.

negative response (F) was caused by addition of 25 μl of antiserum. These data are replotted in Fig. 31 in which the transistor output is shown against total serum concentration. Points at which the antiserum was added lie on the saturation curve, while points where normal serum was added (C, E) are significantly higher. The polarity of the response is in agreement with the higher isoelectric point (more positive charge) of IgG with respect to other major serum proteins. Thus if a preferential binding of IgG takes place the drain current should be higher as observed. It is worth noting that the 100% response occurred within minutes of the addition and that in between the additions the response was remarkably stable.

These results although still preliminary confirm the feasibility of the IMFET. The preparation of an immunochemically reactive surface seems to be the most formidable problem facing this development.

3.7. Reference ISFET

The differential measurement has two main attractions: it can reduce the extraneous interferences such as temperature, light, or common noise and therefore improve the signal-to-noise ratio. It can be also used to

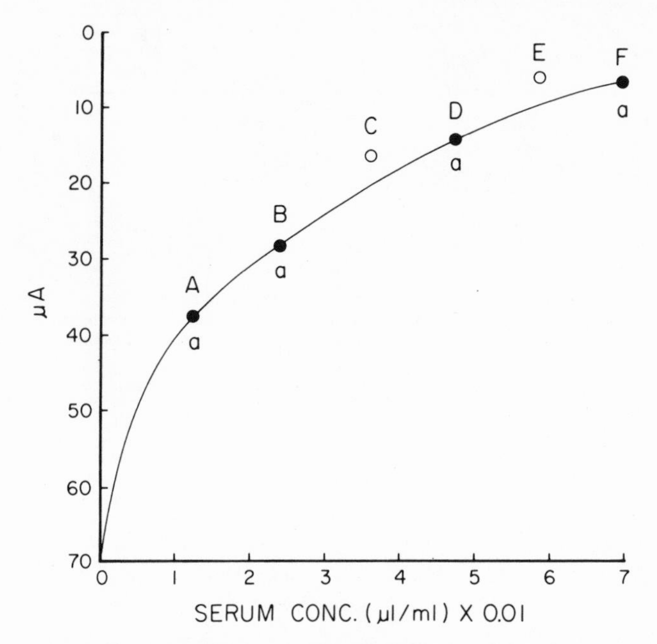

Fig. 31. Data from Fig. 30 replotted as total serum concentration.

eliminate or partially eliminate chemical interferences where a multitude of compounds are present in addition to the measured species of interest.

The CHEMFET, with its single-chip multisensor capability, is ideally suited for this approach. In order to test this concept we have developed an on-chip reference ISFET.[77]

Like the ion-selective electrodes, ISFET's require a reference electrode for a predictable operation. The quality of the measurement, therefore, depends on the performance of both half-cells. In addition, it makes sense to develop a small, rugged ISFET only if a reference electrode of the same size or smaller, can be made. In order to provide a suitable reference electrode for our ISFET's, we have taken two approaches: in the first one we have incorporated a small Ag/AgCl reference electrode in the upper lumen of the catheter[28] which is used as a support for the transistor chip (Fig. 17). The reference electrode compartment was then connected with the sample by a free liquid junction. In the second approach, the H^+ ISFET was placed in the solution by constant pH, which was then connected with the sample solution by a free liquid junction.

Sensor Preparation. The transistor chips were mounted on the distal end of a dual-lumen PVC 6-French catheter and encapsulated with epoxy resin Epon 825. A small well (approximately $500 \times 500 \times 500\ \mu m$) was built

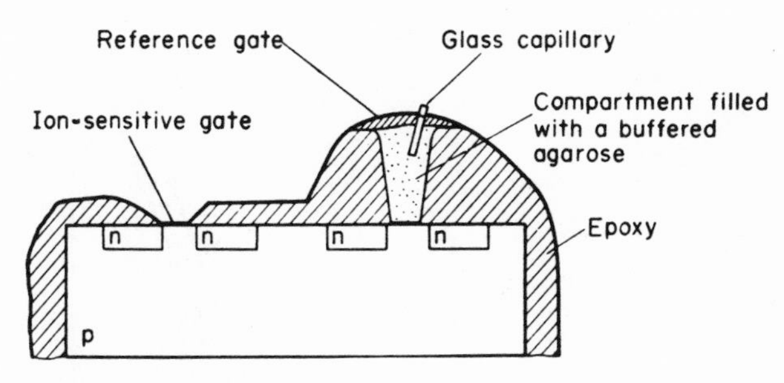

Fig. 32. Reference gate ISFET. (Reprinted with permission from reference 77.)

around one ISFET gate and a pH-sensitive polymeric membrane[59] was cast on *both* gates. Alternatively, bare silicon nitride gates were used. After curing the encapsulant a 1% agarose gel prepared in a suitable buffer was packed into the reference gate well and an approximately 20-μm glass capillary 200 μm long was inserted into the gel. The reference electrode compartment was then closed with Epon 825. After a short practice the preparation of the reference gates takes between 5 and 10 min. A schematic diagram of the gate and the measuring ISFET is shown in Fig. 32. The drain current of both transistors was measured with a simple differential current follower (Fig. 33).

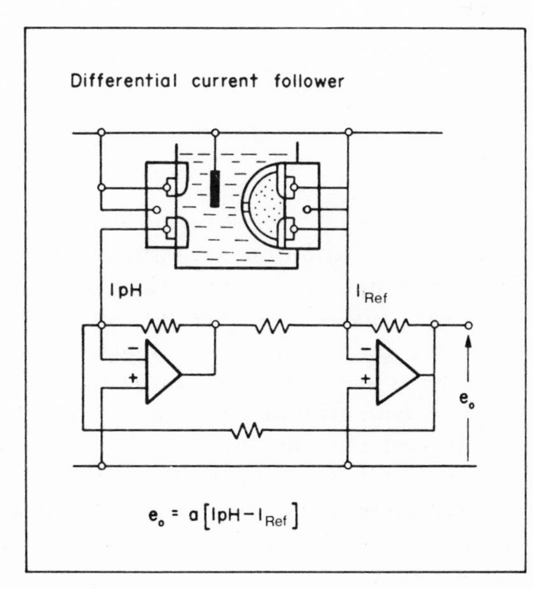

Fig. 33. Differential current follower for measurement of two ISFET's on the same chip (Reprinted with permission from reference 77.)

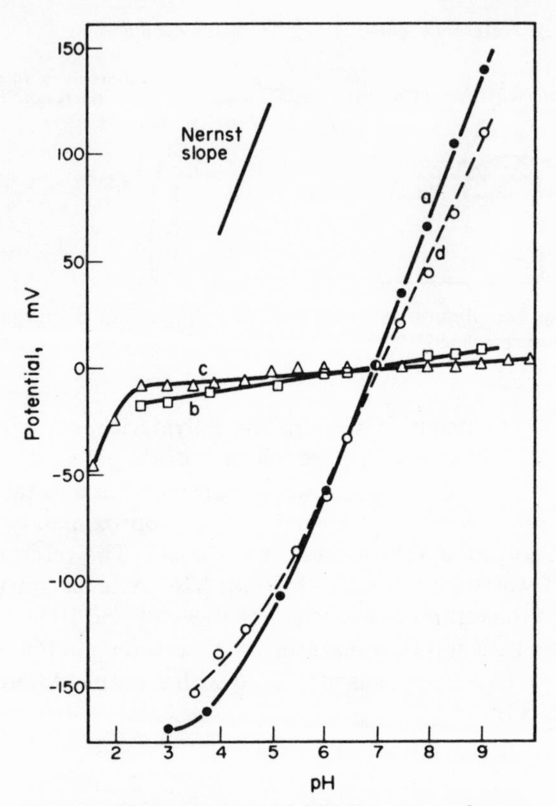

Fig. 34. The response of (a) Si_3N_4 ISFET; reference gate ISFET with (b) 0.15 *M* NaCl and (c) 4 *M* KCl buffered agarose; (d) output of the differential current meter. (Reprinted with permission from reference 77.)

Performance. The performance of the reference gate has been evaluated according to the following criteria: pH response, lifetime, temperature, and noise sensitivity. Figure 34 shows the response of the pH-sensitive gate H^+–ISFET (curve a), reference gate (curves b, c), and the pH response of their difference (curve d). The usable range, pH 4–11, of the pH-sensitive gate is the same as reported by LeBlanc *et al.*[59] The change of the reference gate potential with pH during titration of 0.05 *M* $Na_2B_4O_7$ is linear, 3 mV/pH for internal reference with 0.1 *M* buffer and 1.5 mV/pH for internal reference with 0.1 *M* buffer plus 4 *M* KCl solution. We suspect that the small deviation of the reference gate current at pH values lower than 2.5 and higher than pH 10 (Fig. 34, curve c) is due to the presence of the capillary tip pH potential.

The lifetime of the reference gate was tested by leaving the device continuously in solution and testing the pH response at various intervals. It was found that the performance of the reference gate did not deteriorate upon storage in pH 7.0 buffer solution for minimum of four weeks.

Probably the most attractive feature of the reference gate is its ability to compensate for temperature changes and noise. Figure 35 shows the effect of the change of pH and temperature on the drain current of the pH ISFET (I_{pH} curve) and the difference of drain currents of the pH and the reference gates ($I_{pH} - I_{REF}$ curve). When the solution contact is made via a proper reference electrode (such as Ag/AgCl, satd. KCl‖sample ...) the pH response of the pH gate and the difference is almost identical. However, as the temperature is changed from 37 to 26 °C the drain current of the pH gate changes accordingly, while the difference remains practically unchanged. It can be shown that the change of the liquid junction potential with temperature is very small. Thus the small change (0.008 pH/°C) of the drain current difference can be attributed, at least partially, to the difference between the change of the pH in the reference electrode compartment and the sample.

For the measurements that were discussed up to this point an ordinary stable reference electrode has been used as a contact between the solution and the transistor substrate. However, this is not necessary if we wish to evaluate *only* the differences between the reference and the indicator gate.

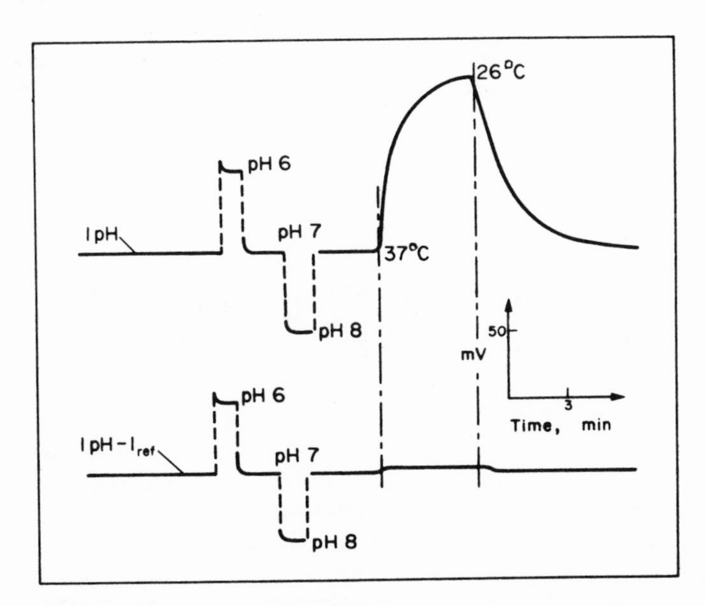

Fig. 35. Effect of temperature on ph ISFET and on the *difference* between pH ISFET and reference ISFET. (Reprinted with permission from reference 77).

Provided that an electrical contact is made between the substrate and the solution, the drain current *difference* is unaffected by the floating potential of this contact. Thus we have used a bare silver wire, an exposed part of the silicon chip, or even a paper clip with no adverse effect on the drain current difference. This point is illustrated in Fig. 36, which shows the response of the drain current difference (lower curve) and the drain current of the pH gate when a noise was "injected" into the solution by touching the solution connection.

We have also experimented with the circuit in which either the reference or the pH gate is operated in a feedback (constant-current) mode. Because this is a preferred mode of operation for a single-gate measurement we expected a better result than with the differential current follower. However, unless the characteristics of the two transistors are perfectly matched (which is rather difficult to achieve) the feedback operation offers no improvement over the differential current measurement.

The general applicability of this approach was tested on the probe which incorporated a K^+ ISFET and the reference gate. The results obtained with this combination were equivalent to the pH reference-gate case.

Because of the small volume of the reference-gate compartment it is preferable to use a true buffer solution with high buffer capacity in this compartment rather than a solution of strong electrolyte such as KCl (for a Ag/AgCl electrode). The high concentration of the neutral salt in the reference gate compartment reduces the liquid junction potential. Because the reference gate can be made relatively easily the only disadvantage of this arrangement is that one gate cannot be used for measurement of other ions.

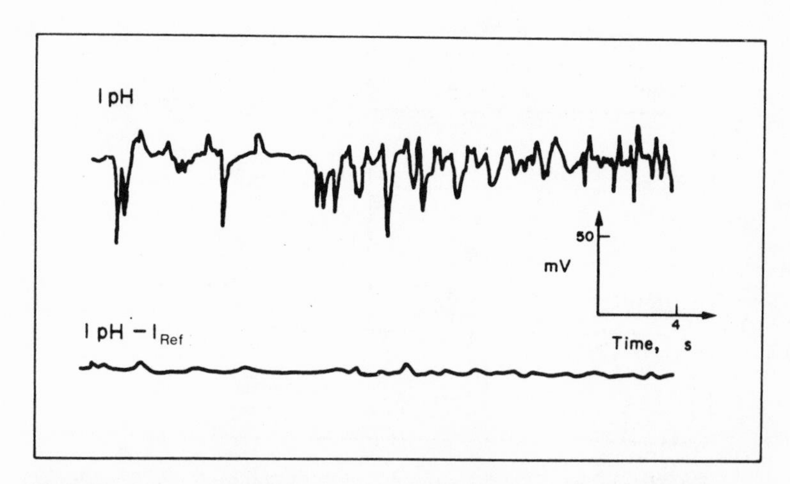

Fig. 36. Noise compensation using reference ISFET. (Reprinted with permission from reference 77.)

This is, however, a relatively small price to pay for an excellent temperature and noise compensation.

3.8. Measuring Circuitry

The fundamental operating mechanism of the ISFET is a change in the drain current in response to a change in the activity of the ions in solution as described by equations (38) and (39), Section 2.3. Maximum sensitivity will occur if the device is operated under conditions that maximize dI_D/dV_a, where $V_a = (RT/z^iF) \ln a^i_j$. To realize this, the operating conditions must be chosen so that they are biased in strong inversion (Section 2.2) and in the unsaturated mode (see Fig. 7, Section 2.2).

Strong inversion conditions result in all of the change in the semiconductor surface charge because of a change in the activity of the ions in solution appearing as a change in the number of mobile charge carriers in the channel. If biased in weak inversion, only a part of this change affects the channel current. The ISFET may be biased in strong inversions by proper setting of the externally applied voltage V_G as defined in Section 2.2.

For a fixed drain voltage, the sensitivity as defined above is greater when the device is operated in the unsaturated mode [equation (38), Section 2.2], as can be seen by inspection of Fig. 8. Unsaturated conditions are realized when

$$V_D < V_G - \overset{*}{V}_T \pm \frac{RT}{z^iF} \ln a^i_2 - E_{REF}$$

However, the operating point must be kept below that at which the sensitivity begins to decrease due to either loss of channel mobility or the effects of parasitic drain resistance as described in Section 2.2.

Once the operating point is chosen, there are two methods of measuring the ISFET response. The applied voltage V_G may be held constant and the change in drain current measured directly, or the applied voltage V_G may be changed in such a way that the drain current is held constant following a change in the activity of the ions in solution. As will be seen, this latter method is preferable.

3.8.1. Constant V_G Operation

Conceptually this is the simplest method. All externally applied voltages are held constant and the change in drain current, ΔI_D, is measured following a change in ionic activity.

In the circuit of Fig. 37, $V_{OUT} = I_D R_{SET}$. The variable resistor, R_{SET}, is a calibration resistor.

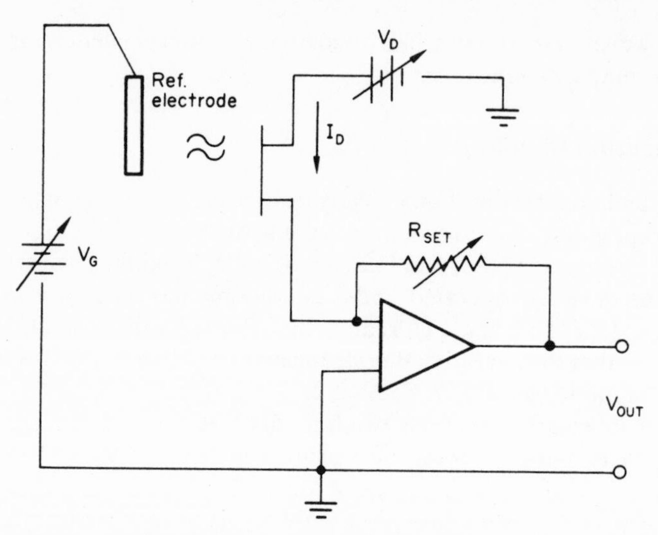

Fig. 37. Schematic of circuit diagram for measurement of drain current at constant V_G.

While this measuring technique is simple, it suffers from the fact that the ISFET is not an ideal device. The drain current equations do not consider all of the parasitic effects, for example, series resistance in the source and drain. These effects cause the relation between drain current and system voltages to be different than the equations predict. As a result the device must be calibrated over its entire operating range and the calibration will be different for every different design. A better measurement method is constant-current operation, as described below.

3.8.2. Constant I_D Operation

By using an operational amplifier in a feedback loop as shown in Fig. 38, the operating point of the ISFET may be held constant. Any change in voltage resulting from a change in ionic activity i appears as a change in the applied potential V_G. As the drain current is held constant, the parasitic effects described above are constant and the change in the potential is read out directly.

In the circuit of Fig. 38, the operational amplifier will maintain the voltage V_{OUT} at whatever value is required to make I_D the same as I_{DSET}. In response to a change in the activity of the measured ions, it will automatically adjust V_G in equation (38) or (39) to balance the change in the term $\pm(RT/z^iF)\ln a_j^i$. Since nothing else changes, the readout is directly in millivolts.

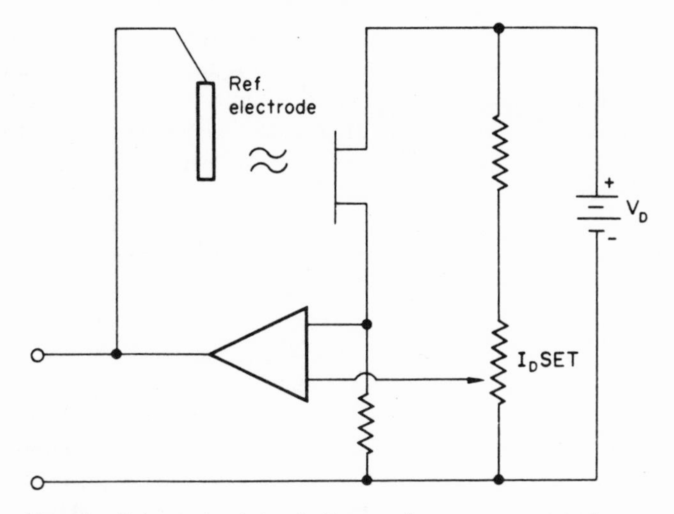

Fig. 38. Schematic of circuit diagram for measurement of ISFET response at constant I_D (feedback).

3.8.3. Differential Measurement

There are numerous applications in which two ISFET devices on the same chip could be used in a differential mode. One device measures the parameter of interest and the other provides a reference signal corresponding to some standard condition. The reference ISFET of Section 3.6 is one example. For these applications, a differential current follower is used as shown in Fig. 33.

4. CONCLUSIONS

It is the purpose of this review to present a common theoretical base for the new type of chemical sensors as well as to summarize the results that have been obtained with these devices. Because of the similarity of ISFET's with ion-sensitive electrodes it is convenient to use the comparison between these two types of sensors. Inevitably the question must be asked, what are the advantages, potential or real, of ISFET's over ISE's?

Because ISFET's are still in their infancy while ISE's have been well tested in many analytical situations, the answer is partially speculative.

First of all, the high resistance of the ion-selective membrane is *in situ* transformed into low output impedance. This means that the signal leads do not have to be heavily shielded in order to minimize electrical interferences. In turn, the probe can be made flexible, which is important for many medical applications. The completely solid state design, particularly the absence of the internal reference solution, make ISFET's light, rugged, and sterilizable. The actual sensing area is very small compared to conventional ISE's although it may never be as small as glass capillary microelectrodes, where the measuring tips are of the order of 1 μm. However, it is well within the state of the art of integrated circuit technology to accommodate tens of transistors on a chip of the size of 1 mm^2. Although the membrane deposition techniques have not yet been developed to the point where various membranes can be applied on several gates in a very small area it is only a matter of time before a *multiprobe* can be made without any increase of the overall *size*. The limiting factor in this direction seems to be the number of leads that can be bonded to the chip. The most promising solution to that problem appears to be the inclusion of a multiplexor and A/D converter on the same chip. Thus an integrated data acquisition/processing unit could be made. As was discussed in Section 2.4 the partial temperature compensation can be achieved by tailoring the transistor characteristics and by the proper selection of the operating point. Another possibility for temperature (and light, if necessary) compensation is to build the compensating elements directly on the sensor chip.

The low fabrication cost of conventional integrated circuits could be achieved for the manufacture of ISFET's as soon as the problems of on-line, automatic encapsulation and membrane deposition are solved.

Finally, many ISE's serve as substructures for enzyme electrodes. In direct analogy, enzyme FET's (or ENFET's) can be envisaged. The particularly attractive feature of FET's in this application is the small size of the active gate area, which could be a significant economical factor in large-scale production of these devices.

There are several unsolved problems. The most pressing one is the automatic encapsulation and automatic membrane deposition. IGFET's are vulnerable to static electricity. This problem is mitigated once the ISFET has been placed in a conducting solution; however, considerable antistatic precautions have to be taken during fabrication and initial testing. Every ISFET reported to date has exhibited a drift, usually equivalent to 1 ± 0.5 mV/hr change of gate voltage. The exact origin of this dtift is not yet known and needs to be further investigated. It is our belief that ISFET's will complement and possibly replace micro ISE's. With the theory of their operation generally understood the remaining problems seem to be technological.

The CHEMFET's with polarized interfaces represent a new dimension in electroanalytical sensors. The electrochemical techniques that are used for study of surface adsorption such as tensammetry are very sensitive to impurities and are not suitable for analytical application. The coupling of the highly specific immunochemical reaction with the surface charge measuring capability of the field effect transistor makes the development of a direct-reading immunochemical probe possible. The elimination of the effect of nonspecific interactions through differential measurement can be easily accomplished. With thousands of antibodies and antigens available the IMFET represents probably the most exciting possibility.

This review covers the literature through January 1979.

ACKNOWLEDGMENT

The partial financial support from the National Science Foundation grant No. CHE 7800637 is greatly appreciated.

Notation

a_j^i	Activity of species i in phase j
C_0	Capacitance per unit area of the gate insulator
d	Thickness of the insulator
E_g	Semiconductor bandgap
E_v	Valence band energy level
E_c	Conduction band energy level
E_F	Fermi energy level
E_i	Intrinsic Fermi energy level
E_0^i	Standard potential of electrode sensitive to species i
E_{REF}	Potential of reference electrode
F	Faraday constant
I_D	Drain current
K_S	Dielectric constant of silicon
L	Channel length
N_A	Volume density of acceptor atoms
q	Charge of electron
Q_B	Charge per unit area in semiconductor surface space-charge region
Q_n	Charge per unit area of electrons in surface inversion layer
Q_S	Total semiconductor surface charge per unit area
Q_{ss}	Intrinsic Si/SiO_2 interface charge
R	Gas constant
T	Absolute temperature
V_D	Drain voltage
V_{DSAT}	Saturation voltage
V_{FB}	Flat-band voltage
V_G	Gate voltage
V_T	Threshold voltage of IGFET
V_T^*	Threshold voltage of ISFET
$V(y)$	Voltage drop along the channel due to I_D
W	Channel width
Z^i	Number of charges of species i
ε_0	Permittivity of free space
Φ_s	Work function of semiconductor
Φ_m	Work function of metal
Φ_{ms}	Metal–semiconductor work function difference (here equals contact potential)
ϕ_F	Fermi level in semiconductor (here equals $\tilde{\mu}_5^e$)
ϕ_j	Inner potential in phase j
ϕ_s	Surface potential
ϕ_B	Barrier potential
μ_n	Effective electron mobility in channel
μ_j^i	Chemical potential of species i in phase j
$\tilde{\mu}_j^i$	Electrochemical potential of species i in phase j
χ	Electron affinity of the semiconductor
χ_i	Electron affinity of the insulator

REFERENCES

1. Zemel, J. N., *Anal. Chem.* **47**, 255A (1975).
2. Afromowitz, M. A., and Yee, S. S., *J. Bioeng.* **1**, 55 (1977).
3. Nagy, K., Fjeldly, T. A., and Johannessen, J. S., in Proceedings of the 153rd Annual Meeting of the Electrochemical Society, Vol. 78, Abstract 108 (1978).
4. Bergveld, P., *IEEE Trans.* **BME-17**, 70 (1970).
5. Matsuo, T., Esashi, M., and Iinuma, K., Digest of Joint Meeting of Tohoku Sections of IEEEJ, October 1971.
6. Bergveld, P., *IEEE Trans.* **BME-19**, 342 (1972).
7. Matsuo, T., and Wise, K. D., *IEEE Trans.* **BME-21**, 485 (1974).
8. Moss, S. D., Janata, J., and Johnson, C. C., *Anal. Chem.* **47**, 2238 (1975).
9. Lundstrom, I., Shivaraman, M. S., Svensson, C., and Lundkvist, L., *Appl. Phys. Lett.* **26**, 55 (1975).
10. Janata, J., and Moss, S. D., *Biomed. Eng.* **11**, 241 (1976).
11. Kelly, R. G., *Electrochim. Acta* **22**, 1 (1977).
12. Zemel, J. N., *Res. Dev.*, April, 38 (1977).
13. Revesz, A. G., *Thin Solid Films* **41**, L43 (1977).
14. Buck, R. P., and Hackleman, D. E., *Anal. Chem.* **49**, 2315 (1977).
15. Cheung, P., Fleming, D. G., Ko, W. H., and Neuman, M. R., eds., Workshop on Theory, Design, and Biomedical Application of Solid State Chemical Sensors, CRC Press, Cleveland, Ohio (1978).
16. Grove, A. S., *Physics and Technology of Semiconductor Devices*, Wiley, New York (1967).
17. Many, A., Goldstein, Y., and Grover, N. B., *Semiconductor Surfaces*, North-Holland, Amsterdam (1965).
18. Lewis, F. A., *The Palladium Hydrogen System*, Academic Press, New York (1967).
19. Lundstrom, I., and DiStefano, T., *Surf. Sci.* **59**, 23 (1976).
20. Bergveld, P., DeRooij, N. F., and Zemel, J. N., *Nature* **273**, 438 (1978).
21. Lundstrom, I., Shivaraman, M. S., and Svensson, C., *Surf. Sci.* **64**, 497 (1977).
22. Chauvet, F., and Caratge, P., *C. R. Acad. Sci. Ser. V* **285**, 153 (1977).
23. Lundstrom, I., Shivaraman, M. S., Stilbert, L., and Svensson, C., *Rev. Sci. Instrum.* **47**, 738 (1976).
24. Shivaraman, M. S., *J. Appl. Phys.* **47**, 5392 (1976).
25. Lundstrom, I., Shivaraman, M. S., and Svensson, C., *J. Appl. Phys.* **46**, 3876 (1975).
26. Janata, J., in *Workshop on Theory, Design, and Biomedical Application of Solid State Chemical Sensors* (P. Chung, D. G. Fleming, W. H. Ko, and M. R. Neuman, eds.), CRC Pres, Cleveland, Ohio (1978).
27. Moss, S. D., Johnson, C. C., and Janata, J., *IEEE Trans.* **BME-25**, 49 (1978).
28. Moss, S. D., Smith, J. B., Comte, P. A., Johnson, C. C., and Astle, L., *J. Bioeng.* **1**, 11 (1977).
29. Mohilner, D. M., in *Electroanalytical Chemistry* (A. J. Bard, ed.), Vol. 1, p. 241, Marcel Dekker, New York (1966).
30. Esashi, M., and Matsuo, T., in Proceedings of the 6th Conference on Solid State Devices, Tokyo (1974), *Suppl. J. Jpn. Soc. Appl. Phys.* **44**, 339 (1975).
31. Esashi, M., and Matsuo, T., *IEEE Trans.* **BME-25**, 184 (1978).
32. Schenck, J. F., in *Workshop on Theory, Design, and Biomedical Application of Solid State Chemical Devices* (P. Chung, D. G. Fleming, W. H. Ko, and M. R. Neuman, eds.), CRC Press, Cleveland, Ohio (1978).
33. Schenck, J. F., *J. Coll. Int. Sci.* **61**, 569 (1977).
34. Fung, D. J., Cheung, P. W., Wong, S. H., Topich, J. A., and Ko, W. H., in Proceedings of the 153rd Annual Meeting of the Electrochemical Society, Vol. 78, Abstract 81 (1978).

35. Wise, K. D., and Weissman, R. H., *Med. Biol. Eng.* **9**, 339 (1971).
36. Cohen, R. M., Huber, R. J., Janata, J., Ure, R. W., and Moss, S. D., *Thin Solid Films* **53**, 169 (1978).
37. Eisenman, G., *Glass Electrodes for Hydrogen and Other Cations*, Marcel Dekker, New York (1967).
38. Ahmed, S. M., in *Oxides and Oxide Films* (J. W. Diggle, ed.), Vol. 1, p. 319 (1972).
39. Tadros, T. F., and Lyklema, J., *J. Electroanal. Chem.* **17**, 267 (1968).
40. Perram, J. W., Hunter, R. J., and White, H. J. L., *Aust. J. Chem.* **27**, 461 (1974).
41. Johannessen, J. S., Spicer, W. E., and Strausser, Y. E., *Thin Solid Films* **32**, 311 (1976).
42. Maguire, H. G., and Augustus, P. D., *J. Electrochem. Soc.* **119**, 791 (1972).
43. Janata, J., Briggs, D., and Davies, G. R., unpublished results.
44. Brouwer, G., *Phys. Lett.* **21**, 399 (1966).
45. Brouwer, G., *J. Electrochem. Soc.* **114**, 743 (1967).
46. Watanabe, T., Fujishima, A., and Honda, K., *Chem. Lett.*, 897 (1974).
47. Janata, J., Blackburn, G., and Jonkman, A., unpublished results.
48. Matsuo, T., and Esashi, M., in Proceedings of the 153rd Annual Meeting of the Electrochemical Society, Vol. 78, Abstract 83 (1978).
49. Koryta, J., *Ion Selective Electrodes*, Cambridge University Press, Cambridge (1975).
50. Buck, R. P., *Electroanalytical Chemistry of Membranes*, Crit. Rev. Anal. Chem. **5**(4), 323 (1975).
51. Aboaf, J. A., *J. Electrochem. Soc.* **119**, 948 (1967).
52. Scott, J., and Olmstead, J., *R.C.A. Review* **26** 357 (1965).
53. Moss, S. D., in *Workshop on Theory, Design, and Biomedical Application of Solid State Chemical Sensors* (P. Chung, D. G. Fleming, W. H. Ko, and M. R. Neuman, eds.), CRC Press, Cleveland, Ohio (1978).
54. Fiedler, U., and Ruzicka, J., *Anal. Chim. Acta* **67**, 179 (1973).
55. Griffiths, G. H., Moody, G. J., and Thomas, J. D. R., *Analyst* **97**, 420 (1972).
56. McBride, P. T., Janata, J., Comte, P. A., Moss, S. D., and Johnson, C. C., *Anal. Chim. Acta* **101**, 239 (1978).
57. Band, D. M., Kratochvil, J., and Treasure, T., *J. Physiol.* **265,** 5P (1977).
58. Brown, H. M., Pemberton, J. P., and Owen, J. D., *Anal. Chim Acta* **85**, 26 (1976).
59. LeBlanc, O. H., Jr., Brown, J. F., Klebe, J. F., Niedrach, L. W., Slusarczuk, G. M. J., and Stoddard, W. M., Jr., *J. Appl. Physiol.* **40**, 644 (1976).
60. Pungor, E., *Anal. Chem.* **39**, 28A (1967).
61. Shiramizu, B. T., Janata, J., and Moss, S. D., *Anal. Chim. Acta* **108**, 161 (1979).
62. Allcock, H. R., *Angew. Chem. Int. Ed.* **16**, 147 (1977).
63. Bray, P. T., Clark, G. C. F., Moody, G. J., and Thomas, J. D. R., *Clin. Chim. Acta* **77**, 69 (1977).
64. Lindner, E., Toth, K., and Pungor, E., *Anal. Chem.* **48**, 1071 (1976).
65. Morf, W. E., Lindner, E., and Simon, W., *Anal. Chem.* **48**, 1596 (1976).
66. Rechnitz, G. A., *Chem. Eng. News*, January, **27**, 29 (1975).
67. Meyerhoff, N., and Rechnitz, G. A., *Science* **195**, 494 (1977).
68. Alexander, P. W., and Rechnitz, G. A., *Anal. Chem.* **46**, 860 (1974).
69. Janata, J., *J. Am. Chem. Soc.* **97**, 2914 (1975).
70. Aizawa, M., Kato, S., and Suzuki, S., *J. Membr. Sci.* **2**, 125 (1977).
71. Wobschall, D., and McKeon, C., *B.B.A.* (*Biochim. Biophys. Acta*) **413**, 317 (1975).
72. Barfort, P., Arquilla, E. R., and Vogelhut, P. O., *Science* **160**, 1119 (1978).
73. Arwin, H. R., Vastra, F., and Lundstrom, K. I., German Patent No. 2643871 (1977).
74. Johnson, C. C., Moss, S. D., and Janata, J., U.S. Patent No. 4,020,380 (1977).
75. Janata, J., and Janata, J., U.S. Patent No. 3,966,580 (1976).
76. Martenson, K., and Mosbach, K., *Biotech. Bioeng.* **14**, 715 (1972).

77. Comte, P. A., and Janata, J., *Anal. Chim. Acta* **101**, 247 (1978).
78. Leistiko, O., Grove, A. S., and Sah, C. T., *IEEE Trans. Electron. Dev.* **ED-12**, 248 (1965).
79. Vadasz, L., and Grove, A. S., *IEEE Trans. Electron. Dev.* **ED-13**, 863 (1966).
80. Grove, A. S., Deal, B. E., Snow, E. H., and Sah, C. T., *Solid State Electron.* **8**, 145 (1965).

SUPPLEMENTARY REFERENCES

Leistiko, O., The selectivity and temperature characteristics of ISFETs, *Phys. Scr.* **18**, 445–450 (1978).

Vlasov, Yu. G., Ion selective field effect transistors (ISFET)—new kind of electrodes for chemical analysis and biomedical studies (in Russian), *Zh. Prikl. Khim.* **52**(1), 3–17 (1979).

Bos, M., Bergveld, P., and Van Veen-Blaauw, A. M. W., The ion sensitive field effect transistor in rapid acid-base titrations, *Anal. Chim. Acta* **109**, 145–148 (1979).

Janata, J., and Huber, R. J., Ion-sensitive field effect transistors, *Ion Select. Electr. Rev.* **1**, 31–79 (1979).

McKinley, B. A., Saffle, J., Jordan, W. S., Janata, J., Moss, S. D., and Westenskow, D. R., *In vivo* continuous monitoring of K^+ in animals using ISFETs, *Med. Instrum.* **14**, 93 (1980).

Zemel, J. N., Chemically sensitive devices, *Surf. Sci.* **86**, 322 (1979).

Chapter 4

A Compilation of Ion-Selective Membrane Electrode Literature

Richard P. Buck, James C. Thompsen, and Owen R. Melroy

INTRODUCTION

This tabular compilation of membrane electrode papers is based on previous, more comprehensive reviews. In fact, the theory and speculation literature has been deleted wherever possible, and the practical articles combined from the five previous biennial reviews. Errors in cataloging can most probably be resolved by consulting the original reviews.

Nomenclature, which ought to come first, has been put at the end, in the last table. Nomenclature is established by IUPAC and shows the idiosyncracies of the working committee. Since these reviews were prepared over a period of ten years, a nomenclature system logical to the writer developed spontaneously. The organization of the electrode field also fell into place naturally. With some regret, the classical electrodes of 0th, 1st, 2nd, and 3rd kind have been omitted. Certainly they are indispensible in their own right as reference electrodes. The electrodes cataloged here are "membrane" electrodes, even though the difference between membrane electrodes with metal backing (all-solid-state examples) and electrodes of 2nd kind may be obscure to some readers. The divisions occur among solid state electrodes,

---

R. P. Buck • Department of Chemistry, University of North Carolina, Chapel Hill, North Carolina 27514. *James C. Thompsen and Owen R. Melroy* • Graduate Students, Department of Chemistry, University of North Carolina, Chapel Hill, North Carolina 27514.

liquid ion-exchanger electrodes, neutral-carrier electrodes, gas, enzyme, and clinical (or bio-) electrodes, and finally classical synthetic organic and inorganic ion-exchange membranes.

In his 1974 review Covington prepared a table of highlights in the development of ion-selective electrodes. This table was appropriate for the types of electrodes thoroughly discussed in the text. However, it omitted many pioneering papers in the glass and synthetic organic and inorganic fixed-site membrane development. Lakshmknarayanaiah extended the table and I have gone considerably further by introducing glass, bioelectrode, and clinical microelectrode history.

The organization of tables in this chapter is as follows.

In Tables 2–15, references are listed chronologically: The capital letter following each reference number denotes the specific reference section at the end of this chapter in which the reference is listed. For example, the first reference in Table 2, Ref. 81B, can be found listed in the "B" reference section, which is a list of references from 1970 to 1971.

The original preparation of these reviews for 1970, 1972, and 1974 was through a page-by-page search of *Chemical Abstracts*. After acquisition of a computer search system for *Chemical Abstract* tapes, searching was done by key words and by authors. In every case literature was acquired until mid-December of odd years. Reviews were organized and written by mid-January. Consequently, an effort at completeness was made, but no pretense of digesting the papers or critical evaluation was made. If these reviews had a purpose, it was to provide a starting place for a proper library search. Surprisingly, I have not received letters from irate authors whose papers were inadvertently omitted; and I always welcomed reprints as reminders.

All of the books written to date have some merit. In every case there is some unique feature. In Table 2 I have listed monographs, edited volumes by "experts," proceedings of conferences, and some lengthy chapters. These are in approximately chronological order. I believe we have seen the end of a rash of general-coverage books and can expect volumes on specific specialty areas in future. Some, hopefully most, reviews are included in Tables 3 and 4. I have not included those written in obscure languages, although these were included in the source reviews. Most of the references here have already appeared in *Analytical Chemistry Annual Reviews*, **42**, 284R; **44**, 270R; **46**, 28R; **48**, 23R; and **50**, 17R. In addition I have included references from *Chemical Abstracts* through December 1978.

Of the solid-membrane electrodes, the LaF_3–fluoride sensor has been most welcome. There have been many hundreds of articles written about its characteristics and applications. This electrode, like the other solid, ion-selective electrodes is based on a material that can be classified as an ion exchanger. Suitable materials are insoluble in ionogenic solvents, although they adsorb and may absorb solvents. The classification "ion exchanger" includes materials such as LaF_3 which exchange ions of the same kind as their constituents. It has been found for AgCl, for example, that exposure of a wafer to radio Ag^+ leads to rapid incorporation of radio Ag^+ in quantities related to the time of exposure. There is no net weight change in the experiment, and pure ion-exchange process is indicated. Exchange rates for both cation and anion need not be rapid. For LaF_3, the fluoride exchange rate is much more rapid than the La^{3+} exchange rate.

For construction of practical devices, the ion-exchange process must be rapid and reversible for one ion or the other. Rapid ion exchange at the electrode surface means that local thermodynamic equilibrium is maintained, and the interfacial potential difference is established by equality of the exchanging ion's electrochemical potential. Rates of stepwise processes, such as adsorption of cations and anions at kink or half-crystal sites, diffusion over the crystal surface, and other processes occurring during growth or dissolution, are not observable at equilibrium. Consecutive chemical reactions exert predictable effects on ion-exchange processes. Concurrent interfacial reactions involving redox species, complexing agents, and metathetical reagents may distort ion-exchange equilibria, and provide new sensors. For example, CN^--sensing electrodes are possible from the corrosive attack of CN^- on AgI and other silver-salt-based membrane electrodes.

An additional requirement of these materials is that they be essentially pure ionic conductors. An electronic component of the solid-membrane conductivity does not cause problems except under unusual conditions involving strong redox reagents. In fact, the presence of some electronic conductivity is beneficial in that a metal-salt contact can be used in place of the inner-reference, ion-exchanging solution. The ability to produce this so-called all-solid-state membrane configuration electrode depends on a reasonable electron conductivity, such that reversible electron exchange can be achieved to create an ohmic-interface device.

The number of solid materials possessing the necessary properties at room temperature is small and restricted to solvent-insoluble salts. Among those considered for selective electrodes are group II and rare earth fluorides; halides of silver, lead, mercury, and thallium(I); sulfides of silver, lead, mercury, zinc, copper(I) and (II), and cadmium; selenides and tellurides of these metals; silver thiocyanate, azide, chromate, cyanide, and phosphate; and bismuth phosphate, lead phosphates, lead sulfate, and others. The metals of these salts are characteristically white, amalgam formers. Their cations possess nearly empty or filled *d* orbitals. Transition metal salts are conspicuously absent. Ions of metals in aqueous solution must be labile to assure rapid ion exchange at the membrane electrode surface. Strong crystal-field-stabilized high-charge ions will not be labile. Thus there are limited numbers of materials suitable for electrode construction for use in aqueous solutions. However, for measurements in nonaqueous and molten salt solvents, other salts may become suitable. The principal necessity of ionic conductivity is general, but the number of materials studied over a wide range of conditions for possible potentiometric application is still small.

The form of materials for solid state electrode applications can be divided into two categories: porous and nonporous, with respect to solvent transport. Organic ion-exchange membranes, precipitate membranes supported in parchment, the older protamine–collodion membranes, oxidized cellophane, and other types of organic membrane with fixed-sites are polyelectrolytes whose solubilities in water are virtually nil. Yet the structures are loose enough in the absence of strong cross-linking to permit solvent absorption. Porous membranes have little practical importance because of limited discrimination among ions of a given charge.

Nonporous materials possessing electrical properties as described include single crystals, disks cast from melts, compressed powders, glasses, and heterogeneous combinations of powders held in hydrophobic polymer binders. These latter membranes are called "supported" and are frequently composed of PVC, polyethylene, or silicone rubber binders. The form selected for an electrode application is based mainly on convenience of fabrication.

The commercial adaptation of low-resistance, permselective, cast-disk, and pressed-pellet membranes made from Ag_2S, AgCl, AgBr (or AgCl–Ag_2S, AgBr–Ag_2S), AgI–Ag_2S, Ag_2S–CuS, PbS, and CdS has provided chemists with new electrodes for the measurement of halide, thiocyanate, sulfide, Ag^+, Cu^{2+}, Cd^{2+}, and Pb^{2+} activities. Although AgCl and AgBr form pressed pellets from powders, AgI tends to crumble when the pressure on the pellet is released. Since soft Ag_2S is more insoluble than any of the other materials, it may be incorporated and serves as a binder. Mixed pellets of CuS, CdS, and PbS with Ag_2S produce electrodes with responses to activities of Cu^{2+}, Cd^{2+}, and Pb^{2+} ions though the typical equilibrium:

$$Cd^{2+} + Ag_2S \rightleftharpoons CdS + 2Ag^+$$

The powders should be prepared by coprecipitation from mixed metal-ion-containing electrolytes. Good responding electrodes require high-temperature and high-pressure pressing.

Pressed-pellet membranes formed from different batches of precipitates can be variable in their response characteristics, ranging from expected ideal Nernstian behavior for component ions to general salt responses without selectivity. Systematic studies of the CuS–Ag_2S and CdS–Ag_2S pressed-pellet responses show that precipitation from weakly acidic solutions of soluble Cd^{2+}, Cu^{2+}, and Ag^+ salts by addition of H_2S or by homogeneous precipitation leads to solids with reversible ion exchange of component cations. In contrast, precipitations made by addition of basic Na_2S presumably induce coprecipitated hydroxides such that log $a_{Cd^{2+}}$ plots vs. potential are not Nernstian. Responses become independent of cation activities at much too high activity values, and monovalent cations show significant interferences. Digestions of the precipitate in acid, and etching of pellet surfaces, improve responses and decrease interferences.

The main topics that have been treated in recent papers include the factors determining lowest-level activity responses in pure systems, role of interferences in terminating low-level responses of cations, nature of selectivity coefficients, and the interplay between nonequilibrium corrosive attack on electrodes and thermodynamic (equilibrium) conversion, origin of time-dependent potential responses, and the problem of ionic/electronic conductor contacts. The classical equilibrium theory of ionic-defect potentiometric responses has been often verified at equilibrium.[18A,19A] However, in mixed electrolytes of composition such that ion exchange occurs, but does not completely convert the underlying surface to a new form (e.g., Br^- attack on AgCl, at low Br^- activities), equilibrium selectivity coefficients do not apply. Hulanicki and Lewenstam [Talanta **23**, 661 (1976); **24**, 171 (1977); and *Proceedings of Ion-Selective Electrodes Conference 1977* (E. Pungor and I. Buzas, eds.), Elsevier, Amsterdam (1978)] have developed and tested a more general model which begins with flux balance during surface attack and ends in thermodynamic equilibrium. Selectivity coefficients vary between the large flux-balance values, depending upon diffusion coefficient ratios, to the much smaller equilibrium solubility product ratio values. In two papers, Liteanu, Hopirtean, and Popescu [*Anal. Chem.* **48**, 2010, 2013 (1976)] establish a statistical basis for the linear response domain (Nernstian or at least linear in log activity). Deviations from linearity at low activities has been a problem all along. Midgley [*Anal. Chem.* **49**, 1211 (1977)] has given mathematical procedures for analyzing low-activity, non-Nernstian responses to diagnose (1) the presence of responsibe species in reagents, (2) the presence of interfering species in the sample, and (3) nonequilibrium between electrode material and solutions as might occur when samples are not presaturated with electrode material. Nonequilibrium processes of the very slow variety frequently show up. Klasens and Goossen [*Anal. Chim. Acta* **88**, 41 (1977)] have also observed potential effects owing to corrosive attack. Yet at equilibrium, electrodes of the second kind and the membrane examples behave identically [C. Harzdorf and H. Keim, *Fresenius' Z. Anal. Chem.* **279**, 263 (1976)]. Baucke's work in Tables 5–9 confirms this result as well.

Exact values of selectivity coefficients remain a problem except in the equilibrium cases. Even then it is not clear experimentally that equilibrium has been achieved. Moody and Thomas have pointed out inconsistencies in published values of selectivity coefficients, and important factors to be considered before reporting values [*Talanta* **18**, 1251 (1971); **19**, 623 (1972)]. Pungor pointed out the corrosive attack and variable selectivity coefficient problem some years ago. In his many review papers, the most reliable selectivity coefficient data are accumulated. Temperature coefficients of electrode responses have not been well documented. Negus and Light [*Instrum. Technol.* **19**(12), 23 (1972)] have made the main contributions.

Part of the selectivity coefficient problem has to do with the slow time response of membrane electrodes. Fast responses are understood from impedance theory and analysis of impedance data. Slow response at noncorrosive surfaces, e.g., step activity changes for a preequilibrated electrolyte, are also known from the diffusion theory of the polarographic method adapted to ion-selective electrodes. However, experiments using rapid-activity steps show several moderately slow time constants [Shatkay, *Anal. Chem.* **48**, 1039 (1976)] and Lindner, Toth, and Pungor [*Anal. Chem.* **48**, 2013 (1976)]. The meaning of these time constants in terms of surface kinetics of ion transfer, or of bulk properties, has not been established. Analysis of various cases has been carried forward by Buffle and Parthasarathy [*Anal. Chim. Acta* **93**, 111 (1977); **93**, 121 (1977)]. Cammann has written out a concurrent ion-exchange theory for several ions of opposite or same charge with finite, small exchange rate kinetics

[*Proceedings of Ion-Selective Electrodes Conference, 1977* (E. Pungor and I. Buzas, eds.), pp. 297–306, Elsevier, Amsterdam (1978)]. Evidence for slow ion-exchange kinetics is difficult to obtain, but step activity experiments show the effect most clearly according to theoretical calculations.

Despite these practical problems, and the analysis of errors [Ebel, S., Glaser, E., and Seuring, A., *Fresenius' Z. Anal. Chem.* **291**(2), 108 (1978)], electrodes are being incorporated into many automatic pieces of analytical equipment [Vandeputte, M., Dryson, L., and Massart, D. L., in *Proceedings of Ion-Selective Electrodes Conference, 1977* (E. Pungor and I. Buzas, eds.), pp. 583–587, Elsevier, Amsterdam (1978); Lingerak, W. A., Bakker, F., and Slanina, J., *ibid.*, pp. 453–462]. Horvai, Domokos, and Pungor have begun to publish an interesting series on the subtle differences and advantages of the various "addition" and "subtraction" methods for internal calibration [*Fresenius' Z. Anal. Chem.* **292**(2), 132 (1978)].

Glass electrodes for pH and for a few monovalent ions (Na^+, K^+, and NH_4^+) represent the highest level of development of all ion-selective electrodes. Furthermore, the development of sodium-selective glasses and "cation"-selective glasses is one of the real achievements of theory applied to practice. These electrodes, when combined with equally good and stable reference electrodes, are capable of precise, thermodynamic measurements. The high standard of performance of junctionless-cell glass electrode measurements sets a goal for performance of the other ion-selective electrodes. Much of the literature in Table 10 is either on applications or examples of improvements. There are many good ideas on salt bridges, liquid junction designs, combination electrode designs, and electrode design. Among the problems currently addressed by researchers on glass electrodes are (1) the theoretical treatment of the role and response of the hydrolyzed surface films on glass electrodes, (2) analysis of the ionic composition profile and the resistivity profile normal to the glass surface, (3) evaluation of internal activity coefficients in the glass, (4) nature of transport through vacancies and accounting for interstitials, (5) evaluation of transport parameters, and (6) the nature of surface and diffuse charge.

It was an Orion liquid-membrane electrode which opened this field of membrane electrodes commercially some years ago. At one time, the years prior to about 1970, development of new liquid ion-exchanger membranes was exceedingly rapid. It seemed that the flexibility in choosing exchangers and supports was so great that electrodes for virtually any ionic species could be made with inherent selectivity. On hindsight, it was true that electrodes for many ions could be made, e.g., anion-sensitive electrodes based on Aliquat salts, but selectivity was not easily achieved. There is still an inability to design selective, charged ion exchangers. There is an absence of practical knowledge of the precise behavior of mediators, interactions between ions and exchangers, interactions between exchanger and organic polymer support, and interactions between ion-exchanger complex and organic polymer supports. Further progress in this field is highly dependent upon lucky guesses and insights, and upon extensive fundamental studies.

Some obvious correlations do exist. Anion selectivities tend to follow the Hofmeister or lyotropic series when the cation exchanger site is merely a charge center without specific interactions with anions. Sensitivities (low-activity-level detection limits) correlate with the ion-exchanger complex solubility in bathing electrolytes. Improvements in Ca^{2+}-selective electrodes have capitalized on use of exchangers with increased oil solubility. Fundamental studies using bi-ionic potential measurements, or back-to-back membrane vs. 2nd kind electrodes, show expected Nernstian responses with good correlations of selectivity coefficients with apparent ion extraction coefficients and with mobilities in the membrane. For reversible, rapid ion exchange (ions of the same charge) the steady state, segmented potential model interprets membrane potentials and interference responses fairly well, in terms of extraction, mobility, ion pairing, or complex formation parameters. Time responses are consistent with the need to reach a diffusion-migration steady state distribution of charged species across the membrane electrode interior. This requirement, except for a few simple cases involving single permeable ions, follows from theory, and distinguishes liquid-membrane electrode responses from solid state responses which can be very much more rapid.

The pioneering work of Simon's group has made neutral-carrier membranes broadly competitive, if not superior in performance, to the other types of responsive membranes. Groups at Cardiff, Newcastle-upon-Tyne, Prague, and Warsaw have also made many contributions and improved performances of electrodes. These membrane electrodes are responsive to cation activities, although one anion system (for HCO_3^-) has been described by Simon. Eisenman has also interpreted some transport effects in lipid bilayers in terms of anion-carried species.

Construction of non-cyclic neutral carriers by Simon's group shows the similarities with natural cyclic carriers. The presence of ether oxygens, carbonyl groups, and sulfur atoms, and the precise placement of these, are crucial to the molecule's ability to form complexes with cations. Their supporting work includes determinations of stability constants (in ethanol) for many alkali and alkaline earth ion-carrier complexes [*Helv. Chim. Acta* **59**, 397 (1976)], use of ^{13}C NMR to show rotational mobility of complexes [*Helv. Chim. Acta* **59**, 2327 (1976)], noninvolvement of C=O in some cases [*Tetrahedron Lett.* **1976**(20), 1709], and ion pairing with lipophilic anions [*Helv. Chim. Acta* **59**, 2407 (1976)]. The more recent reviews by Simon in Table 12-F (1978) are good sources for further reading. However, many other groups are involved in the carrier studies and the book *Topics in Current Chemistry 69*, *Inorganic Biochemistry*, Vol. 2,

p. 91, Springer-Verlag, New York (1977), by W. Burgermeister and R. Winkler-Oswatitsch, covers the extensive literature. Equally relevant are the studies appearing frequently in the biochemical and physiological literature. For example in 1975–1976, numbers of reviews of transport using ion carriers and ion-binding proteins appeared.[35D,45D,46D,52D,73D,74D,84D,119D,154D,161D,162D,203D,205D,217D] Papers on equilibrium properties, such as extraction equilibrium of ions using carriers and complex formation equilibria included Refs. 9D, 43D, 164D. Spectral studies of carrier complexes and ion pairs included Refs. 44D, 47D, 163D, 170D. Measurement of steady state fluxes, currents, and conductances to test the carrier model for Na^+, K^+, and NH_4^+ with actin carriers were by Hladky.[94D,95D]

Organic liquids dissolve salts to varying extents. Occasionally, when used in membrane concentration cell configuration, potentiometric responses are nearly Nernstian to changing cation activities on one side, although carriers have not been added! See Table 1 for classic papers on this effect. These membranes are generally thicker than the Debye length; one then expects potential responses more nearly like constrained liquid junctions because ions of both sign are presumably present and mobile in the organic phase. One would anticipate response slopes to be widely variable and sub-Nernstian with values from positive to negative depending on the relative mobility of cation and anion. However, presence of negative sites (probably neutralized by protons) can confer a region of cation response for bathing electrolytes by permitting anion exclusion. At high activities, changes in response slope, including a maximum in response and slope sign reversal, are possible [*J. Electroanal. Chem.* **82**, 345 (1977); **94**, 59 (1978); **100**, 63 (1979)]. Carriers are oil-soluble complex formers for selected ions. Consequently among their characteristics is the ability to confer some selectivity to the extraction of ions into a membrane phase, whether or not impurity, negative sites are present. Then the simplest view of neutral-carrier electrodes is that near-Nernstian response to a selected ion occurs because the carrier confers selectivity and mobility. The negative site permits uptake of an equal number of cation-carrier complex species, and allows anion exclusion by the conventional Donnan principle. Negative sites may be mobile or fixed in the membrane support.

At an interface, cation-carrier complexes must be relatively labile in comparison with the anion exchange rate. Then cation exchange determines the interfacial potential difference. Otherwise, as in the cases of iron and nickel bathophenanthroline cation complexes, the anion exchange rate is more rapid and the response follows anion activities. There are many other factors addressed in the papers in this table. Among these is the logical effect that oil-soluble anions in a bathing solution will terminate cation response selectivity at lower activities for the more oil-soluble species. Advantage of this effect can be had by incorporating oil-soluble anions, such as picrate, in membranes with carriers. The encroaching anions from the bathing electrolytes are frequently unable to compete and cation response is improved. It is observed and reported in many papers that a particular carrier behaves differently depending on the membrane solvent, the membrane support or the mediator. The specific roles of dielectric constant, complex formation with membrane species, ion-pairing, and higher aggregation species formation are not well known.

The impact of ion-selective electrodes in biology and medicine has been primarily in clinical laboratory measurements. The use of alkali and alkaline earth ion-sensitive electrodes for *in vitro* measurements is well documented. In addition the various enzyme-based electrodes, including gas-sensing electrodes, have made new batch analyses possible. In the long run, continuous monitoring *in vivo* and bed-side care applications will be the areas for major ion-selective electrode applications. The literature in the table following does not emphasize physiological measurements. However, some references to the very interesting microelectrodes using ion-selective tips, and to the open-tip electrodes, are included.

Gas-sensing electrodes and enzyme electrodes are examples of sensitized electrodes or electrodes using interposed chemical reactions. The oldest example is the Stow–Severinghaus pCO_2 electrode which is prepared from an ordinary pH sensing glass electrode. The interposed chemical reaction is the reversible conversion of CO_2 to H_2CO_3 in a thin layer of soluble bicarbonate solution on the electrode surface. The change in pH related to pCO_2 is registered by the electrode. This principle has been applied to the detection and monitoring of many gases that can react to yield a species sensed by the underlying electrode. The pioneering papers in this field have been accumulated in Table 1.

Synthetic ion-exchanger membranes and related homogeneous membrane electrodes have found little use analytically, because, in general, there are only slight discriminations or selectivity differences in the responses to ions of a given charge. There is relatively little additional selectivity for multivalent ions compared to monovalent ions. Solid state membranes tend to show improved selectivity differences for those multivalent ions comprising the membrane salt. The synthetic organic-based anion and cation-exchanger membranes, because of intrinsic structural weakness, swell, imbibe solvent, and lose selectivity based on ionic volume discrimination, unless they are highly cross-linked. Nevertheless, these membranes, often designated as SIEMs, are important for three reasons. They are hydrophobic and nearly inert, and can be used under extreme conditions of acidity and basicity (hostile environments), despite lack of high selectivities. They provide the simplest type of system for testing theories of ion transport in membranes, especially the coupling of ion and solvent transport. They are a basic starting material for experimental desalination schemes using hyperfiltration (reverse osmosis) and related methods.

Although their work is published in foreign journals that are not easily accessible, Russian workers have made progress and extensive application of SIEMs for acidity monitoring,[324B] end-point detection, and activity monitoring.[343B,344B] More recent references are 51D and 148D. An interesting new idea is the use of mediators in solid, fixed-site ion-exchanger membranes to modify selectivites.[179D] The extent of possible experimental applications is not as great as it seems on the surface. SIEMs are very concentrated electrolytes, unless purposely made with low site densities. We have found that many mediator solvents are not miscible with the membrane material, but rather tend to segregate. In Table 14, a list of application papers is provided.

TABLE 1. Highlights in the Development of Membrane Electrodes[a]

Investigator	Active material	Matrix	Selective to	Response slope (mV/decade conc.)	Comments	Ref.[b]
Cremer (1906)					Discovery of glass membrane response	1
Haber and Klemensiewicz (1909)	Thuringian glass	—	H^+	58.1 (20 °C)	Titration curve shapes used. pH not well defined at this time	2
Hughes (1928) MacInnes and Dole (1929, 1930)	Soda-lime glass	—	H^+	Nernstian response observed at pH values less than about 9, depending on conc. of interfering monovalent cations	Basic comparison of glass pH response with standard hydrogen electrode resp. Establishment of a minimum-resistance glass composition	3, 4
Ssokolof and Passynsky (1932)	Lithia-based glasses	—	H^+	Nernstian resp. up to pH 12.5 in absence of Li salt, but in presence of Na salts	Demonstrated Li^+ glasses show minimal Na^+, K^+ error. Provides basis for commercial, high-quality pH glasses	5
Lengyel and Blum (1934)	Alumino- and borosilicate glasses	—	H^+, Na^+	Nernstian slope pH range conspicuously shortened by Na^+ error responses for glasses with added Al and/or B	Established enhanced selectivity for Na^+ by presence of Al and/or B	6
Tendeloo (1936)	Fluorite, CaF_2	—	Ca^{2+}	Linear relation between emf and pCa	Anderson (ref. 8 below) found that it did not work as Ca selective	7

Kolthoff and Sanders (1937)	Silver halide disks	—	Cl^- Br^- I^-	57.9 57.4 52	AgI not affected by Cl^-, Br^-; $KMnO_4$ had no effect	9
Marshall (1939)	Natural zeolites (chabazite, apophyllite)	—	M^+ M^{2+}	Chabazite M^+ 40–50, M^{2+} 20–30 Apophyllite M^+ 49–58, M^{2+} 21–30	—	10
Marshall and co-workers (1941–1942)	Clay (montmorillinite, bentonite)	—	K^+ NH_4^+		Interference from Na but not from Ca, Mg	11, 12
Sollner and co-workers (1943–1954)	Collodion (oxidized in 1 *M* NaOH and dried)	—	Li^+, Na^+, K^+, NH_4^+, Mg^{2+}		Serious interferences, useful in titrations	13–15
	Collodion (treated with protamine sulfate and dried)		F^-, Cl^-, Ac^-, IO_3^-, NO_3^-, ClO_3^-, ClO_4^-		Serious interferences. First anion-responsive electrode	
Wyllie and Patnode (1950)	Commercial cation-exchange resins	Polymethyl methacrylate; polystyrene	Na^+	51–56		16
Sinha (1953–1955)	Commercial cation- and anion-exchange resins	Polystyrene	Cations, anions	—	Estimation of activities of cations and anions; acid–base titrations	17
Caldwell (1954)	Glass microelectrode	—	H^+		First microelectrode for pH. Used microbulb configurations	18
Woermann *et al.* (1956)	Ion-exchange resin containing dipicrylamine groups	Resin pressed to form a membrane with binder	K^+	—	Poor selectivity to K^+ over Na^+	19

continued overleaf

TABLE 1 *(Continued)*

Investigator	Active material	Matrix	Selective to	Response slope (mV/decade conc.)	Comments	Ref.[b]
Tendeloo and Krips (1957)	Calcium oxalate and other Ca salts	Paraffin + nonionic detergent on gauze	Ca^{2+}	—	Acceptable but criticized by Shatkay (ref. 21 below)	20
Eisenman, Rudin, and Casby (1957)	NAS 11-18 glass	—	Na^+	59 from pNa 1 to about 6	Devised a composition for highest-quality pNa responses. Low interference from K^+	22
Stow, Baer, and Randall (1957); Severinghaus and Bradley (1958)	Membrane, water; membrane, buffer	pH glass electrode	CO_2	Near-Nernstian	Classical CO_2 sensor for bioapplications	23, 24
Gregor and co-workers (1957–1964)	Alkaline earth and iron(III) stearates and/or palmitates	Multilayers formed between edges of cracked glass plate	Ca^{2+}, Ba^{2+}, Mg^{2+}, Sr^{2+}, Fe(III)	—	Fairly selective to alkaline earth ions. Used with Na and K solutions to estimate the activities of alkaline earth ions. Very difficult to fabricate the electrodes	25–28
Parsons (1958)	Commercial cation and anion exchangers	Polystyrene	Na^+	58 in the range pNa 1–3	K^+ interferes; used in titrations	29
Fischer and Babcock (1958)	$BaSO_4$, $BaCrO_4$	Paraffin without gauze	Ba^{2+}, SO_4^{2-}	—	Not selective to cations or anions	30
Hinke (1959)	Glass microelectrode	—	Na^+, K^+		First microelectrode for Na^+, K^+ monitoring	31

Tendeloo and Krips (1959)	Potassium tetraphenyl borate	Polystyrene + gauze	K^+	—	Not selective	32
Tendeloo and Van der Voort (1960)	Calcium stearate	Paraffin as above	Ca^{2+}	—	No response to K; stronger response than CaC_2O_4 electrode	33
Cloos and Fripiat (1960)	CaC_2O_4	Paraffin and detergent	Nonspecific	—	Shows memory effect	34
Pungor and Hollos-Rokosinyi (1961)	AgI	Paraffin	I^-	—	KCl does not interfere	35
Ilani (1963)	Organic liquids	Millipore filter	K^+ over Na^+ only for the liquid toluene + butanol	52 for log K^+ between 0.7 and 1.8	Membrane resistance high	36
Sollner and Shean (1964)	Amberlite La-2 (lauryl trialkylmethyl amine salts	Benzene, xylene, nitrobenzene	Cl^-, CNS^-	58 for KCl	—	37
Pungor *et al.* (1964)	Ion-exchange resins	Silicone rubber	SO_4^{2-}, Cl^-, OH^-, H^+, K^+, Zn^{2+}, Ni^{2+}	—	Not specific but selective to valence type	38
Pungor *et al.* (1964–1965)	$BaSO_4$, AgI	Silicone rubber	SO_4^{2-}, I^-	$K_2SO_4 = 24–30$ $KI = 50–60$	Phosphate interferes but not 0.1 *M* KCl	38, 39
Pungor *et al.* (1965)	Silver halides, manganese(III) phosphate aluminum oxine nickel dimethyl-glyoxime	Silicone rubber	Ag^+, X^-, PO_4^{3-}, Al^{3+}, Ni^{2+}	—	—	40
Morazzani-Pelletier and Baffier (1965)	Cobalt phosphate nickel dimethyl-glyoxime MnC_2O_4 NiC_2O_4	Collodion paraffin	Co^{2+}	20–23 in cobalt solutions	Poor responses to SO_4^{2-} porous membranes, highly permeable to KCl	41

continued overleaf

TABLE 1 *(Continued)*

Investigator	Active material	Matrix	Selective to	Response slope (mV/decade conc.)	Comments	Ref.[b]
Geyer and Syring (1966)	TiO_2, Fe_2O_3, SnO_2	Polyethylene	H^+, OH^-	—	Studied titrations	42
	ZrO_2	Polypropylene	Na^+		—	
	Al_2O_3	Paraffin				
	K_2SiF_6	Agar or paper	SiF_6^{2-}, K^+			
	$Ag_4Fe(CN)_6$	Agar	Both ions			
	$PbWO_4$	Paraffin				
Bonner and Lunney (1966)	Dinonylnaphthalene-sulfonate salts	Nitrobenzene + *o*-dichlorobenzene	Na^+, NH_4^+, Ca^{2+}	Less than theoretical	Useful range 3×10^{-1}–3×10^{-3} *M*	43
	Aliquat 336 (tricaprylmethyl ammonium chloride)	*o*-Dichlorobenzene				
Frant and Ross (1966)	LaF_3	—	F^-	58	OH^- interferes	44, 45
	Ag_2S	—	Ag^+, S^{2-}	59.1	Tested and characterized by Hseu and Rechnitz (ref. 46 below)	
Ross (1967)	Calcium didecyl phosphate in di-*n*-octylphenyl phosphonate	Filter	Ca^{2+}	Theoretical	H^+, Zn^{2+} interfere	47
Shatkay and co-workers (1967)	CaC_2O_4	Paraffin + nonionic detergent + gauze	Ca^{2+}	15–20	Not specific; not completely permselective	21, 48
	Theonyltrifluoro-acetone	Polyvinyl chloride + tributyl phosphate	Ca^{2+}	27–28	Very selective; compared well with the commercial electrode	

Coetzee and Freiser (1968)	Aliquat 336 in 1-decanol	Millipore filter	Anions	50–58	Depending on the ionic form, electrodes selective to particular anions	49
Schultz *et al.* (1968)	Calcium diethylhexyl phosphonate	Collodion	Ca^{2+}	28–29	Comparable with Orion liquid membrane. Evaluated by Rechnitz and Hseu (ref. 51 below)	50
Pioda *et al.* (1969)	Valinomycin in diphenyl ether	Filter	K^+	58.3	Highly selective for K^+ over Na^+	52
Guilbault and Montalvo (1969)	Enzyme, buffer	Cellophane or polymer gel on cation-sensitive glass electrode	Urea	sub-Nernstian	Operates 2 weeks in 10^{-4}–10^{-2} M region	53
Higuchi, Illian, and Tossounian (1970)	Organic liquids	PVC	Oil-soluble organic cations	Approaching Nernst	Mechanism not known	54
Liteanu and Hopirtean (1970)	Organic liquids	PVC	H^+ primarily	Sub-Nernstian	Suitable for acid–base titration monitoring	55
Moody, Oke, and Thomas (1970)	Calcium didecyl phosphonate in dioctylphenyl phosphonate	PVC	Ca^{2+}	Nernstian pCa 4.3–1	Long life	56
Ruzicka and co-workers (1970–1972)	Suitable salt in organic liquid	Porous graphite	Particular cation or anion concerned	—	This work resulted in the construction of selectrodes (refs. 58 and 59 below) selective to specific cations and anions	57

continued overleaf

TABLE 1 *(Continued)*

Investigator	Active material	Matrix	Selective to	Response slope (mV/decade conc.)	Comments	Ref.[b]
Cattrall and Freiser (1971)	Calcium didecyl phosphate in di-*n*-octylphenyl phosphonate	Platinum wire coated with PVC containing the electroactive material	Ca^{2+}	Theoretical	Found very promising	60
Levins (1971)	Polyethylene glycols as carriers with tetraphenylborate in *p*-nitroethylbenzene		Ba^{2+}	Nernstian pBa 5–1	Excellent selectivity over Ca and Mg	61
Amman, Pretsch, and Simon (1972)	Synthetic, non-cyclic carrier for Ca^{2+} in *p*-nitroethylbenzene	—	Ca^{2+}	Nernstian pCa 4–1	Excellent selectivity over Mg, Na	62
Rechnitz *et al.* (1977)	Membrane, buffer bacteria	Ammonia electrode	Arginine	Sub-Nernstian	5×10^{-5} to 1×10^{-3} *M* 2–20 day life	63

[a] After format for A. K. Covington, *Crit. Rev. Anal. Chem.* **3**, 355 (1974) and N. Lakshminarayanaiah, *Membrane Electrodes*, Academic Press, New York (1976).
[b] Numbers in this column refer to the references listed on pages 193 and 194.

1. M. Cremer, *Z. Biol.* **47**, 562 (1906).
2. F. Haber and Z. Klemensiewicz, *Z. Phys. Chem.* (*Leipzig*) **67**, 385 (1909).
3. W. S. Hughes, *J. Chem. Soc.* **1928**, 491.
4. D. A. MacInnes and M. Dole, *Ind. Eng. Chem. Anal. Ed.* **1**, 57 (1929); *J. Am. Chem. Soc.* **52**, 29 (1930).
5. S. I. Ssokolof and A. H. Passynsky, *Z. Phys. Chem. Abt. A* **160**, 366 (1932).
6. B. Lengyel and E. Blum, *Trans. Faraday Soc.* **30**, 461 (1934).
7. H. J. C. Tendeloo, *J. Biol. Chem.* **113**, 333 (1936); *Rec. Trav. Chim.* **55**, 227 (1936).
8. R. S. Anderson, *J. Biol. Chem.* **115**, 323 (1936).
9. I. M. Kolthoff and H. L. Sanders, *J. Am. Chem. Soc.* **59**, 416 (1937).
10. C. E. Marshall, *J. Phys. Chem.* **43**, 1155 (1939).
11. C. E. Marshall and W. E. Bergman, *J. Am. Chem. Soc.* **63**, 1911 (1941); *J. Phys. Chem.* **46**, 52, 325 (1942).
12. C. E. Marshall and C. A. Krinbill, *J. Am. Chem. Soc.* **64**, 1814 (1942).
13. K. Sollner, *J. Am. Chem. Soc.* **65**, 2260 (1943).
14. C. W. Carr and K. Sollner, *J. Gen. Physiol.* **28**, 119 (1944).
15. H. P. Gregor and K. Sollner, *J. Phys. Chem.* **50**, 88 (1946); **58**, 409 (1954).
16. M. R. J. Wyllie and H. W. Patnode, *J. Phys. Chem.* **54**, 204 (1950).
17. S. K. Sinha, *J. Indian Chem. Soc.* **30**, 529 (1953); **31**, 572, 577 (1954); **32**, 55 (1955).
18. P. C. Caldwell, *J. Physiol.* (*London*) **126**, 169 (1954).
19. D. Woermann, K. F. Bonhoeffer, and F. Hellferich, *Z. Phys. Chem.* (Frankfurt am Main) **8**, 265 (1956).
20. H. J. C. Tendeloo and A. Krips, *Rec. Trav. Chim.* **76**, 703, 946 (1957).
21. A. Shatkay, *Anal. Chem.* **39**, 1056 (1967).
22. G. Eisenman, D. O. Rudin, and J. U. Casby, *Science* **126**, 831 (1957).
23. R. W. Stow, R. F. Baer, and B. F. Randall, *Arch. Phys. Med. Rehabil.* **38**, 646 (1957).
24. J. W. Severinghaus and A. F. Bradley, *J. Appl. Physiol.* **13**, 515 (1958).
25. H. P. Gregor and H. Schonhorn, *J. Am. Chem. Soc.* **79**, 1507 (1957); **81**, 3911 (1959).
26. H. Schonhorn and H. P. Gregor, *J. Am. Chem. Soc.* **83**, 3576Z (1961).
27. H. P. Gregor, A. C. Glatz, and H. Schonhorn, *J. Am. Chem. Soc.* **85**, 3926 (1963).
28. J. Bagg and H. P. Gregor, *J. Am. Chem. Soc.* **86**, 3626 (1964).
29. J. S. Parsons, *Anal. Chem.* **30**, 1262 (1958).
30. R. B. Fischer and R. F. Babcock, *Anal. Chem.* **30**, 1732 (1958).
31. J. A. M. Hinke, *Nature* **184**, 1257 (1959), Suppl. No. 16.
32. H. J. C. Tendeloo and A. Krips, *Rec. Trav. Chim.* **78**, 177 (1959).
33. H. J. C. Tendeloo and F. H. Van der Voort, *Rev. Trav. Chim.* **79**, 639 (1960).
34. P. Cloos and J. J. Fripiat, *Bull. Soc. Chim. France* 423 (1960).
35. E. Pungor and E. Hollos-Rokosinyi, *Acta Chim. Acad. Sci. Hung.* **27**, 63 (1961).
36. A. Ilani, *J. Gen. Physiol.* **46**, 839 (1963).
37. K. Sollner and G. M. Shean, *J. Am. Chem. Soc.* **86**, 1901 (1964).
38. E. Pungor, J. Havas, and K. Toth, *Acta Chim. Acad. Sci. Hung.* **41**, 239 (1964).
39. E. Pungor, K. Toth, and J. Havas, *Hung. Sci. Instrum.* **3**, 2 (1965).
40. E. Pungor, J. Havas, and K. Toth, Z. Chem. **5**, 9 (1965).
41. S. Morazzani-Pelletier and M. A. Baffier, *J. Chim. Phys.* **62**, 429 (1965).
42. R. Geyer and W. Syring, *Z. Chem.* **6**, 92 (1966).
43. O. D. Bonner and D. C. Lunney, *J. Phys. Chem.* **70**, 1140 (1966).
44. M. S. Frant and J. W. Ross, Jr., *Science* **154**, 1553 (1966).
45. M. S. Frant and J. W. Ross, *Eastern Analytical Symposium*, Plenum Press, New York, (1968).
46. T. Hseu and G. A. Rechnitz, *Anal. Chem.* **40**, 1054 (1968).
47. J. W. Ross, Jr., *Science* **156**, 1378 (1967).

48. R. Bloch, A. Shatkay, and H. A. Saroff, *Biophys. J.* **7**, 865 (1967).
49. C. J. Coetzee and H. Freiser, *Anal. Chem.* **40**, 2071 (1968).
50. F. A. Schultz, A. J. Petersen, C. A. Mask, and R. P. Buck, *Science* **162**, 267 (1968).
51. G. A. Rechnitz and T. M. Hseu, *Anal. Chem.* **41**, 111 (1969).
52. L. A. R. Pioda, V. Stankova, and W. Simon, *Anal. Lett.* **2**, 665 (1969).
53. G. C. Guilbault and J. G. Montalvo, *Anal. Lett.* **4**, 283 (1969).
54. T. Higuchi, C. R. Illian, and J. L. Toussounian, *Anal. Chem.* **42**, 1674 (1970).
55. C. Liteanu and E. Hopirtean, *Talanta* **17**, 1067 (1970); *Rev. Roum. Chim.* **15**, 749 (1970).
56. G. J. Moody, R. B. Oke, and J. D. R. Thomas, *Analyst* **95**, 910 (1970).
57. J. Ruzicka and J. C. Tjell, *Anal. Chim. Acta* **49**, 346 (1970).
58. J. Ruzicka and C. J. Lamm, *Anal. Chim. Acta* **54**, 1 (1971).
59. J. Ruzicka, C. J. Lamm, and J. C. Tjell, *Anal. Chim. Acta* **62**, 15 (1972).
60. R. W. Cattrall and H. Freiser, *Anal. Chem.* **43**, 1905 (1971).
61. R. J. Levins, *Anal. Chem.* **43**, 1045 (1971).
62. D. Ammann, E. Pretsch, and W. Simon, *Anal. Lett.* **5**, 843 (1972); *Tetrahedron Lett.* (24), 2473 (1972).
63. G. A. Rechnitz, R. K. Kobos, S. J. Riechel, and C. R. Gebauer, *Anal. Chim. Acta* **94**, 357 (1977).

TABLE 2. Books and Articles on Ion-Selective Electrodes and Closely Related Topics (in Approximately Chronological Order)

Author	Title	Reference
Durst (ed.)	*Ion Selective Electrodes*	81B
Moody and Thomas	*Selective Ion Sensitive Electrodes*	244B
Lavallee, Schanne, and Hebert (eds.)	*Glass Microelectrodes*	195B
Buck (Weissberger, Rossiter, eds.)	*Potentiometry: pH Measurements and Ion Selective Electrodes*	46B
Cammann	*Das Arbeiten mit Ionenselectiven Elektroden* (two editions)	73C
Geddes	*Electrodes and the Measurement of Biological Events*	134C
Eisenman (ed.)	*Membranes—A Series of Advances*	114C
Pungor (ed.)	"Silicone Rubber Based Graphite Electrodes"	377C
Tenygl	"Ion-Selective Electrode Analysis," in *Int. Rev. Sci., Ser. One*	444C
Lakshminarayanaiah	"Membrane Phenomena," in Specialist Periodical Reports, *Electrochemistry*	266C
Koryta	*Ion-Selective Electrodes*	110D
Bailey	*Analysis with Ion-Selective Electrodes*	11E
Baiulescu and Cosofret	*Applications of Ion Selective Membrane Electrodes in Organic Analysis*	12E
Bloch and Lobel (Meares, ed.)	"Recent Developments in Ion-Selective Membrane Electrodes," in *Membrane Separation Processes*	28E
Buck	*Electroanalytical Chemistry of Membranes*	33E
Lakshminarayanaiah	"Membrane Phenomena," in Specialist Periodical Reports, *Electrochemistry*	170E

Author	Title	Reference
Tenygl	"Ion-Selective Electrode Analysis," in *Int. Rev. Sci. Phys. Chem. Ser. Two*	263E
Kessler, Clark, Lübbers, Silver, and Simon (eds.)	*Ion and Enzyme Electrodes in Biology and Medicine*	155E
Pungor (ed.)	*Ion-Selective Electrodes*, Proceedings of the Second Matrafured Conference, 1976	226E
Fuchs	*Ion-Selective Electrodes in Medicine*	96E
Lakshminarayanaiah	*Membrane Electrodes*	171E
Cavagnaro	"Ion Selective Electrodes," citations part 1 and 2	41F, 42F
Pungor (ed.)	*Ion Selective Electrodes*, Proceedings of the Budapest Conference, September 1977	194F
Vesely, Weiss, and Stulik	*Analysis With Ion-Selective Electrodes*	262F
Freiser (ed.)	*Ion Selective Electrodes in Analytical Chemistry*, Vol. I	68F

TABLE 3. Reviews of Principles and Applications (in Approximately Chronological Order)

Author	Reference
1968–1969	
Andelman	2A
Deelstra	35A
Johansson	76A
Kaminski	77A
Lambert	87A
Oehme	110A
Pungor	116A
Pungor and Toth	117A
Vanleugenhaghe	152A
Zinser	167A
1970–1971	
Cammann	51B, 52B, 53B
Covington	67B
Durst	82B
Florence	101B
Liteanu	207B
Pearson	287B
Simon	347B, 351B
Taubinger	370B
Ishibashi	162B
Moody and Thomas	243B
Moody, Oke, and Thomas	245B
Oehme	266B
Rechnitz	300B

continued overleaf

TABLE 3 (*Continued*)

Author	Reference
Pungor and Toth	296B
Marton and Pungor	221B, 222B
Pungor and Toth	297B, 298B, 299B
Toth and Pungor	374B
1972–1973	
Koryta	248C
Baumung	40C
Clerc and Pretsch	82C
Clerc, Kahr, Pretsch, Scholer, and Wuhrmann	83C
Fiori and Formaro	125aC
Hozumi	199C
Hulanicki	201C
Ijsseling	204C
Oehme	341C
Omang	248C
Pretsch, Scholer, Kahr, and Wuhrmann	374C
Pungor and Toth	376C, 378C–381C, 452C, 453C
Sapio and Braun	400C
Wilson	484C
1974–1975	
Moody and Thomas	141D
Covington	38D, 39D
Szczepaniak	194D
Sollner	191D
Riande	166D
1976–1977	
Koryta	165E
Pick	222E
Simon, Pretsch, Ammann, Morf, Gueggi, Bissig, and Kessler	253E
Mussini	201E
Freiser	93E
Thomas	265E
Durst	76E
1978	
Aizawa, Suszuki	3F
Antson and Suntola	6F
Bates and Robinson	10F
Brunfelt	30F
Buck	31F–33F
Childs	44F

Author	Reference
Ito and Ueno	112F
Kambara and Kataoka	123F
Kivalo and Virtanen	131F
Moody and Thomas	158F, 160F
Oehme	180F
Pataki	183F
Pick	185F–188F
Pungor, Toth, and Nagy	198F, 199F
Pungor	197F
Pungor and Toth	195F
Rock	204F
Thomas	245F
Weiss	261F

TABLE 4. Specific Application Reviews for Ion-Selective Electrodes[a]

Topic	References
Industrial processes	27B, 204B, 205B, 2C, 179C, 368C, 375C, 420C, 64E, 289E, 243F, 232F, 83F
Electroplating	103B, 368C, 449C
Toxicology and industrial hygiene	48B, 196E, 199E, 270F
Water and air quality and pollution	3B. 56B, 223B, 309B, 310B, 352B, 403B, 277C, 291C, 294C, 313C, 399C, 433C, 475C, 476C, 490C, 40E, 174E, 224E, 225E, 269E, 6D, 37D, 97D, 129D, 137D, 249F, 64F, 75F, 201F, 200F, 137F
Sea water	90C, 474C, 227E, 285E, 288E
Continuous monitoring	91C, 127C, 229C, 390C, 391C, 70F, 139F, 169F, 168F, 196F, 167F, 62F
Biological and biochemical applications	10C, 48C, 74C, 170C, 219C, 259C, 270C, 436C, 75E, 194E, 195E, 233E, 255E, 266E, 275E, 152F, 159F, 224F

[a]See also Table 5 for further examples and applications by electrode type.

TABLE 5. Properties and Applications of Fluoride-Selective Electrodes

Topic	References
A (1968–1969)	
1. Potentiometric response to pF in water and aqueous ethanol K_{sp} values for fresh LaF_3 and EuF_3	93A
2. Fluoride titrimetry with tetraphenylantimony	112A

continued overleaf

TABLE 5 *(Continued)*

Topic	References
3. Fluoride titrimetry with La^{3+}	3A
4. Fluoride titrimetry with Th^{4+}	91A
5. Automated potentiometric titrations of F^- using titrants Al^{3+}, La^{3+}, Y^{3+}, Ce^{3+}, or Th^{4+}	59A
6. Coulometric titration of F^- with La^{3+}	34A
7. Coulometric titration of F^- with Al^{3+}	107A
8. Kinetic analysis of F^- using LaF_3 electrode	126A
9. Continuous F^- monitor	5A
10. Microtitration of F^- using Al^{3+}	75A
11. Nonaqueous titration of Li^+ using F^-	6A
12. Direct potentiometry of F^- in aqueous solutions with various acids and inert salts	10A, 39A, 55A, 104A, 153A, 158A, 159A
13. Precision techniques for F^-, standard addition	7A
Precision techniques for F^-, analate addition	38A
14. Acidity effects and HF speciation	143A, 151A
15. Association constants for F^- complexes of Sn(II), Th(IV), and V(IV)	57A, 109A
16. Potentiometric responses of BiF_3	45A
Potentiometric responses of CaF_2	(99A)
17. Electrode design patent	138A
18. Heterogeneous, supported fluoride electrode	96A
B (1970–1971)	
1. Responses to F^- of LaF_3 doped successively with five rare earth fluorides	382B
2. Responses of LaF_3 to F^- at extreme dilution	20B
3. Responses of LaF_3 to F^- in 1 *M* NaCl	388B
4. Microelectrode design using LaF_3	80B
5. $Ca^{2\pm}$ response to mixed electrodes LaF_3–CaF_2 (10–50% CaF_2)	90B
6. F^- response unaffected by γ-dose rates less than 15,000 rads/min	187B
7. Reports of metal-contacted LaF_3 electrodes	91B, 93B
8. Titration of fluoride	4B, 26B, 87B
9. Potentiometry of F^-	96B
Potentiometry of F^- and SiF_6^{2-}	171B
10. Titration of Al(III)	24B
11. L-Chymotrypsin indirect potentiometry	88B
12. Activity coefficients in NaCl–NaF mixtures	49B
13. Stabilities of F^- conplexes of Cd, Mg, Ni, Zn, Tl(I), Ag	37B
14. Ionization constant of aq. HF at 25 °C	283B
15. Stability constants at 25 °C for HF, HF_2^-, AlF_{1-4} species	23B
16. Stability constants at 25 °C of HF and HF_2^- in 1–4 *M* $NaClO_4$	182B

Topic	References
17. Stability constants at 5, 25, and 45 °C of HF and ThF_{1-4} species	22B
18. Stabilities of thorium fluoride species	261B
19. Solubility of species in Pb(II)–F^- system	39B
20. Stabilities of fluoride species of Mg, Ca, Sr, and Ba	38B
21. Stability constants of MgF^+ and CaF^+	85B
22. Stability constants of MgF^+	109B
23. Stabilities of mixed species $BF_2(OH)_2^-$ and $BF(OH)_3$	247B
24. Stability constants of fluoroborate species $BF_n(OH)_{4-n}^-$	119B
25. Stabilities of Nb(V)–OH^-–F^- species	260B
26. Potentiometry of F^- in geological materials	62B, 95B, 159B, 160B, 295B, 394B
27. Potentiometry of F^- in soils	178B, 379B
28. Potentiometry of F^- in natural waters	177B, 389B, 390B
29. Potentiometry of F^- in potable waters	285B
Potentiometry of F^- in fluoridated water supplies	64B
Potentiometry of F^- in Austrian water	326B
30. Standard addition titration of F^- in sea water	5B
31. Potentiometry of F^- in plants	167B, 196B, 380B, 402B
32. Potentiometry of F^- in biological materials	401B
33. Potentiometry of F^- in vegetation and gases	44B, 272B
34. Potentiometry of F^- in air	31B, 202B
35. Potentiometry of F^- in preservatives and treated wood	71B, 383B
36. Potentiometry of F^- in organic and organometallic compounds using closed flask combustion	102B, 157B, 286B, 335B, 342B, 373B
37. Potentiometry of F^- in organic and organometallic compounds using hydrolysis	312B
38. Potentiometry of F^- in urine	57B, 257B, 377B
39. Potentiometry of F^- in serum	108B, 353B
40. Potentiometry of F^- in beverages	94B
41. Potentiometry of F^- in plaque and tooth deposits	32B, 354B
42. Potentiometry of F^- in difficultly soluble phosphates, and in the presence of phosphate species	43B, 78B, 289B, 333B, 378B
43. Potentiometry of F^- in insoluble aluminum, calcium, and silicon compounds	271B
C (1972–1973)	
1. Review of mechanisms and applications	211C
2. Studies of varying solubilities of LaF_3 in relation to applications	121C
3. Response mechanism studied by ^{18}F transport	427C
4. Similarities between glass and LaF_3 membranes	467C
5. Nonspecific response of LaF_3 membrane electrodes	493C
6. Microdetermination of F^- using Gran plots	413C
7. Studies of potentiometric titrations of F^-	81C, 117C, 175C, 442C

continued overleaf

TABLE 5 *(Continued)*

Topic	References
Measurement of F^-	
in dental pastes	299C
in enamel	269C
in glass	30C, 338C
in Cr plating baths	128C
in wood preservations	264C
in sugar cane	289C
in drinking water	45C, 65C
in natural water	473C
in clays	488C
in aluminum processing	351C, 435C, 448C
in welding fluxes and coating	52C
in phosphate processing	112C, 168C, 255C
in silicate rocks	222C
in pickling baths	116C, 118C
in Pu-containing oxides	292C
in HF production	256C
in urine	88C, 151C, 456C
in biological samples	167C, 246C
in plants and vegetation	25C, 200C, 220C, 221C, 446C
in organics containing B and P	393C
in household products	406C
in wine	103C, 325C
in organics	178C
9. Determination of activities in multicomponent systems	276C
10. Determination of dissociation constant of HF (student experiment)	95C
11. Determination of formation constants of U(VI), V(IV)–F^- complexes	7C
12. Determination of formation constants of Nb(V)–F^- complexes	268C
13. Determination of formation constants 1st row divalent transition metal ion–F^- complexes	53C
14. Determination of formation constants Al(III)–F^- complexes	6C
15. Indirect determination of Si, Al, Fe, Ca, and Mg	331C
16. A continuous monitor for F^-	218C
17. Use with furnace as gas chromatograph detector	243C
18. Use as a membrane for reference electrode in F^- melts	62C
19. An electrode for sensing Ca^{2+} and Mg^{2+} using CaF_2 and/or MgF_2 and LaF_3	122C
20. An improved LaF_3–EuF_2 membrane electrode	26C
21. Properties of LaOF as a solid electrolyte	359C

Topic	References
D (1974–1975)	
1. F^- electrode using sintered LaF_3, EuF_3, and CaF_2	90D
2. LaF_3—discovery of response to monofluorophosphate	17D
3. Procedures for determining fluoride in airborne dusts	96D
4. Establishment of F^--selective ISE as an official AOAC method for feeds	202D
E (1976–1977)	
1. Ag-bonded LaF_3 electrodes	177E, 178E, 273E
2. Patent on the Radelkes LaF_3 electrodes	68E
3. Extended F^- detection limits using nonaqueous solvents	261E
4. Use of Tiron as an Al^{3+} masking agent in F^- titrimetry	262E
5. Impedonic measurements and time responses of LaF_3 electrodes	188E, 280E
6. Round-robin F^- analyses results	143E, 279E
7. F^- in urine	203E
8. F^- in blood	254E
9. F^- in seawater	166E
10. F^- by standard addition technique	231E
11. F^- in microencapsulation processes	179E
F (1978)	
1. Mechanistic study of LaF_3 membranes, based on impedance	153F
2. Increased sensitivity in detection of F^- and Cl^-	105F
3. Fluoride profiles and fluoride depletion in the surface layers of LaF_3 electrodes	170F, 65F
4. Shaped crystals for macro- and microelectrodes	88F
5. Trace fluoride in reagent as a source of error	97F
6. Fluoride electrode based on BiF_3	63F
7. Determination of F^- in organoaqueous solvents	71F, 146F
8. Use of buffers for F^- determination in the presence of Al(III)	125F
9. F^- determination in water analysis	121F
10. Automated F^- monitoring in water	147F, 184F
11. Determination of F^- in soils	244F
12. Determination of F^- in milk	227F
13. Determination of F^- in steelworks	24F
14. Use of F^- in gas monitoring	106F, 258F
15. Analysis of trace F^- in nearly pure beryllium	191F
16. Analysis of F^- in various ores	1F, 92F, 104F, 89F, 250F, 269F

TABLE 6. Solid State Halide-Sensing Electrodes

Topic	References
A (1968–1969)	
1. Use of heterogeneous, supported AgI in titrimetry	32A
2. Direct potentiometry of $S_2O_3^{2-}$ and CN^- using heterogeneous, supported AgI electrodes	60A, 63A, 160A
3. Use of supported iodide electrodes in equilibrium studies of BiI_3 and I_3^-–I^-	23A
4. Use of heterogeneous, supported AgCl, AgBr electrodes including titrations	61A, 62A
5. Construction of a heterogeneous halide electrode	98A
6. Use of supported halide electrodes in automatic instruments	97A
7. Use of solid state membrane electrodes in titrimetry (78A), in potentiometry (154A) and in continuous monitoring (89A)	
B (1970–1971)	
1. Titration of Cl^- in plant extracts; $AgNO_3$ titrant	188B
2. Titration of Cl^- in natural waters; $AgNO_3$ titrant	97B
3. Differential potentiometry, 1–350 ppb in high-purity waters	100B
4. Titration of Cl^- in soil extracts	150B
5. Potentiometric determination of Cl^- in milk	251B
6. Microdetermination of Cl^- and N_3^-	334B
7. Potentiometric determination of Br^- in natural waters	288B
8. Comparison of pressed pellet with heterogeneous, supported membrane electrodes of AgI and Ag_2S	40B
9. Titration of Hg(II) using I^- electrode and standard NaI titrant	275B
10. Potentiometric determination of I^- and IO_3^- using AgI membrane electrode	281B
11. Potentiometric determination of I^- in organic materials, using hydrolysis	282B
12. I^- measurements in reacting systems using I^--sensing electrode	400B
13. Ruzicka–Lamm "electrodes" using halides and sulfides on hydrophobized carbon rods	319B, 320B
14. Titration of Cl^-, Br^-, I^- in mixtures, CN^- and SCN^- using standard $AgNO_3$ titrant	26B
15. Titrations of 2.5–60 μg of individual halide ions; 0.002 *M* $AgNO_3$ into approx. 75% HOAc solvent	184B
16. Potentiometric determination of Cl^- and Br^- in natural waters using corresponding electrodes	395B
17. Titration of halide mixtures using I^--selective electrode	176B
C (1972–1973)	
1. Use of Cl^- electrode to study Cd(II) and Zn(II) chloro-complexes	69C
2. Potentiometric determinations of anion in presence of interferences with Cl^--selective electrode	171C

Topic	References
3. Cl^--sensitive electrode using hydrous zirconium oxide in silicone rubber	443C
4. Determination of Cl^- in pesticides	458C
5. Determination of Cl^- in sweat	437C
6. Continuous monitoring of Cl^- in natural water	333C
7. Determination of Cl^- in soils and soil extracts	140C, 254C, 414C
8. Determination of Cl^- in sea water	343C
9. Determination of Cl^- in silicate rocks	174C
10. Determination of Cl^- in milk	177C
11. Determination of Cl^- in sugar industry	296C
12. Monitoring of Cl^- in refinery control	371C
13. Determination of Cl^- in pulping liquors	355C
14. Determination of Cl^- in acetonitrile	273C
15. Determination of Cl^- in brewing water	454C
16. Estimation of Br^- in plasma	104C
17. Applications of Br^--sensitive electrodes in	164C, 165C
18. Determination of Br^- in soft drinks	455C
19. Following Br^- in oscillating chemical reactions	247C
20. Determination of SCN^- with AgBr membrane electrode	245C
21. Response effects of polymorphic forms of AgI in AgI–Ag_2S membrane electrodes	147C
22. New model for the response of CN^- at I^--sensing electrode	59C
23. Response of the I^--selective electrode	107C
24. Response of the I^--selective electrode in concentrated aqueous electrolytes	324C
25. Use of I^- electrode to detect Au(III)	477C
26. Use of I^- electrode to determine Bi(III)	68C
27. Potential of, and salt diffusion in, parchment-supported AgI membranes	41C, 42C
28. Comparison of responses of commercially available CRYTUR electrodes with heterogeneous membranes and with 2nd kind electrodes	29C
29. Comparison and selectivity of single-crystal and heterogeneous-membrane electrodes	403C
30. Activity measurements in high ionic strength aqueous solutions	24C
31. Determination of azide with an AMEL(Italian)-selective electrode	58C
32. Determination of SO_3^{2-} with I^--selective electrode	206C
33. Applications in microanalysis of volatile organics (Cl^- and Br^-)	373C
34. Potentiometric measurements in nonaqueous media with Br^-- and I^--selective electrodes	124C
35. Determination of alkaloid halides using halide-sensitive ion electrodes	228C
36. Mercurimetric determination of halides and pseudohalides	423C
37. Applications to monitoring of reaction kinetics	242C
38. Potentiometric titration of Na pentacyanonitrosyl ferrate	422C

continued overleaf

TABLE 6 *(Continued)*

Topic	References
D (1974–1975)	
1. Establishment of Cl^- responses of a solid state ion-selective electrode in 10–100% iso-PrOH for $10^{-7} < Cl^- < 10^{-1}$ *M*	105D
2. Chloride electrode from HgS–Hg_2Cl_2 with Hg contact	115D
3. Chloride electrode from $AgCl$–Ag_2Te or Ag_2S–Hg_2S–Hg_2Cl_2	196D
4. Exploration of suspension effects and interferences in ion-selective electrode responses to Cl^- and Ca^{2+} in aqueous soil suspensions	83D, 112D
5. Iodide electrode from AgI–Ag_2Se	115D
6. AgI–Ag_2S electrode, study of Nernstian CN^- response from $2 < pCN < 6$ at pH 13	122D
7. Iodide electrode, Ag_3IS from 1:1 mixture of AgI–Ag_2S molded ~100 kg/cm^2 heated to just below mp of AgI in N_2	91D
8. Iodide electrode, reproducible preparation by spray of soluble metal solutions into precipitating agent. Product is an aerosol of definite size	16D
9. Iodide electrode response, identifies deviations in dilute I^- due to unstable, more soluble γ-AgI mixed with β-AgI. Sensitivity to light was measured	208D
10. Analysis of natural waters for I^- and CN^-	181D
11. AgCl, AgBr, AgI electrode studies of CN^- responses, including cations which form complexes (Cd^{2+}, Ni^{2+}, and Zn^{2+}). Attack of AgCl is rapid and produces unacceptable sensitivity to stirring	132D
E (1976–1977)	
1. Precise comparison of Ag/AgCl and AgCl membrane potentiometric responses	21E, 22E
2. Improved membrane claimed by melting Ag_2S–$AgCl$	292E
3. Cl^--sensitive micropipet tip electrodes using *in situ* generated AgCl	9E
4. Use of Cl^--sensitive electrode for formolysis kinetics in nonaqueous solvents	26E
5. Analysis of Cl^- in high-purity water and heavy water using Hg_2Cl_2–HgS electrode	244E
6. Br^--selective electrode from Hg_2Br_2–HgS	271E
7. Demonstration of upper response limits of AgI-based electrodes by dissolution	215E–219E
8. Micro- and submicrodeterminations of thiocarbonyl and thiol groups using I^- electrode	119E
9. Responses of silicone rubber-supported AgI membranes with regard to surface morphology changes	182E
10. Halide-selective electrodes made from thin films of AgX or Ag_2S; Ag_3PO_4 on $Ag_{19}I_{15}P_2O_7$ responsive to HPO_4^{2-}; other examples in patent	276E–278E

Topic	References
11. Halide-selective AgX film on pressed powder of Ag and on Ag foil	169E
12. Halide-selective AgX film on glass with metal contact	8E
13. Halide-selective (Br^-, Cl^-, SCN^-, CN^-) electrodes based on Hg(I) salts and Hg_2S. Comparison with corresponding silver-based electrodes	242E
14. Solid state electrodes based on Ag or Hg sulfide, selenide, and telluride matrices. Includes electrodes for halides and heavy metal cations. Transition metal electrodes are unsatisfactory, as expected	243E
15. Direct measurement of halide interferences with halide-sensitive electrode	71E–73E
16. Improving halide detection limits using alcohol–water mixtures	30E, 151E
17. AgX–Ag_2S mixtures on Ag; PbS and Pb; Ag_2Te–MnTe–Mn	27E
F (1978)	
1. Review of solid state membrane electrodes	78F
2. Theory and practice of Cl^- determination with ion-selective electrodes	242F
3. Errors in Cl^- analysis induced by the presence of Fe(III)	17F
4. Sweat testing for cystic fibrosis	25F
5. Composition for a Cl^- electrodes AgCl : Ag_2S in epoxy resin	236F
6. Potentiometric determination of Cl^- in water	107F
7. Test to show lack of interference from SO_4^{2-} in Cl^- determination	74F
8. Test of AgCl and AgI electrodes in mixed solvents, DMF, DMSO, and dimethylacetamide	35F
9. Coulometric method for calibrating a Br^--selective electrode	79F
10. AgI-, Ag_3SBr- and Ag_3SI-based electrodes show Nernstian responses when chemical treatment removes or neutralized adsorbed ions	252F, 254F
11. Microdetermination of Hg(II) using I^--selective electrode	251F
12. Preparation of Ag-based electrodes using salts and metallic Ag	181F
13. Construction of pressed pellet halide-sensing electrodes	8F
14. Electrical properties of mixed AgX–Ag_2S electrode compositions	209F
15. Analysis of multi-ion-containing solutions using electrodes	256F
16. Halide determinations in natural waters	127F

TABLE 7. Solid State Cyanide Responses[a]

Topic	References
1. Potentiometry of CN^-, indirect determination via Ag^+ in equilibrium with $Ag_2(CN)_2$	99B
2. Titration of CN^- using standard Ni^{2+} solution	368B
3. Potentiometry of CN^- in waters, wastes, and biological materials	65B
4. Potentiometry of CN^- to determine cyanogenic glucosides, using hydrolysis, acid distillation	33B
5. Potentiometry of CN^- in forage	116B
6. Influence of pH on the response of a CN^--sensing electrode	302C
7. Diffusion barrier model for the CN^--sensing electrode	120C
8. Determination of rhodanase with a CN^--selective electrode	203C
9. Cyanide determination after distillation	360C
10. Cyanide determination in the presence of chelating agents	181C, 244C
11. Cyanide determination in mineral waters	478C
12. Cyanide determination in solutions	154C
13. Apparatus for free and combined CN^-	290C
14. Determination of CN^- in cigarette smoke	468C
15. Determination of CN^- using standard addition	129C
16. Potentiometric titration of CN^- and Cl^-	89C
17. Determination of CN^- in waste waters	392C
18. Determination of CN^- in flowing systems	126C
19. Ag_2S electrode responses to CN^-, I^-, Hg^{2+}, (Hg_2^{2+}), and H^+	159D
20. AgX- and Ag_sS-Au electrode responses to component ions and to CN^-	205aD
21. Cyanide response applications of Ag_2S electrodes	50E
22. Continuous monitoring of HCN with gas-permeable membrane and Ag^+ electrode	61F
23. Potentiometric monitoring of CN^-	13F
24. CN^- monitoring in sea water	130F
25. Determination of CN^- in presence of sulfur compounds	171F

[a] See also entries in Table 6(A–E).

TABLE 8. Solid State, Mainly Sulfide-Sensing, Electrodes

Topic	References
A (1968–1969)	
1. Response evaluation of Ag_2S membrane electrodes	68A, 92A
2. Application of Ag_2S electrode in titrimetry with Ag^+	24A
3. Direct potentiometry with Ag_2S electrodes	13A
4. Complex chemistry study of SnS_3^{2-} using Ag_2S electrodes	68A
B (1970–1971)	
1. Ag_2S membrane electrode	105B
2. Application of Ag_2S electrode	81B

Topic	References
3. Evaluation of an Ag_2S electrode	236B
4. Titration of sulfide using 10^{-4}–10^{-1} *M* Na plumbite; CN^- interferes; Cl^-, Br^-, I^-, SO_3^{2-}, and SCN^- do not interfere	256B
5. Titration of sulfide using $AgNO_3$ titrant; $S_2O_3^{2-}$ interferes	124B
6. Titration of thiols; Hg^{2+} interferes	123B
7. Titration of nanogram amounts of S^{2-} using Pb^{2+} in 1 *M* NaOH, 1.5 *M* N_2H_4 solution presaturated with PbS	356B
8. Titration of sulfide in pulping liquors using silver ion titrant and back titration. Direct potentiometry also is possible with known addition method	364B
9. Potentiometry of sulfide in lime-sulfide solutions	231B
10. Determination of NH_4 diethyl dithiophosphate in flotation mill solutions	313B
C (1972–1973)	
1. Metal selenides, tellurides, and sulfides as materials for ion-selective electrodes	465C
2. Microelectrodes based on materials used in membrane ion-selective electrodes	99C, 172C
3. Comparison of three methods for sulfide titration using ion-selective electrode end-point detection	205C, 336C
4. Ag_2S ceramic electrode as chromatographic detectors	285C
5. Field of existence of Ag_2S (II)	54C
6. Studies on thioacetimide with sulfide-selective electrodes	354C
7. Determination of thiourea and *p*-urazine	258C
8. Potentiometric titration of trace sulfur in petroleum	260C
9. Potentiometric titration of thiols	361C
10. Sulfide determination in solids	11C
11. Sulfide determination in slurries	1C
12. Sulfide determination in organics	424C
13. Use of sulfide-sensing membrane to study *o*-acetylserine sulfhydrylase	335C
14. Micro- and semimicrodetermination of thiols	412C
15. Responses of a series of Ag_2S membranes showing varying E^0 values with methods of preparation	466C
16. Selectivity coefficients of sulfide-sensitive electrodes	407C, 408C
17. Durable, two-layer construction for ion-selective electrodes	31C
18. Determination of S^{2-} and S_2^{2-} with sulfide-sensitive electrodes	438C
19. CRYTUR—single-crystal electrodes, selectivities, and other properties	3C
20. Applications of CRYTUR electrodes	432C
21. Properties of the $Cu_{1.4}Se$ CRYTUR electrode	464C
D (1975–1976)	
1. Demonstration of titrimetry of H_2S with $CdCl_2$ monitored by solid state ion-selective electrode in nonaqueous solvent (aimed at petroleum fraction analysis)	172D

continued overleaf

TABLE 8 (*Continued*)

Topic	References
2. Sulfide electrode: Ag_2S–Ag_2Se	115D
3. Sulfate electrode: Ag_2Se–$PbSe$–$PbSO_4$	115D
4. Ag_2S electrode response; Nernstian pS^{2-} from $1 < pS^{2-} < 7$ at pH 14	158D
5. Solid state electrodes based on sulfides and halides; instead of using Ag_2S, 5–95 mol % Cu_2S is used	210D
E (1976–1977)	
1. Determination of xanthates by potentiometric titration and Ag_2S electrode	116E
2. Sulfite-sensitive electrode from melted AgCl, Hg_2Cl_2, and Ag_2S	104E
3. Electrode for detection of H_2S using Ag_2S, Ag/AgCl (ref.), air gap, and porous Teflon membrane	291E
4. Potentiometric determination of sulfite with Hg_2Cl_2–HgS electrode	272E
F (1978)	
1. Construction of a Ag_2S-electrode on Ag foil	128F
2. Patent on a Ag_2S electrode structure	142F
3. Potentiometric monitoring of dithiooxamide during Ag^+ titration	218F
4. Prereduction system for determination of a variety of sulfur compounds (including SO_4^{2-}) as sulfide with an electrode	46F
5. Use of electrode in control of sulfidation of oxidized copper ores	119F, 118F

TABLE 9. Solid State Electrodes for Divalent Ions and Miscellaneous

Topic	References
A (1968–1969)	
1. Heterogeneous, supported membranes containing sulfates, phosphates, and carbonates of divalent metal ions, as proposed cation activity sensors	17A
2. A heterogeneous, supported Ca^{2+} sensor	82A
3. Use of PbS–Ag_2S electrodes in sulfate titrimetry	128A
4. Use of CuS–Ag_2S electrodes in chelometric titrimetry of Cu^{2+}	119A
5. Response of a heterogeneous Ca^{2+} sensor	118A
B (1970–1971)	
1. Metal-sulfide–Ag_2S electrodes	104B
2. AgSCN–Ag_2S electrode patent	104B
3. Hot-pressed PbS, Cu_2S–Ag_2S electrodes	151B
4. Testing of Cu_2S–Ag_2S electrode	152B
5. Cu(II) tetracyanoquinodimethan-radical salt electrode	340B

Topic	References
6. Patents on construction	92B, 127B, 155B, 224B, 268B, 291B, 349B
7. Titration of Cu(II) with EDTA and stability of complexes, using Cu-selective electrode	139B
8. Pb(II) titrations in aqueous media, nonaqueous media and in biological systems, using Pb-selective electrode	304B
9. Potentiometric determination of Ag^+ in 10^{-7} M range	249B
10. Potentiometric determination of Ag^+ to follow adsorption of Ag^+	83B
11. Titration of Zr(IV), Fe(III), Th(IV), Hg(II), Sm(III), La(III), and Ca(II) with EDTA using Cu(II)–EDTA intermediate and Cu-selective electrode	21B
12. Titration of Cu(II), Ni(II), and Zn(II) with 1,10-phenanthroline, tetraethylenepentamine and EDTA	314B
13. Back titration of Al(III) with Cu(II) diaminocyclohexanetetraacetic acid and Cu-selective electrodes	362B
14. Titration of nitriloacetic acid using Cu(II) titrant and Cu-selective electrode	301B
15. Pb^{2+} detection for lead poisoning	303B
16. Interferences and titration applications using Cd^{2+} and Cd-sensitive electrode in water and in DMSO	42B
17. Microtitration of phosphate using $Pb(ClO_4)_2$ titrant, pH 8.25–8.75, Pb-selective electrode	336B
18. Semimicrotitration of oxalate using $Pb(ClO_4)_2$ titrant in 40% *p*-dioxane–water, pH 3.5–10.5 with Pb-selective electrode	337B
C (1972–1973)	
1. Response vs. electrode preparation for CdS–AgS electrodes	303C
2. Storage of sulfide-based electrodes in reducing media	367C
3. Effects of surface preparation on electrode response	226C
4. Patented sulfate-selective electrode based on $BaSO_4$ in silicone rubber	401C
5. Potentiometric titration of SO_4^{2-} in mineral and sea water with Pb^{2+}-selective electrode	304C
6. Responses of a natural Cu_2S electrode to sulfide	56C
7. A polythene-supported Ag_2S–CdS electrode	303C
8. A silicone rubber-supported Ag_2S–CdS electrode	195C
9. Responses of a synthetic Cu_2S electrode to Cu^{2+}	404C
10. Use of a synthetic Cu_2S electrode to study Cu(I) complexes in acetonitrile	176C
11. Preparation and responses of CuI electrodes	337C
12. Potentiometric determination of tetraphenylborate with Ag^+ using a halide-responsive electrode	421C
13. Responses of mixed sulfide electrodes for Cu^{2+}, Cd^{2+}, Pb^{2+}, Hg^{2+}, Ni^{2+}, CO^{2+} and Zn^{2+}	16C
14. Cu^{2+} determination in water by standard addition	426C

continued overleaf

TABLE 9 (*Continued*)

Topic	References
15. Potentials across $Co_2Fe(CN)_6$-supported membranes; properties of Mn, Co, Ag, Cd, ferrocyanide membranes	352C, 417C
16. Determination of silver in ZnS, ZnSe, CdS, and CdSe	207C
17. Determination of silver in fiixing baths	471C
18. Flow-through electrode based on Cu^{2+}, Cd^{2+}, and Pb^{2+}-sensing compositions	447C
19. Potentiometric titrations of SO_4^{2-}, Ba^{2+}, and Sr^{2+}, using Pb^{2+}-selective electrode	169C
20. Potentiometric titrations using Cu^{2+}- and Pb^{2+}-selective electrodes	349C
21. Optimization of compositions for PbS–Ag_2S and CdS–Ag_2S electrodes	144C
22. Ag^+-sensitive electrode made from Ag_2Se, Ag_2Te	187C
23. Cd^{2+}, Zn^{2+}, Cu^{2+}, Pb^{2+}-sensitive electrodes from 24 mol % metal telluride with 75 mol % Ag_2S	186C, 188C, 189C, 190C
24. Responses of PbSe, PbTe–Ag_2S electrodes	193C
25. Mn^{2+}-, Co^{2+}-, Ni^{2+}-, and (Cr^{3+})-sensitive electrodes from 25 mol % metal telluride and 75 mol % Ag_2S	183C, 184C
26. Cr^{3+} is detected with Ni^{2+} electrode	185C, 191C
27. Response of a Cd^{2+}-selective ceramic electrode	194C
28. Response of a Pb^{2+}-selective ceramic electrode	192C
29. Behavior of CuS–Ag_2S electrodes	226C
30. Effect of Cu^{2+} on algal growth	425C
31. Behavior of PbS-, PbSe-, PbTe-based electrodes in buffered Pb^{2+} solutions	293C
32. A silicone-based Cs–12 molybdophosphate Cs^+-sensitive electrode	85C
33. Potentiometric determination of Cs^+ using the former electrode	84C
34. An epoxy-based Cs^+-sensitive electrode	86C
35. An epoxy-based Tl^+-sensitive electrode using Tl(I) molybdo- or tungstophosphate	87C
36. A silicone-based Cu(II)-sensitive electrode	363C
37. A collodion-paraffin membrane selective to Cu^{2+} and Ni^{2+}	314C
38. A silicone-based, $Cu_{2-x}S$, Cu^{2+}-sensitive electrode	365C
39. Applications of the previous electrode	367C
40. Ag_2S ceramic electrodes—statistical study of Ag^+ response	282C
41. A polythene-based Pb^{2+}-sensitive electrode	301C
42. Ag_2S ceramic electrodes—influence of preparation conditions on Ag^+ response	372C
43. "Selectrode" applications	167aC, 267aC, 395aC, 395bC
44. Sulfate-sensing electrode based on PbS–$PbSO_4$–Cu_2S–Ag_2S and on PbI_2	319aC, 388aC, 169aC

Topic	References
D (1974–1975)	
1. Responses of Ag_2S-based electrodes in terms of preparation conditions	103D, 130D, 153D
2. Air oxidation and leaching effects on Ag_2S-based and/or AgI-based electrodes	23D, 41D, 173D, 190D, 208D
3. SO_4^{2-}-electrode response determination from 2×10^{-4} to 2×10^{-2} M	11D
4. Orion Cu^{2+} electrode response; Cl^- in test solutions causes irregular response and "tarnish" on electrode surface	40D
5. Orion Cu^{2+} study of responses using Cu^{2+} buffers and converse experiments to determine Cu^{2+} activities and stability constants for complexing agents	147D
6. AgI–Ag_2S electrode responses to Hg^{2+} solutions	160D
7. Glassy $PbCrO_4$ and $PbCrO_4$ in silicone rubber. Metal contacts, Pt and Hg, used; claims responses to CrO_4^{2-}. Claim interferences by SO_4^{2-} and HPO_4^{2-}; electrodes appear sensitive to H^+ and cations	78D
8. TlI–AgI electrodes respond indirectly to Tl^+ through Ag^+. Selectivity coefficients for many cations are reported. I^- is an interferent	68D
9. Ag_2HgI_4 electrode gives linear response to pAg^+, pHg^{2+}, and pI^- from 1 to 6.5	67D
10. AgI and AgI–Ag_2S electrode response study over a wide activity range	208D
11. Ag_2S electrode response characterization from $0 < pAg < 7$ by dilution, and $4.3 < pAg < 23.5$ by precipitation and complexation	121D
12. Cd, Pb, Cu, with Ag_2S; 5-μm powders of Cd, Pb, Cu sulfides treated with aqueous $AgNO_3$ to yield metathetical surface reactions. Later pulverized and pressed into electrodes	79D, 80D, 81D
13. Cu^{2+} electrode patents on $Cu_{2-x}S$, $x = 0$–2.1	92D, 93D, 204D
14. Cd^{2+} electrode from CdSe–Ag_2S	146D
15. Cd^{2+} electrode from CdS–Ag_2S–Bi_2O_3 (25–60–15 wt %)	82D
16. Pb^{2+} electrode from single-crystal PbS	36D
17. Hg^{2+} electrode from Ag_2HgI_4 in epoxy binder	69D
18. Tl^+ electrode from TlI–AgI (3/7 wt. ratio)	70D
19. Various ions, PTFE binder for various solids, among these $AlPO_4$, $BaSO_4$, and Cd_3PO_4	151D
E (1976–1977)	
1. Production of Cu^{2+} at $Cu_{1.8}Se$-, CuS–Ag_2S-, and Cu^{2+}-sensing "Selectrodes," as a function of ambient O_2 and pH	191E
2. Response of CuS–Ag_2S electrodes to chelating compounds	210E
3. Effects of Cl^- and F^- on calibration curves of Cu^{2+}-sensing electrodes	192E, 209E
4. Responses of three commercial Cd^{2+}- and Pb^{2+}-sensing electrodes	161E, 162E

continued overleaf

TABLE 9 (*Continued*)

Topic	References
5. Response character of TCNQ radical anion electrodes	247E
6. Titration of benzyldithiocarbamate with Cu^{2+}, monitored with a Cu^{2+}-sensing electrode	55E
7. Extraction rates monitored by Cu^{2+}-sensing electrode	160E
8. Use of Cu^{2+}-selective electrodes in chelometric titrations–back titrations	281E
9. Microcell (25 μl) Cu^{2+}-selective electrode with standard addition	282E
10. Use of Cu^{2+}-selective and Cd^{2+}-selective electrodes to monitor EDTA titration of VO^{2+}	202E
11. Response of Ag_2S electrodes to Cu^{2+}; use in complexometric titrimetry	127E
12. Single-crystal Cu_2S and a study of responses to Cu^{2+} and Cu(I) activities	139E
13. Copper telluride $Cu_{1.75}Te$ as a Cu^{2+} sensor	105E
14. Ternary and quaternary compounds of S, Se, and Te with Cu(II), Hg(II), Mn(II), Cd(II), and Pb(II) and Ge, Sn, and In; studied as possible ion-selective electrodes	65E
15. From melts (Cd, Ag)S functions as Cd(II) sensor; (Zn, Ag)S fails to sense Zn^{2+} activities	293E
16. Potassium zinc ferrocyanide in silicone rubber responds to K^+ activities. Selectivity over Na^+ is about the same as cation glass electrodes	232E
17. Cs and Tl tungstoarsenates in Araldite resin are suggested as selective electrode components. Calibrations are improved in mixed solvents	181E
18. $CaWO_4$ and $BaMoO_4$ are reported as responsive to Ca^{2+} and Ba^{2+} activities	284E
19. Ag–heavy metal oxides, fired on glass substrates	109E
20. Determination of sulfate by Pb^{2+} titration: (a) in 80% isopropanol with commercial Pb^{2+}-sensitive electrode; (b) using single-crystal PbS electrode	236E, 117E
21. Another study of SO_4^{2-} responses of $BaSO_4$ in PVC: claims approx. Nernstian response from 10^{-4} to 2×10^{-1} *M*	260E
F (1978)	
1. Compositions and construction of Cu(II)-, Cd(II)-, and Pb(II)-sensing electrodes using Ag_2S–MS	7F
2. All-solid-state configuration preparation from fused materials, including use of halides, sulfides, and selenides	214F, 223F
3. Study of CdS–Ag_2S over wide composition range with point of view of doping effects on responses	174F, 175F
4. Construction by chemical means and vapor deposition of electrodes in flat form on Ag_2S, Ag_3SBr, or Ag_3SI supports	252F

Topic	References
5. Series of papers on physical–chemical properties of Cu(II)- and Pb(II)-sensing electrodes with relation to potentiometric and titration responses	93F–96F, 255F
6. Patent of copper-sensitive electrodes using selenides	248F
7. Description of copper-sensitive electrode using sulfides	260F
8. Patent and paper on a phosphate-sensitive composition using Ag_3PO_4	179F, 238F
9. Patent on phosphate- and arsenate-sensing electrodes based on Pb compounds	241F
10. Patent on an arsenate-sensing electrode based on Ag compounds	43F
11. Sulfates responses of homogeneous $PbSO_4$–PbS–Ag_2S and heterogeneous $BaSO_4$ in PVC	156F
12. Flat-form Cu and Hg-chalcogenide electrodes using conducting base	173F
13. Determining Pb(II) in polluted soils (264F) and complex formation of Cu(II) with soil fulvic acids (28F)	
14. Use of Cu(II)-sensing electrode for Cu in Si (239F), and Cu(II) in water (108F)	
15. Determination of xanthates (9F) and tartrate (59F)	
16. Hg(II)-sensitive electrode from $AgPO_3$–HgI_2 glassy system	20F
17. Ni(II)-sensitive electrode based on nickel dimethylglyoxime	155F
18. Patent on Mg^{2+} sensor from MgF_2, SiO_2, and binder	129F
19. Stearic and Ca stearate on Pd-hydride and Pd as a Ca^{2+} sensor	56F
20. Sr^{2+}-sensitive electrodes from $Sr_3(AsW_{12}O_{40})_2$	116F
21. Second-type on all-solid-state membrane electrodes for Ag and Cu from vapor sulfided wires of corresponding metals	257F

TABLE 10. Glass Electrodes

Topic	References
A (1968–1969)	
1. Practice of pH determination	56A
2. Blood pH determination	147A
3. pH standards and uses	8A, 31A
4. Porous glass and porous ceramic potentiometric responses	1A, 65A, 84A, 142A
5. Micro-open-tip electrodes	129A
6. Micro-pH, -pNa, -pK electrodes	79A, 156A
7. Origins of pH glass electrode potentials and selectivities	14A, 20A, 40A, 150A
8. Double-layer effects and time constants of glass electrodes	21A, 22A, 76aA, 120A, 149A, 166A
9. Low-Na^+-error pH glass compositions	12A, 100A, 101A, 146A

continued overleaf

TABLE 10. *(Continued)*

Topic	References
10. Glass pH electrodes for high-temperature use	36A, 46A, 85A
11. Hydrophobized pH glass electrode surfaces	155A
12. Sodium-selective glass compositions	41A, 42A, 53A, 123A, 145A
13. pNa^+ potentiometry in high-purity water	161A
14. pNa^+ potentiometry in natural waters	64A, 86A
15. pNa^+ potentiometry in salt solutions	111A
16. pNa^+ potentiometry in suspensions	133A
17. pNa^+ potentiometry interpretation especially activity coefficients	88A, 102A, 130A, 131A, 136A, 148A
18. "Cationic" glass compositions	48A
19. Glass surface modifications	26A, 27A, 122A
20. "Cationic" glass applications in nonaqueous solvents	33A, 103A, 139A
21. "Cationic" glass in titrimetry using Ag^+	30A, 47A
22. "Cationic" glass in titrimetry: liquid junction elimination, heat of mixing studies, and polyelectrolyte activity measurements	80A, 90A, 106A, 125A
B (1970–1971)	
1. Hydrolyzed surface film development and composition, on pH glass electrodes	6B, 75B, 76B
2. Alkaline errors of pH glass electrodes	174B, 227B, 228B, 229B
3. Precision pH titrations and calibration technique in NaCl solution	146B, 316B
4. Solvent effects in pH measurements:	
MeOH, DMF, HOAc	163B
DMF	172B
aqueous 1 M $NaClO_4$	280B
aqueous corrosive ions phosphate, arsenate, tungstate, molybdate, and borate	118B
liquid ammonia	25B, 68B, 170B
5. Techniques for:	
pH glass electrode storage	107B
fabric coating	252B
heat sterilizing	173B
pressurization	186B, 233B, 360B, 381B
manufacture	168B
6. Design patents	149B, 216B, 230B, 232B, 262B, 323B, 348B, 366B, 384B, 391B
7. Coatings to prevent pH bulb fouling	175B
8. Junction designs	61B, 156B, 213B, 259B, 290B, 338B, 363B, 371B
9. Design of interior reference electrodes for pH glass bulbs	13B
10. Na^+- and K^+-sensing electrode responses	242B, 248B, 359B
11. Na^+- and K^+-sensing glass compositions	36B, 45B, 126B, 169, 357B

Topic	References
12. Na^+-electrode applications in nearly pure water (boiler water)	84B
13. Na^+-electrode application in active transport studies	265B
14. Na^+-electrode application in MeOH	164B
15. Na^+-electrode application in blood	258B, 273B
16. Na^+-electrode application in determination of activity coefficients	158B
17. "Cationic" K^+ glass applications	77B, 147B, 276B
18. Glass microelectrodes	195B
19. Glass microelectrode fabrication and responses	50B, 58B, 59B, 214B, 284B, 404B, 406B
C (1972–1973)	
1. Characterizing the surface layer composition of pH glass electrodes by ion profile analysis, ion exchange rates, and radio tracers	32C, 97C, 98C, 141C, 182C, 230C, 231C, 232C, 320C, 383C, 480C, 481C, 482C
2. Response time and reproducibility of pH glasses in nonaqueous solvents	310C, 322C, 328C
3. New pH glass compositions	50C, 51C, 494C
4. A new model for pH response	66C
5. Precision pH measurement techniques	111C, 227C
6. Statistical standardization procedure	387C
7. Suspension effect	77C, 318C, 319C
measurements in high-ionic-strength solutions	39C, 356C, 357C
measurements in sea water	173C
8. pH response of a Na^+-selective glass	491C
9. Evaluation of steady and time responses of "cation"-sensitive glasses	9C, 362C
10. Intramucosal gastric acid measurements (microelectrodes)	485C
11. *In vivo* pH and pNa in the myocardium (microelectrodes)	463C
12. pH electrode for nanoliter samples (microelectrodes)	233C
13. Direct intracellular pH measurement (microelectrodes)	350C
14. Analyses of parameters affecting glass microelectrodes	457C
15. Capillary glass electrode for small blood volumes	46C
16. Electrical resistance of glass capillary electrodes	440C
17. Continuous monitoring of blood pH in the heart coronary sinus	133C, 315C
18. Apparatus for pH determination of gastric contents	47C
19. Telemetric measurements of gastric acid	234C
20. Measurement of gastric juice acidity	295C
21. Design and use of micro pH electrodes	135C–138C, 257C, 462C
22. Measurement of alkaline reserve in crustaceans	105C
23. Measurement of soil pH	405C, 418C
24. Study of the acidity region of molten $KHSO_4$	469C
25. Determination of acidity of aluminosilicate catalysts	250C

continued overleaf

TABLE 10. (*Continued*)

Topic	References
26. Alkali determination in the aluminum industry	261C, 450C, 459C
27. Compact electrode system for environmental use	479C
28. Patents on pH electrode (and system) designs	109C, 123C, 265C, 311C, 398C, 441C
29. pH glass composition patent	309C
30. Seals for pH glass electrodes	156C, 157C
31. pH meter for concentration readout	312C
32. Instrument for direct measurement of glass electrode resistance	251C
33. Table of pH and nanoequivalents for pH 7–8	262C
34. Methods for establishing and measuring the isoelectric potential (i.e., a constant potential value at all temperatures for glass electrodes vs. Ag/AgCl ref.)	241C
35. Following sodium ion activity in active transport in multicellular membranes	332C
36. Potentiometric determination of Na^+ in whole blood	297C
37. Sodium ion analysis in clinical medicine	358C
38. Intracellular sodium activity and sodium pump in small neurons	445C
39. Sodium and potassium activities in muscle	20C
40. Analysis of sodium and potassium in corn and sugar beet tissues	76C
41. Analysis of sodium in soda-industry products	344C
42. Monitoring tablet dissolution	305C
43. Analysis of sodium in ceramics	339C
44. Continuous determination of sodium in high-purity water	108C
45. Potentiometric analysis of sodium in river and mineral waters	461C
46. Rapid potentiometric analysis of sodium in drug solutions	249C
47. Determination of sodium in waste waters	386C
48. Sodium determination in mill liquors	253C, 274C, 329C
49. Analysis of saline soils	434C
50. Studies of sodium tartrate complexes	113C
51. pH-adjusting solution for pNa measurements	115C
52. Determination of potassium in inorganic and organic substances	8C
53. Determination of sodium and potassium activities in dilute solutions	72C, 236C
54. Increasing sensitivity of a sodium electrode by Ag^+ pretreatment	492C
55. Response of sodium silicate glasses to univalent and divalent cations	369C, 370C
56. A potassium-sensitive glass electrode composition	75C
57. Use of Cs/Cs^+ glass electrodes	15C

Topic	References
D (1974–1975)	
1. Hydrolysis layer ionic profile analysis for pH glass electrodes	7D, 12D, 13D
2. Identification of surface resistance by impedance measurements on pH glass electrodes	176D
3. Correlations of electrode response quality with durability and properties of hydrolyzed layers on pH glass electrodes	212D, 213D, 214D
4. Extension of transport theory for deriving potentiometric response function of glass electrodes	30D, 31D, 49D, 117D
5. Cation error of pH glasses	33D
6. pH errors of Na^+-responsive glass electrodes, and selectivity coefficients for K^+, Ag^+, and NH_4^+	215D, 216D
7. New glass composition studies relating to control of pH or pM responses	15D, 157D, 178D
8. pH studies of competitive equilibria	148D, 180D
9. pNa study of Na thiosulfate complexes	177D
10. Construction and electrode design patents	19D, 56D, 72D, 139D
E (1976–1977)	
1. Further studies and experimental analysis of the composition of the hydrolyzed-leached layer on pH glass electrodes	10E, 20D, 23E, 24E
2. Analysis of the glass electrode membrane potential	251E
3. Transient responses of pH and "cation"-sensitive glass electrodes	1E
4. Long-term drifts of glass electrode responses	29E, 148E
5. Acid errors of pH glass electrodes	34E, 49E, 187E
6. Unusual pH-sensitive structures using supported minerals with metallic contacts	66E, 67E
ion-sensitive glass on metal backing	85E
ion-sensitive glass on metal film and insulator backing	259E
7. Demountable combination pH electrode with built-in thermistor	167E
8. All-solid-state glass electrode-configuration bulb with fused ionic conductor inner electrolyte	110E
9. Sodium ion-selective composition of Al_2O_3 and Ta_2O_5	290E
10. Na β-alumina as basis for Na^+-selective electrode	87E, 92E
11. Comparison of Na^+ analysis in foods by AA and ion-selective electrodes	241E
12. Determination of Na^+ in alumina	267E
13. Comparison of K^+ analyses by four different and unusual electrodes: minerals in supports, K^+-selective glass particles in paraffin, $KBPh_4$ or K dipicrylamine, etc., in paraffin and an oxidized carbon rod. The latter was preferred for monitoring titrations of K^+ with BPh_4^-	101E

continued overleaf

***TABLE* 10 (*Continued*)**

Topic	References
14. NH_4^+-responsive glass compositions based on sodium-aluminosilicates (vanadates)	128E
15. Microelectrode for pH (spear, internal capillary, and recessed tip)	121E
16. Microbulb pH electrodes	62E
17. Elongated tube combination pH electrode suitable for catheter applications	137E
18. Flat pH-sensitive electrode for skin and other uses	145E
19. Tip potentials and resistances of microelectrodes	108E
F (1978)	
1. Transport processes in glass membrane electrodes, emphasizing surface layer effects	14F, 21F
2. Accuracy test	
for glass membrane electrodes	215F
for acid errors	34F
3. Combination electrodes patent	26F
4. A submicrometer glass-membrane pH microelectrode	192F
5. Construction patent for pH sensor, reference electrode, and sealed electronic package	19F
6. Hybrid (glass film on metal conductor on insulator substrate) construction of a glass electrode sensor	2F
7. Patent on	
glass electrode composition and construction	89F
glass electrode with electrolyte film for gas analysis	27F
8. Sodium transport study using glass membrane with sodium amalgam contact	67F
9. Highly sodium-selective glass compositions, manufacture, and responses	23F
10. Automated analyzer for sodium and potassium in whole blood	60F
11. Measurement of sodium in albumin solutions	157F
12. Application of sodium-selective glass with Ag^+ response to CN^- titrimetry	4F
13. Patent on Na-selective glass	22F
Applications in monitoring Na^+ of infusion solutions	226F
14. Application of Na-sensing electrode for thermodynamic measurements in *n*-heptanol and *n*-octanol solutions of NaI	182F
15. Application of K^+-selective glass in analysis of ores and optical glass compositions	15F

TABLE 11. Liquid Membrane Electrodes—Cation Sensitive

Topic	References
A (1968–1969)	
1. Ca^{2+}-selective using thenoyltrifluoroacetone	11A
2. M^{3+}-, M^{4+}-sensing electrodes using dinonylnaphthalene sulfonic acid	58A
3. Construction, composition patents	127A, 132A
4. Use of Ca^{2+}-electrode in kinetic measurements, and in titrimetry	105A, 121A, 163A, 164A, 1651
5. Use of Ca^{2+}-electrode in activity measurements	25A, 43A, 44A, 70A, 134A, 135A, 137A
6. Applications in milk analysis	83A
7. Evaluation of a solidified Ca^{2+} electrode	118A
B (1970–1971)	
1. Patent on a Cu(II) sensor	315B
2. Construction patent on a liquid electrode	115B
3. Cation determination using dithizonate as ion exchanger	318B
4. Cation determination using diethyldithiocarbonate and dipicrylamine	317B
5. Na^+ determinations using K^+ (*p*-chloro)tetraphenylborate	17B
6. An electrode using dithizone in PVC with dipentylphthalate as mediator	60B, 361B
7. Dithizone as ion exchanger in plastic membranes	189B, 190B, 191B, 192B
8. Responses of the Orion Pb^{2+} liquid membrane electrode	191B
9. Electrodes based on polymer plasticizers with low site content	148B, 206B, 208B, 209B, 210B, 211B, 212B
10. Micro-liquid-membrane electrodes	267B, 386B
11. Design of a liquid membrane, phosphate-sensing electrode	129B
12. Characterization of the Orion Ca^{2+} electrode and related electrodes	9B, 10B, 120B, 397B
13. A Ca^{2+}-sensitive electrode using PVC matrix support	241B
14. Patent on liquid ion exchanger for a Ca^{2+} electrode	398B
15. Patent covering the thenoyltrifluoroacetone Ca^{2+} sensor	35B
16. Clay and conventional synthetic fixed-site membrane electrode	197B, 278B, 279B
17. Ca^{2+}-electrode response to cyclohexylammonium ion from 1–1000 m*M*	63B
18. Activity coefficients in $CaCl_2$–$MgCl_2$–$SrCl_2$ mixtures using Ca^{2+} electrode	199B
19. Activity coefficients in $CaCl_2$–$MgCl_2$–NaCl mixtures using Ca^{2+} electrode	200B

continued overleaf

TABLE 11 *(Continued)*

Topic	References
20. Activity coefficients in $CaCl_2$ using Ca^{2+} electrode	341B
21. Activity coefficients in $CaCl_2$–NaCl, $NaNO_3$, KCl, or KNO_3 using Ca^{2+} electrode	8B
22. Solubility of $CaSO_4 \cdot \frac{1}{2}H_2O$ measured using Ca^{2+} electrode	372B
23. Ca tri- and tetrametaphosphate complex stabilities using Ca^{2+} electrode	393B
24. Ca albumin aggregation study using Ca^{2+} electrode	166B
25. EDTA titration of Ca^{2+}, Mg^{2+} using Ca^{2+} electrode	137B
26. Automatic EDTA titration using Ca^{2+} and Mg^{2+} using Ca^{2+} electrode	365B
27. Ca^{2+} measured in venous human blood	41B, 141B, 144B, 201B, 246B, 321B, 322B, 328B
28. Measurement of Ca^{2+} in soil extracts	86B, 98B
29. Measurement of Ca^{2+} in glass batches	183B
30. Measurements of Ca^{2+} interactions with montmorillonite	339B
31. Mg^{2+}–carbonate species stabilities by use of Ca^{2+}-sensing electrode	254B
C (1972–1973)	
1. Application of extraction parameters to interpretation of liquid membrane electrode responses	23C, 36C, 110C, 217C, 223C, 308C
2. Digital simulation of monovalent cation errors of a Ca^{2+}-electrode	67C
3. Current-voltage character of a Ca^{2+}-responsive electrode	132C, 281C
4. Sources of error in Ca^{2+} measurements	70C
5. New optimized compositions for Ca^{2+}-responsive electrodes	94C, 152C, 153C, 307C, 394C
6. Application of Ca^{2+} electrodes in titrimetry	202C
7. Measurements of calcium in serum and plasma	60C, 130C, 131C, 263C, 278C, 382C, 384C 419C, 431C
8. Applications in sugar beet industry	180C
9. Applications to detergent builders	49C
10. Application of Ca^{2+} and divalent ion electrodes in water analysis	235C, 298C, 300C, 346C, 347C
11. Electrode with Ba^{2+}- and Sr^{2+}-selective responses	275C, 306C
12. Electrode for selective response to -onium ions	409C
13. Electrodes for measuring choline and applications to analysis of acetocholineesterase and organophosphate pesticides	33C–35C, 37C
14. Electrodes sensitive to vitamins B_1 and B_6	212C
15. Electrode sensitive to potassium using ion exchanger in plasticized polymer membrane	38C

Topic	References
16. Electrode sensitive to potassium using potassium tetra-*p*-chlorophenyl borate in PVC	102C, 430C
17. Microelectrode selective to potassium activities	489C
18. Applications of liquid ion exchange microelectrodes for potassium activity measurements	234C, 237C–240C, 470C, 472C
D (1974–1975)	
1. Precision and accuracy of Ca^{2+} electrodes for thermodynamic applications	28D
2. Selectivity coefficient test using dipicrylamine membranes sensitive to alkali and tetraalkylammonium ions	104D, 218D
3. Liquid ion exchanger membranes in PVC supports	140D
4. Testing Ca^{2+} response mechanisms using $^{45}Ca^{2+}$ tracer	200D
5. Chiroselective membrane examples	116D
6. K^+, Ag^+ response electrode using $NaBPh_4$ in cyclohexanone	99D
7. K^+-responsive electrode using $KBPh_4$ in $PhNO_2$	198D
8. K^+-responsive electrode using $KBPh_4$ in bis(2-ethylhexyl)adipate and $PhNO_2$	10D
9. Ca^{2+}-responsive electrode influenced by surfactant, both anionic and cationic	124D
10. K^+-responsive electrode using $KBPh_4$ in *o*-NO_2 toluene	64D, 65D
11. Ni^{2+}, Cd^{2+}, Pb^{2+} responsive electrode using bis(*O*,*O*′-diisobutyldithio phosphato)-M(II) in PhCl	135D
E (1976–1977)	
1. Li^+-responsive electrode uses only (sic) *n*-decanol	123E
2. Na^+-responsive electrode uses monensin; microelectrode construction	168E
3. K^+-responsive microelectrodes for bioapplications	283E
4. Double-barreled K^+- or Cl^--selective electrode for bioapplications	98E
5. Cs^+-selective electrodes (BPh_4^- in nitroaromatic solvents)	25E, 52E
6. Ca^{2+} Orion exchanger (92–20–02) in PVC matrix demonstration of Ca^{2+} permeability using radioisotopes	57E, 59E
7. Ca^{2+}-di-*n*-octyl phenylphosphate in PVC; responses as functions of added mediators; selectivity coefficients	58E
8. Ca^{2+}-selective fixed-site membrane electrode using grafted alkyl phosphate sites on PVC, poly(vinyl alcohol), copolymer matrix	153E
9. Ca^{2+}-selective electrode optimization using di-*n*-decyl phosphate in various tri-*n*-alkyl phosphates. Long-chain alkyl groups are preferred	100E

continued overleaf

TABLE 11 (*Continued*)

Topic	References
10. Use of Ca^{2+}-responsive electrode for determining Ca^{2+}-complex formation constants for tartrate, malate, succinate, malonate, maleate, and fumarate	154E
11. Applications of Ca^{2+}-responsive electrodes for serum-ionized calcium in whole blood	95E, 118E, 238E, 239E
12. Hardness response ($Ca^{2+} + Mg^{2+}$), attributed to a solid, phosphanated polyethylene film electrode	250E
13. Ni^{2+}-response electrode using 10^{-3} *M* $(Et_4N)_2(NiL_2)$, L = 5,6-dithiobenzo-7,8-phenazine in nitrobenzene	176E
14. Zn^{2+}-responsive polymeric membrane electrode resembling a Ca^{2+}-sensing electrode	107E
15. Pb^{2+}-responsive membrane electrode based on Pb diisobutyl dithiophosphate	184E
16. Pb^{2+}-responsive PVC membrane electrode using diethyldithiocarbamate in tetra-hydrofuran	257E
17. Tl^{+}-responsive membrane electrode based on *O,O'*-didecyldithiophosphate in chlorocyclohexane	258E
18. Ag^{+}-responsive membrane electrode using a complicated chelating thiotrizaine in $CHCl_3$	14E
19. Cu^{2+}-responsive membrane electrode using pyrrolidine dithiocarbamate in $CHCl_3$	17E
20. Cu^{2+}-responsive and Hg^{2+}-responsive membrane electrode using diphenylthiocarbazide, Ruhemann's purple, and salicylaldoximate in $CHCl_3$	13E, 15E, 16E, 54E
21. Hg^{2+}-selective electrode application to EDTA titrimetry	130E
22. PVC electrodes responsive to Hg^{2+}, Zn^{2+}, Co^{2+}, and Ni^{2+} based on oil-soluble complex salts $M^{2+}(NL_4)^{2-}$, $L = SCN^-$	234E
23. Coated-wire electrode responsive to Ca^{2+}	237E
24. Coated-wire electrode responsive to NH_4^+ and to K^+	133E, 134E
F (1978)	
1. Construction patent for liquid ion exchanger membranes	222F
2. Role of site mobility and ion pairing in determining potentiometric selectivity of membrane electrodes	234F
3. Improved Ca^{2+}-responsive electrodes using special, new, oil-soluble bis-phosphates	161F, 126F, 51F
4. Radiotracer studies of Ca, Sr, and Ba ion transport in membrane electrodes	115F
5. Use of solid contacts with PVC-supported Ca^{2+}-sensing, ion exchanger electrode	109F
6. PVC supported Ca^{2+}-sensitive electrode for continuous monitoring applications	144F

Topic	References
7. Patent on Ca^{2+} sensor composition using a bis-phosphate	208F
8. Flow injection method for Ca^{2+} in serum and water	85F
9. Uranyl-sensitive electrode composition	154F
10. Hg(II)-sensitive electrode using a chelate in $CHCl_3$	48F
11. A Cu(II)-responsive diethylhexyl phosphate electrode	73F
12. Mg^{2+} responsive liquid membrane based on surfactant anion salts in 1-decanol	87F
13. Attempt to generate Cu-, Ag-, Au-, and Hg-responsive membranes from uncharged organic extractants in PVC	40F
14. Study of the K^+ response of the dipicrylaminate exchanger electrode	101F
15. Cs^+- and Tl^+-sensitive liquid membrane electrodes using tetrakis (*m*-trifluoromethylphenyl)borates	47F
16. A pyridinium-sensitive electrode using the molyboarsenate salt	229F
17. Cationic dye-sensitive electrode based on tetraphenylborate and 12-tungstosilicate salts	66F
18. A pyrocatechol violet, ion-selective electrode with applications to chelatometry	124F
19. Construction and studies of Chloramine-T and picrate-sensitive electrodes	82F, 135F, 136F
20. Drug-responsive membrane electrodes:	
Chlorpheniramine	69F
dibenyldimethylammonium	166F
vitamin B_6	72F
ephedrine	213F
novocaine	103F
as triazines	49F
six quaternary pharmaceuticals	50F
21. Electrode response to cationic surfactants and determination of critical micelle concentrations	230F

TABLE 11a. Liquid Membrane Electrodes—Anion Sensitive

Topic	References
A (1968–1969)	
1. Anion-selective system using methyltricaprylylammonium salts	28A, 29A
2. Selectivity coefficient method test using liquid anion sensor	144A
3. Response of a NO_3^- sensor	115A
4. Application of a NO_3^- sensor in soil analysis	108A
5. Responses of a ClO_4^- sensor	69A

continued overleaf

TABLE 11a *(Continued)*

Topic	References
6. Application of a ClO_4^- electrode	
in mercury complex chemistry	66A
in tetraphenylarsonium titrimetry	4A
B (1970–1971)	
1. Patent on electrodes sensitive to anions using tris-phenanthroline Ni(II) and Co(II)	226B
2. Application of BF_4^- electrode in agriculture	55B
3. Use of R_4N^+ Cl^- in benzene as Cl^- sensor	329B, 330B, 332B
4. Use of tetradecyltrimethylammonium picrate as a sensor for titrimetry of oil-soluble anions and for determining critical micelle concentrations	110B, 111B, 112B, 125B
5. Determining heavy metals as anion Cl^- complexes	331B
6. Determining amino acid anions	226B
7. Determining HPO_4^{2-}	253B
8. Determining ClO_4^-	369B
9. Determining I^- (Br^- and NO_3^- interfere)	121B
10. Determining salicylate	145B
11. Determining Br^-, I^-, NO_3^-, and ClO_4^-	277B
12. Determining NO_3^- using solidified electrodes	2B, 29B
13. Nitrate analysis using a fluoride reference electrode	218B, 219B
14. Titration of nitrate with diphenyl thallium(III) sulfate; perchlorate interferes	72B
15. NO_2, NO in ppm in air using O_2 oxidation at 300 °K	73B
16. Nitrate in plants, soils, waters; buffer used	235B
17. Nitrate in soils	69B, 98B, 215B, 269B, 274B
18. Nitrate in plant tissues; comparison with colorimetric methods	11B, 54B, 165B
19. Nitrate in water; comparison of detection limits with colorimetric methods	1B, 47B, 174B, 180B, 194B, 217B, 220B
20. Nitrate in food	396B
21. Titration of nitric acid on oleum using standard NaOH	308B
22. Nitrate in nitrite using hydroxylamine	113B
23. Continuous monitoring of nitrate	234B
C (1972–1973)	
1. Time responses of anion-sensitive liquid membrane electrodes	389C
2. Dependence of selectivity of NO_3^- over Cl^- on organic salt concentration in a nitrate electrode	100C
3. NO_3^--selective electrode using Orion liquid in PVC	101C
4. NO_3^--selective electrode composition patent	486C
5. NO_3^--selective membrane electrodes	143C, 145C, 146C, 149C
6. Determination of nitrate and nitrite in mixtures	327C

Topic	References
7. Determination of nitrate in mineral waters	317C
8. Determination of nitrate in cotton petioles	61C
9. Determination of nitrate in plants	19C
10. Application of Cl^- electrode for salt in cheese	385C
11. Iodide-selective electrode composition	210C
12. Perchlorate selective electrodes and responses	22C, 148C, 208C, 330C
13. Determination of boron with BF_4^--selective electrode	483C
14. Bicarbonate-selective electrode compositions	150C, 487C
15. Effect of structure of carboxylate anions on responses	345C
16. Electrodes selective to maleic and phthalic acids	213C
17. Electrode selective to trifluoroacetate	214C
18. Electrode selective to thiocyanate	216C
19. Electrode selective to aromatic sulfonates	209C, 215C
20. Electrode selective to bisulfite	43C
21. Electrode selective to picrate	21C
22. Liquid ion exchange membrane assemblies	252C, 323C
23. Liquid microelectrode construction	429C
D (1974–1975)	
1. Aliquat–nitrobenzene membrane biionic potentials	63D
2. Selectivity coefficient testing of liquid membranes	53D
3. Studies on phosphate-selective membranes	150D
4. Use of organo-lead and organo-thallium salts in membrane electrode compositions	184D
5. Analysis of natural water for NO_3^-	211D
6. Procedure for precision (±4%) determination of boron using BF_4^- electrode	113D
7. Response of ClO_4^- electrode to IO_4^-, and automatic reaction rate method for vicinal glycols, based on time to consume a fixed amount of IO_4^-	48D
8. Electrode responsive to Br^- using tetradecylammonium $HgBr_3^-$ in tributylphosphate	111D
9. Electrode responsive to I^-, SCN^- using tetraphenylarsonium, and crystal violet in $PhNO_2$, 1.2-dichloroethane, and $CHCl_3$	106D
10. Electrode responsive to ClO_4^- using Nb(V) oxinate in $CHCl_3$	195D
11. Electrode responsive to NO_3^- using nitronnitrate in $PhNO_2$	197D
12. Electrode responsive to NO_3^- using tetraalkyl phosphonium NO_3^-	188D
13. Electrode responsive to I^- using tetraoctylphosphonium iodide	171D
14. Electrode responsive to NO_3^- using tetradecylammonium NO_3^- in PhCl	133D

continued overleaf

TABLE 11a *(Continued)*

Topic	References
15. Hydroxyacetate-, benzoate-, *p*-OH benzoate-, salicylate-, cyclohexylsulfonate-, *o*-benzosulfimide-sensitive electrode using 10% methyl trioctyl salts in 1-decanol	185D
16. Electrode responsive to surfactant anions, dodecyl-, tetradecyl-, and hexadecyl sulfates, using oil-soluble cations in *o*-dichlorobenzene	20D, 21D, 22D
17. Electrode responsive to CO_3^{2-} using 1% tricaprylmethylammonium chloride in trifluoroacetyl-*p*-butyl benzene	85D, 87D
18. Electrode responsive to $HSeO_3^-$ using 3,3′-diaminobenzidine in hexane	128D
19. Electrode sensitive to SO_4^{2-} using trioctylamine and trilaurylamine in a variety of hydrophobic solvents	134D
20. Electrode responsive to Mo(V) as $MoO(SCN)_5^-$ using this anion in 2:3 $PhNO_2$: *o*-$PhCl_2$ solvent	58D
21. Patent on anion-responsive electrodes using alkylammonium salts	62D
22. Patent on orthophosphate-sensing electrode using Ag(I) thiourea derivatives	75D
23. Electrode responsive to NO_3^- using trialkylbenzylammonium chloride in benzylchloride	71D
E (1976–1977)	
1. PVC NO_3^--responsive electrode compositions: various tetraalkylammonium nitrates in dibutylphthalate	206E
2. NO_3^--responsive electrodes using Gentian violet or tetraphenyl phosphonium bromide in nitrobenzene or in tetrachlorethane	135E, 136E
3. NO_3^- determination using electrodes and colorimetry compared	248E
4. NO_3^- determination in difficult systems such as HNO_3–HF	39E
5. NO_2^- determination compared with official AOAC method	249E
6. ClO_4^--responsive electrodes based on methylene blue in nitrobenzene	131E, 149E
7. PVC-ClO_4^--responsive electrode based on tris(bathophenanthroline)iron(II)	43E
8. Titrimetry of ClO_4^- using PhAsCl or Ph_4PCl	245E
9. Applications of IO_4^-(ClO_4^-)-responsive electrodes in monitoring analytical reactions involving IO_4^-	78E, 79E
10. ReO_4^-, $SbCl_6^-$, $TiCl_4^-$, $AuCl_4^-$, and anionic dye-responsive systems based on cationic dyes	88E–91E
11. ReO_4^-(ClO_4^-)-responsive membrane electrodes using quaternary phosphonium salts in nitrobenzene	274E

Topic	References
12. BF_4^--responsive membrane electrodes based on Ph_4PBr in tetrachlorethane	175E
13. ClO_3^--responsive quaternary alkylammonium-based electrodes	99E, 122E
14. CrO_4^{2-}-sensitive membrane electrodes based on crystal violet in nitrobenzene	106E, 129E
15. $FeCl_4^-$-sensitive membrane electrodes using triphenylpyrilium cation in dichloroethane	132E
16. $AuCl_4^-$-sensitive membrane electrodes using complex extracting agent	82E, 103E
17. $AuCl_4^-$- and $Au(CN)_4^-$-responsive electrodes based on Ph_4As^+	41E, 42E
18. Fixed-site, quaternary ammonium anion exchangers, wetted with nitrobenzene, show similar selectivities as corresponding liquid membranes	147E
19. Coated-wire electrodes using PVC and chlorocomplexes of Hg(II), Cu(II), Zn(II), and Cd(II)	45E–48E
20. Simple monocarboxylic acid (C_1–C_4), benzoic, and substituted benzoic acid responses of quaternary ammonium and metal ligand cationic-site membrane electrodes	124E, 185E, 211E
21. Phenobarbital-responsive coated-wire membrane electrode	44E
22. Chromazural-S- and Eriochrome-Black-T-responsive membrane electrodes	149aE, 228E
23. 2,4-Dichlorophenoxyacetic acid, and 8-quinolinol-5-sulfonate-responsive membrane electrodes	120E, 256E
24. Bile salt anions (taurocholate, taurodeoxycholate)-responsive membrane electrodes	102E
25. Dodecylsulfate-ion responsive membranes based, respectively, on:	
alkyldimethylbenzylammonium cation	223E
rosaniline	186E
ferrous orthophenthroline	5E, 51E
Closely related dodecylbenzenesulfonate electrode used *o*-phenthroline cobalt(III)	
F (1978)	
1. Crystal Violet as ClO_4^-, SCN^-, or I^- in nitrobenzene was subject for a selectivity coefficient study	265F
2. Use of serial dilution to establish stability with time of a liquid anion-sensing electrode	143F
3. Observation of a streaming potential effect with liquid membrane electrodes	76F
4. 2-Aminoperimidine sulfate with Aliquat 336 in PVC (trifluoroacetyl-*p*-butylbenzene as mediator) provided a SO_4^{2-}-sensitive electrode	12F

continued overleaf

TABLE 11a *(Continued)*

Topic	References
5. Studies by the Leningrad School on ion exchangers and mediators for halide-sensitive liquid membrane electrodes	148F, 151F, 176F
6. Solvent effect on response of Br^--sensing electrode using trinonylactadecylammonium cation as exchanger	231F
7. Br^-- and Cl^--responsive electrodes using tetradecylammonium cation and HgX_3^-	165F
8. Quaternary ammonium SCN^- electrode responsive to SCN^-	216F
9. ClO_3^--sensitive electrode using very large, oil-soluble dyelike cations as exchangers	134F
10. Perchlorate, thiocyanate, tetrafluoroborate and nitrate responsive electrode studies using triphenylmethane-type cation dyes, quaternary ammonium cations, Ni and Fe cationic complexes, ferricenium cation	77F, 98F, 100F, 138F
11. Determination of boron using BF_4^- electrode	110F
12. Nitrate electrodes based on quaternary ammonium, phosphonium cations, and Ni-cation complexes	18F, 149F, 221F, 263F, 240F
13. Flow injection analysis and water analysis for nitrate	84F, 246F
14. Electrode compositions responsive to perrhenate, $AuCl_4^-$, $Au(CN)_2^-$, and $FeCl_4^-$	268F, 178F, 267F, 217F, 99F
15. Analytical uses of a picrate-responsive electrode	54F, 81F
16. Design and compositions for nicotinate-responsive electrodes	37F, 38F, 39F
17. Use of cetylpyridinium, tetradecylammonium and triphenylmethane dyes as exchangers for construction of organic anion-sensitive electrodes	102F, 150F
18. Fundamental studies of responses of surfactant anion-sensing electrodes	52F, 53F, 177F, 247F
19. Anion-sensing electrodes based on "Neosepta"	206F

TABLE 12. Neutral-Carrier-Based Electrodes

Topic	References
B (1970–1971)	
1. Pioneering paper on neutral carriers giving selectivity of K^+ over Na^+ in potentiometric response	358B
2. Cellulose acetate holding valinomycin or gramicidin in di-Me phthalate mediator (K^+); monensin in ethylhexyldiphenylphosphate (Na^+); nonactin in di-Bu sebacate (NH_4^+)	179B
3. Simon patent on actin homologs, gramicidin, and/or valinomycin	350B

Topic	References
4. Orion K^+-selective electrode description and Beckman K^+ electrodes	106B, 225B
5. Experimental evaluation of K^+ neutral carrier selectivity over Na^+	89B, 193B, 294B
6. Application of K^+-carrier electrode:	
in kidney tubules	181B
in blood plasma or serum	237B, 293B, 376B, 399B
in cells	325B
7. Application of K^+-carrier electrodes:	
in KCl activity measurement	49aB
in K^+ adenosine triphosphate equilibria studies	306B
in soil analyses	12B
8. NH_4^+-selective carrier electrode using actins	66B, 185B, 327B
9. Micro-K^+-electrode	385B
10. Ba^{2+}-carrier electrode using Igepal CO-880	198B, 345B
C (1972–1973)	
1. Potentiometric K^+ selectivities correlated with complex formation constants	388C
2. Molecular level parameters involved in carrier selectivity	326C
3. Ion pairing in K^+-crown ether species	439C
4. Substituted crown ethers for K^+-selective electrodes	396C, 397C
5. Noncyclic neutral carriers for di- and monovalent cations	12C–14C
6. Properties of silicone-rubber-supported K^+ carrier electrodes	364C, 366C
7. Patents on Beckman solid-neutral-carrier electrode systems	92C, 93C
8. Patents on carrier electrodes giving supports and solvents	64C, 430C, 451C
9. Application of K^+ carrier electrodes:	
in biological systems	57C, 196C, 197C, 272C, 288C, 395C
in sea water	17C
in soil	27C, 402C
10. Membranes which respond to ion activities and may involve ion exchange or neutral carriers or both	63C, 80C, 198C, 267C, 279C, 280C, 283C, 284C
D (1974–1975)	
1. Responses of neutral-carrier-based electrodes in terms of carrier types, mediators, and interferences	25D, 38D, 110D, 143D–146D
2. Studies on chiral neutral carriers	152D, 199D
3. Synthesis of new macrocyclic ether carriers	156D
4. "Bioelectrodes," a review of electrodes mainly based on neutral carriers	50D
5. Electrode for K^+, based on macrocyclic polyether in PVC	193D

continued overleaf

TABLE 12 *(Continued)*

Topic	References
6. Electrode for K^+, based on valinomycin on PVC + dibutylphthalate	153D
7. Electrode for K^+, based on valinomycin in PVC + dioctyladipate or dioctylphthalate	54D
8. Electrode for Na^+ based on new synthetic carrier in PVC plus mediator	3D
9. pH response of tricresylphosphate in PVC	98D, 100D
10. Response of R_4N^+ compounds using dioctylphthalate in PVC	88D, 89D
11. Dextromethorphan response using *N,N'*-dimethyloleamide in PVC	88D, 89D
12. Sr^{2+}-response of a Sr^{2+}–polyethyleneglycol carrier complex	14D
13. pH response to a butanol membrane in a fritted glass support	138D
14. pH response to membranes made from 14 aqueous immiscible organic liquids in fritted glass supports	123D
15. K^+ response of valinomycin or dibenzo 18-crown-6 in PVC with several plasticizers	131D
E (1976–1977)	
1. Studies on synthesis and properties of carriers for selective electrodes	36E, 37E, 38E, 159E, 180E
2. Studies on anions and mechanisms involving anions in membranes containing cation carriers	35E, 146E, 152E, 198E, 220E, 264E
3. Ionophores for alkali and alkaline earth-sensitive electrodes, preparation, testing in PVC with mediators *o*-nitrophenylether or dibutylsebacate	3E
4. Synthesis of neutral ionophores for Ca^{2+} membrane electrodes	252E
5. Neutral ionophore for liquid membrane electrodes with high selectivity for Na^+ over K^+	112E
6. Neutral carrier for Ba^{2+}	113E
7. Poly(propylyl oxide) neutral carriers for Ba^{2+} with tetraphenylborate in PVC and dioctylphenylphosphonate mediator	140E–142E
8. Antibiotic A23187 in nitrobenzene, held in cellulose ester membrane as electrode	56E
9. Nitrogen-linked (through pyromellitic diamide) cyclic polyether (crown types) for ion binding	94E
10. Potentiometric titration study of alkali metal ion cryptate stabilities in water	63E
11. Synthetic ligand selectivities for Ca^{2+} and Na^+ used in lipid bilayers to enhance conductance, were consistent with selectivities in thick membranes	2E
12. A review of development and characterization of synthetic carriers	197E
13. Microelectrode for K^+, based on valinomycin and applications to plasma analysis	183E, 208E

Topic	References
14. K^+-selective valinomycin-based electrode in polycarbonate, siloxane copolymer	172E
15. K^+-selective crown ether electrode, PVC and dipentylphthalate mediator	221E
16. Anion interference characterization and means for partial removal. Electrodes use valinomycin and crown ethers in PVC. Improved responses were found by using alkyl chain phthalate mediators	235E
17. Identity of CRYTUR K^+-sensitive electrode as valinomycin-based, using dipentylphthalate	246E
18. Anion interference removal for $KVal^+$ electrode uses low dielectric constant solvent and fixed sites (negative) on surface-coated support	173E
19. Optimizing solvent mediators for Na^+-, K^+-, and Ca^{2+}-responsive polymeric electrodes	83E, 84E
20. Neutral-carrier Li^+-responsive PVC membrane electrode and microelectrode	111E, 267E
21. Neutral-carrier Ca^{2+}-responsive PVC membrane electrode and microelectrode	4E, 207E
22. Application of ion-selective electrodes to study lipid bilayer behavior a review with 135 references	7E
F (1978)	
1. Review of neutral-carrier, ion-selective electrodes	162F, 225F
2. Recent developments in the field of neutral-carrier-based ion-selective electrodes	164F
3. Selective transport processes in artificial membranes	226F
4. Applications of crown ethers	220F
5. Designing and synthesis of carriers for ion-selective membrane electrodes	16F, 190F
6. Effect of anions on carrier transport	45F, 140F
7. Response time measurements	141F
8. Transport properties of neutral-carrier membranes	163F
9. Exchange kinetics at potassium-selective liquid membrane electrode interfaces	36F
10. Determination of K^+ in small samples, including whole blood	90F, 91F
11. Titration of potassium binding to membrane proteins	86F
12. Determination of K^+ in sea water	207F
13. Automated determination of alkali metal ions	205F
14. Poly(propylene glycol) adducts of alkali and alkaline earth ions as sensors on components	113F, 114F
15. Siloxane polymers with mediators for measuring H^+ and Li^+	210F
16. Nonactin-based NH_4^+-sensing electrode	233F
Tetranactin-based NH_4-sensing electrode	237F

TABLE 13. Gas, Enzyme, and Clinical Electrodes

Topic	References
A (1968–1969)	
1. Blood pH determination	147A
2. Conversion of pH glass electrodes to pCO_2 electrodes	37A, 124A, 162A
3. Conversion of pH glass electrodes to urea sensors	54A
B (1970–1971)	
1. Review of enzyme and other sensitized electrodes	128B, 250B, 302B
2. pCO_2 electrodes, calibration, and applications	7B, 14B, 30B, 114B, 135B, 138B, 142B, 143B, 153B, 311B
3. Conversion of pH glass electrodes to pNH_3 sensors	28B, 270B
4. Immobilized enzyme potentiometry and amperometry	79B
5. Urea enzyme electrode	240B
6. Urease enzyme electrode	238B, 239B
7. Application of a urea enzyme electrode for blood and urine	131B
8. L- and D-amino enzyme electrodes	130B, 132B, 133B
9. Amygdalin-sensing electrode	305B
10. Use of ion-permeable membranes as electrode coatings	392B
11. Membrane electrode, responsive to acetylcholine cation	15B, 16B, 18B, 19B
C (1972–1973)	
1. Review of enzyme and sensitized electrodes	18C, 160C, 342C
2. Ammonia-sensing electrodes using pH or pNH_4^+ sensors	28C, 71C, 106C, 139C, 142C, 271C, 460C 316C, 321C
3. Urea-enzyme electrode	163C
4. D-amino acid and asparagine electrode	158C
5. 1-phenylananine electrode	162C
6. Glutamine electrode	159C
7. Penicillin electrode	353C
8. Urea and L-tyrosine electrode	161C
9. Electrode monitoring of cholinesterase	96C
10. Electrode monitoring of rhodanase	287C
11. Electrode monitoring of β-glucosidase	286C
12. Electrode monitoring of blood glucose	166C, 340C
D (1974–1975)	
1. Use of Ag_2S electrodes to monitor low-level anion activities of biological interest	204aD
2. Use of Ag_2S electrodes to monitor proteins	1aD
3. SMAC system for direct potentiometric determination of Na^+ and K^+ in blood serum	165D, 206D
4. Electrodes for monitoring Na^+, K^+, and Ca^{2+} in blood serum	118D

Topic	References
5. Potentiometric system for monitoring Ca^{2+} movement across biological membranes	126D
6. Fluoride determination in plasma for use in clinical laboratories. Based on slope determination by dilution and standard addition	61D
7. Fluoride determination in biological materials: separation techniques and analysis	207D
8. Analysis of NO_3^- and NO_2^- reductases using NO_3^-- and NH_4^+-sensing electrodes	101D
9. Serum CO_2 by a continuous-flow method using a CO_3^{2-}-sensing electrode	86D
10. I^- permeability of lipid bilayers followed with ISE's	182D
11. Cl^- analysis in blood serum using an anion exchange membrane (SIEM)	127D
12. Enzymic cholesterol determination using I^- electrode	155D
13. Construction and testing of enzyme electrodes	34D
14. Use of NH_3 electrode for determining creatinine and serum urea	125D, 201D
15. Use of Ag_2S electrodes for protein monitoring and antibody–antigen precipitin reaction monitoring	1D, 2D
16. Improved penicillin-sensing enzyme on glass electrode	42D
17. Systems for urea analysis	76D, 77D
18. Systems for L-aminoacid analysis	149D
19. Systems for cholesterol analysis	66D
20. Miniature probe containing sensors for CO_2, O_2, H^+, K^+, or Ca^{2+}	29D
21. Gas-sensing electrodes for CO_2, NH_3, Et_2NH, SO_2, NO_2 based on pH glass membranes; H_2S, HCN, HF, and Cl_2 based on Ag salt membranes	169D
22. Ammonia sensor based on "selectrodes" with HgS, Ag_2S, or Ag_2S–CuS tips	4D
23. Ammonia, SO_2 and "NO_x" sensors based on pH glass membranes. Also a continuous analyzer. Recent applications to boiler feed water, fresh water, blood plasma, Kjeldahl digests, and effluents are referenced	8D
24. Ammonia electrode construction patent	5D
25. Orion patents on ammonia sensor, HCN sensor	60D, 167D
26. CO_2 sensors of Severinghaus-type	59D, 189D
27. CO_2 sensor combined with second pH sensor	114D
28. Completely air-gap sensors	55D, 174D, 175D
29. Circulating system for monitoring Na^+, K^+, Mg^{2+}, and Cl^-	32D
30. Orion Patent: Two streams, one monitored and one containing known concentration of reference ion are mixed. Two selective electrodes, one monitoring and one sensing reference ion, provide a difference signal	168D
31. Foxboro Patent: Two streams, sample and reagent mix. Treated sample and fresh reagent pass on two sides of a selective electrode	120D

continued overleaf

TABLE 13 *(Continued)*

Topic	References
32. Systems for Cl^- and F^- in surface and underground water	18D
33. Systems for monitoring dissolved H_2S making use of air-free NaOH solution	27D
34. Cyanide monitor in ore treatment. Study of suitability of Ag_2S, Cu_2S, and Ni_3S_2 as electrodes, especially sensitivity to xanthates	26D
35. Cyanide monitoring using flowing solution, treated coulometrically with Ag^+. Standard Ag^+ is a second solution. Each is fed to opposite sides of a Ag_2S membrane and difference potential measured.	57D
36. Behavior of CuS–Ag_2S as a micro-flow-through electrode for low-activity Cu^{2+} monitoring	24D
37. Feasibility of monitoring Cl^-, Ca^{2+}, and Mg^{2+} in sea water by successive dilutions. Author is optimistic.	192D
38. Feasibility of monitoring sulfides in sea water	142D
39. Feasibility of monitoring Cu^{2+} in sea water	102D
40. Use of split crystals to monitor test and reference streams by differential potentiometry	209D
41. Flow-through NO_3^- sensing electrode with 50 μl dead volume, applied to monitoring NO_3^- and NO_2^- effluents from ion exchange columns	183D
42. Systems for hydrogenating sulfur-, Cl-, and F-containing effluents from GC columns to give H_2S, HCl, or HF, which are absorbed in electrolytes and detected with small dead volume ISE cells	107D, 108D
E (1976–1977)	
1. pH electrodes based on Pd/PdO, and CO_2 sensors based on these electrodes coated with CO_2-permeable polymer	205E
2. Miniaturization of coated wire sensors	86E
3. Microelectrodes for K^+, Na^+, and Cl^-	32E
4. Four-barrel microelectrode	156E
5. All-solid-state *in vivo* K^+ electrode	60E
6. Patent on ion-measuring system	193E
7. Catheter K^+ electrodes	18E, 19E, 270E
8. Surface skin measurement electrodes	30E, 31E
9. Ca^{2+} serum analyses by flow-through electrodes	95E, 97E
10. Ca^{2+} serum analyses by direct dip using CO_2-controlled and atmospheric pH	238E–240E
11. Blood-parameter analyzers	77E, 125E
12. CO_2 and NH_3 measurement using electrodes	212E, 229E, 286E
13. For 5′-adenosine monophosphate (AMP)	214E
14. AMP-metal complexation and thermodynamic quantities	163E
15. AMP-*d*-fructose-1,6-diphosphatase binding	230E

Topic	References
16. Automated analysis of adenosine deaminase using enzymes and NH_3-sensing electrode	126E
17. Electron transfer electrodes (Pt, Au, C) using lactate dehydrogenase and ferricyanide for lactate analysis	74E
18. Electron transfer enzyme-covered Pt electrode for determining blood lactates via cytochrome b_2 and ferricyanide	115E
19. Lactose, sucrose, and maltose analysis using pO_2 electrodes and appropriate enzymes	53E
20. For creatinine (urine and serum)	189E
21. For L-arginine and L-lysine using decarboxylase- and CO_3^{2-}-sensitive electrode	268E
22. For L-asparagine in conjunction with enzyme and NH_3-sensing electrode	287E
23. For L-phenylalanine in conjunction with enzyme and NH_3-sensing electrode	138E
24. For urea in whole blood using enzyme with NH_3-sensing electrode	213E
25. For uric acid with immobilized uricase and pCO_2 electrode	150E
26. For SO_4^{2-} by amperometric interference of 4-nitrocatechol oxidation on Pt. Uses 4-nitrocatechol sulfate and arylsulfatase. PO_4^{3-} also inhibits the oxidation	61E
27. For phosphate: two enzyme systems using amperometric monitoring	114E
28. For nitrite using nitrite reductase (to NH_3) and NH_3-sensing electrode	157E
29. Construction patents for immobilized enzyme electrodes	81E, 158E, 204E
30. Acetylcholine–ATP binding using a liquid ion exchanger selective for acetylcholine	164E
31. Dibenzyl-dimethylammonium-selective electrode using tetraphenylborate in dichloroethane as liquid ion exchanger	200E
32. Hapten-responsive liquid ion exchanger using trimethylphenylammonium ion sites. Application in immunochemistry	190E
33. Immunomonitoring using indirect reactions and release of marker reagent	70E
34. Patent on Janata's "Immunoelectrode"	144E
35. Further studies on protein responses of silver sulfide membrane electrodes	69E
F (1978)	
1. Bioelectrochemical sensor review	172F
2. Determination of NH_3, using gas-sensing ion electrodes	57F, 111F, 132F
3. Patent on amperometric CO analyzer using gold-plated polymer membrane	29F
4. Microelectrodes for NH_3 and CO_2	193F

continued overleaf

TABLE 13 *(Continued)*

Topic	References
5. A zero-current chronopotentiometric method for gas analysis using membranes	212F
6. A PVC-based electrode for DDA^+, dibenzyldimethylammonium cation	219F
7. Glucose and L-amino acid enzyme electrodes	145F
8. Application of enzyme electrodes to enzyme analysis	117F
9. Continuous analysis of urea	211F
10. Membrane electrode prepared with living bacterial cells	202F
11. Bacterial electrode for L-aspartate	133F
12. Bacterial electrode for glutamine	203F
13. Bacterial electrode for lysozime enzyme	58F
14. Immobilized antibody membrane for antigen protein couples to an oxygen electrode using catalase	235F
15. Review on enzyme and microorganism-based electrode	122F
16. Kinetic creatinine determination in urine with picrate electrode	55F
17. Continuous monitoring of Na^+, K^+, Ca^{2+}, and β-D-glucose using electrodes	209F
18. Announcement of the flat-form differential and direct format electrodes	5F, 11F
19. Patent on electrode arrangements	80F
20. Automated immunoassay with a silver sulfide electrode	20F

TABLE 14. Classical Synthetic Organic and Inorganic Ion Exchange Membranes

Topic	References
A (1968–1969)	
1. Applications in titrimetry, as electrodes	71A, 72A, 73A, 74A
2. Membrane potential equation, tests, and applications in corrosive (HF/media)	9A, 51A, 81A
3. Applications in:	
complex ion studies	50A
nonaqueous potentiometry	49A
salt release titrimetry	15A
4. Relation of bulk polymer and membrane physical properties	52A
5. Examples of:	
parchment-impregnated membranes AgI	140A, 141A
alkaline earth sulfates, oxalates, selemates, and tellurates	67A, 94A
arsenates	95A
heavy metal ferro- and ferricyanides	113A, 114A, 157A

Topic	References
B (1970–1971)	
1. Electrochemistry of ion exchange resins	117B
2. Soviet studies of synthetic ion exchange membranes in:	
strong acids	324B
end-point detection aqueous and nonaqueous solutions	343B, 344B
dye-concentration determination	263B, 264B
3. Complexing resin membranes for cations. Complexing resin membranes for anions.	29B, 74B, 122B
4. Responses of poly(methylmethacrylate) membrane electrode to Ca^{2+} and Mg^{2+}	197B
5. Synthetic membranes for:	
anion activities	292B
cation activities	367B
6. K^{+}-selective, cellulose acetate membrane	34B
7. Tests of synthetic-membrane potentials	70B, 134B, 154B, 161B, 255B, 307B, 345B, 346B, 355B, 375B, 387B
8. Applications of synthetic cation exchange membrane electrodes	136B, 140B, 203B
C (1972–1973)	
1. Surfactant measurements	411C, 416C
2. Membrane with interior, but no exterior sites	224C, 225C
3. Acid and monovalent ion responses	4C, 119C, 334C, 415C, 428C
4. Divalent ion responses	55C, 78C, 79C
5. Anion responses	5C, 44C
6. Synthetic pH-sensing membrane. Synthetic Cl^{-}-sensing membrane.	155C, 410C
7. Cellulose ion exchange paper responses	387C
D (1974–1975)	
1. Synthetic membranes for use as electrodes in HF pickling baths for H^{+} monitoring	51D, 148D
2. Incorporation of divalent-ion complex former to modify fixed-site membrane selectivity	179D
3. Charge and sites in cellulosic membranes	109D, 186D
4. Studies aimed at making anionic surfactant-selective synthetic membranes	21D, 136D
F (1978)	
1. Amberlite IR 120 in epoxy-based membrane electrode for cation analyses Na, K, Li, and Ca	120F

TABLE 15. Potentiometry and Ion-Selective Electrode Nomenclature[a]

The following list of terms, their definitions, and abbreviations is based on current usage consistent with recommendations of IUPAC Divisions of Analytical and Physical Chemistry.[1–5] Two exceptions are marked with an asterisk.

General

Potentiometry, differential potentiometry, potentiometric titration, differential, derivative, and other variant potentiometric titrations. See definitions 4.1–4.7 of reference 6 below.

Galvanic cell organization and potentials defined, illustrated with recommended symbols[7]

Equilibrium between two charged, simple phases, definitions and symbols.[8]

Activities, activity coefficients, concentrations, electrolyte solutions and related quantities, definitions, and symbols.[5,9] In this topic, use a_A, a_B, a_C, c_A, c_B, and c_C for species activities and concentrations.

The quantity pH; see reference 10 below.

Specific

Blocked interface, polarizable interface—no charge carriers cross the interface, e.g., AgX/insulator of NaF(aq.)/Hg.

Calibration curve—cell voltage, emf, or potential difference (equivalent terms for E) as ordinate vs. pa_A ($pa_\pm$) or pc_A, etc. as abscissa with activities increasing to right carriers—generally neutral, hydrophobic ionophores, e.g., valinomycin. IUPAC also includes liquid ion exchangers cell organization; for example:

Cu (left)	External reference electrode: typically mercury–calomel or silver–silver chloride in fixed activity Cl^-	Salt bridge	External monitored solution	Membrane, reversible and permselective to cations or anions	Inner filling solution containing a fixed activity or permselective ion and fixed Cl^- activity	AgCl; Ag; Cu (right)

Combination electrode—a complete cell in one body including sensor and reference half-cells

Crystalline, primary, homogeneous electrode—"solid" electrode, e.g., single-crystal, pressed pellet, or melted pellet AgCl, or $AgCl–Ag_2S$

Crystalline, primary, heterogeneous electrode—"solid" electrode, e.g., AgCl powder in PVC, or an AgCl Selectrode, AgCl/hydrophobized graphite

Detection limit, drift, hysteresis—see reference 3 below

Electrode kind, type, or class (equivalent terms): *zeroth—redox only, e.g., $Pt/Fe(CN)_6^{4-}$, $Fe(CN)_6^{3-}$; first—one ion-crossing interface, e.g., Ag/Ag^+; second—two involved ion-crossing interfaces, e.g., $Hg/Hg_2Cl_2/Cl^-$; three—three involved ion-crossing interfaces, e.g., $Pb/PbC_2O_4/CaC_2O_3/Ca^{2+}$

Electrodes, membrane—half-cell including phase-separating electrolyte solutions such that ordinary charge-transport is modified and potential differences are created, e.g., 2-interface ion-exchanger phase with inner reference as in cell (above). All-solid-state configurations omit inner filling solution and connect common metal wire to salt.

Electrodes, reference—a non-polarizable half-cell of known or reproducible potential difference, e.g., usually second kind electrodes such as $Ag/AgCl/Cl^-$

Mediator—a solvent added to ion-exchanger membranes to modify ion selectivity coefficients

Nonblocked interface, nonpolarizable interface—rapid, reversible ion exchange, potential difference-defined interface

Noncrystalline, primary electrodes—general term for porous, supported (e.g., Millipore filter, glass frit) and nonporous (e.g., PVC) liquid ion exchanger, neutral carrier, solvent-mediated membrane electrodes. May include mediators and gelling agents. Also defines glass membrane electrodes.

Potentiometric selectivity coefficient—$k^{pot}_{A,B}$ used in potentiometric response equation

Potentiometric response equation—historically a form closely connected with Horovitz, Nicolsky, and Eisenman. It is vaguely a form of Goldman–Hodgkins–Katz equation or Nernst equation. When response to only one ion is involved, and is ideal, i.e., obeys the following equation, the response is called "Nernstian."

$$*E = \text{const.} + 2.303RT/F \log a_A^{1/z_A} + (k^{pot}_{A,B} a_B)^{1/z_B}$$

E is cell voltage (SI unit V). The response for the two ion activities indicated here is the voltage, potential difference, or emf found by E_{Cu}(right) − E_{Cu}(left) and this quantity has thermodynamic significance in the absence of liquid junction potential differences.

Sensitized ion-selective electrodes—gas-sensing and enzyme substrate electrodes. An interposed chemical reaction is involved with conversion of sensed substance or intermediate to a product ion detected by the underlying electrode.

---

[a]The superscript numbers in parentheses correspond to the following references:

1. IUPAC, "Classification and Nomenclature of Electroanalytical Techniques," *Pure Appl. Chem.* **45**, 81 (1976).
2. IUPAC, "Manual of Symbols and Terminology," Appl. III Electrochemical Nomenclature, *Pure Appl. Chem.* **37**, 501 (1974).
3. IUPAC, "Recommendation for Nomenclature of Ion-Selective Electrodes," *Pure Appl. Chem.* **48**, 127 (1976).
4. IUPAC, *Manual of Symbols and Terminology*, Butterworth, London (1973).
5. IUPAC, "Proposed Terminology and Symbols for the Transfer of Solutes from One Solvent to Another," *Inf. Bull.*, No. 34 (August 1974).
6. Reference 1, p. 86.
7. Reference 2, pp. 504–506; reference 4, pp. 9–10, 27–28.
8. Reference 2, pp. 507–508.
9. Reference 2, pp. 510–511; reference 4, pp. 36–40.
10. Reference 4, pp. 29–30.

"A" REFERENCES (1968–1969)

1. Hair, I., and Altug, M. L., *J. Phys. Chem.* **71**, 4260–4263 (1967).
2. Andelman, J. B., *J. Water Pollut. Contr. Fed.* **40**, 1844–1860 (1968).
3. Anfalt, T., and Jagner, D. *Anal. Chim. Acta* **47**, 483–494 (1969).
4. Baczuk, R. J., and DuBois, R. J., *Anal. Chem.* **40**, 685–689 (1968).
5. Babcock, R. H., and Johnson, K. A., *J. Am. Water Works Assoc.* **60**, 953–961 (1968).
6. Baumann, E. W., *Anal. Chem.* **40**, 1731–1732 (1968).
7. Baumann, E. W., *Anal. Chim. Acta* **42**, 127–132 (1968).
8. Beck, H. W., Bottom, A. E., and Covington, A. K., *Anal. Chem.* **40**, 501–505 (1968).
9. Bercik, J., and Hladky, Z., *Proc. Conf. Appl. Phys. Chem. Methods Chem. Anal. Budapest* **1**, 142–151 (1966).
10. Bock, R., and Strecker, S., *Fresenius' Z. Anal. Chem.* **235**, 322–334 (1968).
11. Bloch, R., Shatkay, A., and Saroff, H. A., *Biophys. J.* **7**, 865–877 (1967).
12. Bobrov, V. S., Kalmykova, L. P., and Shult's, M. M., *Elektrokhimiya* **4**, 1322–1327 (1968).

13. Bock, R., and Puff, H. J., *Fresenius' Z. Anal. Chem.* **240**, 381–386 (1968).
14. Boksay, Z., and Csakvari, B., *Ann. Univ. Sci. Budap. Rolando Eotvos Nominatae Sect. Chim.* **10**, 125–134 (1968).
15. Botre, C., Marchetti, M., Borghi, S., and Memoli, A., *Biochim. Biophys. Acta* **183**, 249–252 (1969).
16. Brand, M. J. D., and Rechnitz, G. A., *Anal. Chem.* **41**, 1185–1191 (1969).
17. Buchanan, E. B., Jr., and Seago, J. L., *Anal. Chem.* **40**, 517–521 (1968).
18. Buck, R. P., *Anal. Chem.* **40**, 1432–1439 (1968).
19. Buck, R. P., *Anal. Chem.* **40**, 1439 (1968).
20. Buck, R. P., *J. Electroanal. Chem. Interfacial Electrochem.* **18**, 363–380 (1968).
21. Buck, R. P., and Krull, I., *J. Electroanal. Chem. Interfacial Electrochem.* **18**, 381–386 (1968).
22. Buck, R. P., and Krull, I., *J. Electroanal. Chem. Interfacial Electrochem.* **18**, 387–399 (1968).
23. Burger, K., and Pinter, B., *Hung. Sci. Instrum.* **8**, 11–14 (1966).
24. Burroughs, J. E., and Attia, A. I., *Anal. Chem.* **40**, 2052–2053 (1968).
25. Butler, J. N., *Biophys. J.* **8**, 1426–1433 (1968).
26. Carter, N. W., Ger. Patent 1,281,183 (24 October 1968).
27. Carter, N. W., U.S. Patent 3,415,731 (10 December 1968).
28. Coetzee, C. J., and Freiser, H., *Anal. Chem.* **40**, 2071 (1968).
29. Coetzee, C. J., and Freiser, H., *Anal. Chem.* **41**, 1128–1130 (1969).
30. Covington, A. K., and Lilley, T. H., *Phys. Chem. Glasses* **8**, 88–91 (1967).
31. Covington, A. K., Paabo, M., Robinson, R. A., and Bates, R. G., *Anal. Chem.* **40**, 700–706 (1968).
32. Csakvari, B., and Meszaros, K., *Hung. Sci. Instrum.* **11**, 9–13 (1968).
33. Csakvari, B., Torros, K., and Gomory, P., *Ann. Univ. Sci. Budap. Rolando Eotvos Nominatae Sect. Chim.* **9**, 115–117 (1967).
34. Curran, D. J., and Fletcher, K. S., *Anal. Chem.* **41**, 267–273 (1969).
35. Deelstra, H., *Ind. Chim. Belge* **34**, 177–182 (1969).
36. Dolidze, V. A., Khutsishvili, A. N., and Tsutskiridze, Z. I., *Izobret. Prom. Obraztsy Tovarnye Znaki* **45**, 94 (1968).
37. Dubowski, K. M., *Clin. Pathol. Serum Electrolytes* **1966**, 124–142.
38. Durst, R. A., *Mikrochim. Acta* **1969**, 611–614.
39. Edmond, C. R., *Aust. Miner. Develop. Lab. Bull.* **7**, 1–14 (1969).
40. Eisenman, G., *Ann. N.Y. Acad. Sci.* **148**, 5–35 (1968).
41. Eisenman, G., U.S. Patent 3,450,604 (7 June 1969).
42. Eisenman, G., and Ross, J. W., U.S. Patent 3,356,595 (5 December 1967).
43. Frant, M. S., *Anal. Chem.* **40**, 457 (1968).
44. Frant, M. S., *Anal. Chem.* **40**, 458 (1968).
45. Frant, M. S., U.S. Patent 3,431,182 (4 March 1969).
46. Geisler, H., *Chem. Tech.* **20**, 233–234 (1968).
47. Gerchman, L. J., and Rechnitz, G. A., *Fresenius' Z. Anal. Chem.* **230**, 265–271 (1967).
48. Geyer, R., and Chojnacki, K., *Wiss. Z. Tech. Hochsch. Chem. "Carl Schorlemmer" Leuna-Marseburg* **9**, 325–331 (1967).
49. Gordievskii, A. V., Filippov, E. L., and Shterman, V. S., *Tr. Mosk. Khim. Tekhnol. Inst.* **58**, 105–109 (1968).
50. Gordievskii, A. V., Filippov, E. L., Shterman, V. S., and Krivoshein, A. S., *Zh. Fiz. Khim.* **42**, 1998–2004 (1968).
51. Gordievskii, A. V., Kiryushov, V. N., Tevlina, A. S., and Novikova, S. P., *Mosk. Khim.-Tekhnol. Inst.* **51**, 206–211 (1966).

52. Grebemyuk, V. D., Pevnitskaya, M. V., and Gnusin, N. P., *Zh. Prikl. Khim.* **42**, 578–584 (1969).
53. Guha, S. K., and Bhattacharyya, A. P., *Trans. Indian Ceram. Soc.* **26**, 136–141 (1967).
54. Guilbault, G. C., and Montalvo, J. G., *Anal. Lett.* **2**, 283 (1969).
55. Guth, J. L., and Wey, R., *Bull. Soc. Fr. Mineral. Cristallogr.* **92**, 105–107 (1969).
56. Hadert, H., *Glas-Instrumen.-Tech. Fachz. Lab.* **13**, 209–214 (1969).
57. Hall, F. M., and Slater, S. J., *Aust. J. Chem.* **21**, 2663–2667 (1968).
58. Harrell, J. B., Jones, A. D., and Choppin, G. R., *Anal. Chem.* **41**, 1459–1462 (1969).
59. Harzdorf, C., *Fresenius' Z. Anal. Chem.* **245**, 67–70 (1969).
60. Havas, J., Huber, M., Szabo, I., and Pungor, E., *Hung. Sci. Instrum.* **1967**, 1923.
61. Havas, J., Papp, E., and Pungor, E., *Acta Chim.* **58**, 9–21 (1968).
62. Havas, J., Papp, E., and Pungor, E., *Magy. Kem. Foly.* **74**, 292–297 (1968).
63. Havas, J., Toth, K., Szabo, I., and Pungor, E., *Proc. Conf. Appl. Phys. Chem. Methods Chem. Anal. Budapest* **1**, 159–165 (1966).
64. Hawthorn, D., and Ray, N. J., *Analyst (London)* **93**, 158–165 (1968).
65. Hersh, L. S., *J. Phys. Chem.* **72**, 2195–2199 (1968).
66. Hirsch, R. F., and Portock, J. D., *Anal. Lett.* **2**, 295–303 (1969).
67. Honig, E. P., Hengst, J. H., and Hirsh-Ayalon, P., *Ber. Bunsenges Phys. Chem.* **72**, 1231–1242 (1968).
68. Hseu, T. M., and Rechnitz, G. A., *Anal. Chem.* **40**, 1054–1060 (1968).
69. Hseu, T. M., and Rechnitz, G. A., *Anal. Lett.* **1**, 629–640 (1968).
70. Huston, R., and Butler, J. N., *Anal. Chem.* **41**, 200–202 (1969).
71. Ijsseling, F. P., and Van Dalen, E., *Anal. Chim. Acta* **40**, 421–430 (1968).
72. Ijsseling, F. P., and Van Dalen, E., *Anal. Chim. Acta* **43**, 77–87 (1968).
73. Ijsseling, F. P., and Van Dalen, E., *Anal. Chim. Acta* **45**, 121–132 (1969).
74. Ijsseling, F. P., and Van Dalen, E., *Anal. Chim. Acta* **45**, 492–505 (1969).
75. Jaselskis, B., and Bandemer, M. K., *Anal. Chem.* **41**, 855–857 (1969).
76. Johansson, G., *Kem. Tidskr.* **81**, 6–12 (1969).
76a. Johansson, G., and Norberg, K., *Electroanal. Chem. Interfacial Electrochem.* **18**, 239–250 (1968).
77. Kaminski, R. K., *Instrum. Technol.* **16**, 41–45 (1969).
78. Katz, D. A., and Mukherji, A. K., *Microchem. J.* **13**, 604–615 (1968).
79. Khuri, R. N., Agulian, S. K., and Harik, R. I., *Pfleugers Arch. Gesamte Physiol.* **301**, 182–186 (1968).
80. Kim, J. J., and Rock, P. A., *Inorg. Chem.* **8**, 563–566 (1969).
81. Kiryushov, V. N., Gordievskii, A. V., and Kolesnikov, G. S., *Zavod. Lab.* **34**, 795–798 (1968).
82. Komarova, N. A., and Becklemesheva, N. V., *Izobret. Prom. Obraztsy Tovarnye Znaki* **44**, 130 (1967).
83. Kraemer, R., and Lagoni, H., *Milchwissenschaft* **24**, 68–70 (1969).
84. Krischer, C. C., *Z. Naturforsch.* **B24**, 151–155 (1969).
85. Kryukov, P. A., *Izobret. Prom. Obraztsy Tovarnye Znaki* **44**, 124 (1967).
86. Kuhlann, E. J., *Power* **111**, 69–71 (1967).
87. Lambert, J. L., *Adv. Chem. Ser.* **73**, 18–26 (1968).
88. Lanier, R. D., and Johnson, J. S., *U.S. Office Saline Water Res. Develop. Progr. Rept.* **302**, 15–17 (1968).
89. Lee, T. G., *Anal. Chem.* **41**, 391–392 (1969).
90. Leszko, M., *Zesz. Nauk. Uniw. Jagiellor. Ser. Nauk Mat.-Przyr. Mat. Fiz. Chem.* **12**, 163–169 (1967).
91. Light, T. S., and Mannion, R. F., *Anal. Chem.* **41**, 107–111 (1969).

92. Light, T. S., and Swartz, J. L., *Anal. Lett.* **1**, 825–836 (1968).
93. Lingane, J. J., *Anal. Chem.* **40**, 935–939 (1968).
94. Liteanu, C., and Mioscu, M., *Rev. Roum. Chim.* **13**, 209–218 (1968).
95. Liteanu, C., Mioscu, M., and Popescu, I. *Rev. Roum. Chim.* **13**, 569–572 (1968).
96. Macdonald, A. M. G., and Toth, K., *Anal. Chim. Acta* **41**, 99–106 (1968).
97. Malissa, H., and Jellinek, G., *Fresenius' Z. Anal. Chem.* **245**, 70–75 (1969).
98. Mascini, M., and Liberti, A., *Anal. Chim. Acta* **47**, 339–345 (1969).
99. Materova, E. A., Grinberg, G. P., and Evstifeeva, M. M., *Zh. Anal. Khim.* **24**, 821–824 (1969).
100. Matsushita Electric Ind. Co., Ltd., Fr. Patent 1,507,834 (29 December 1967).
101. Matsushita, H., and Furuta, S., *Chubu Kogyo Daigaku Kiyo* **3**, 154–159 (1967).
102. Matsushita, H., and Furuta, S., *Chubu Kogyo Daigaku Kiyo* **4**, 121–127 (1968).
103. McClure, J. E., and Reddy, T. B., *Anal. Chem.* **40**, 2064–2066 (1968).
104. Mesmer, R. E., *Anal. Chem.* **40**, 443–444 (1968).
105. Mukherji, A. K., *Anal. Chim. Acta* **40**, 354–356 (1968).
106. Murray, R. C., and Rock, P. A., *Electrochim. Acta* **13**, 969–975 (1968).
107. Muto, G., and Nozaki, K., *Bunseki Kagaku* **18**, 247–252 (1969).
108. Myers, R. J. K., and Paul, E. A., *Can. J. Soil Sci.* **48**, 369–371 (1968).
109. Noren, B., *Acta Chem. Scand.* **23**, 931–942 (1969).
110. Oehme, F., *Dechema (Deut. Ges. Chem. App.) Monogr.* **62**, 137–152 (1968).
111. Onishi, K., *Bunseki Kagaku* **16**, 1152–1155 (1967).
112. Orenberg, J. B., and Morris, M. D., *Anal. Chem.* **39**, 1776–1780 (1967).
113. Panday, K. K., and Agrawal, V. K., *Indian J. Chem.* **6**, 48–49 (1968).
114. Panday, K. K., and Agrawal, V. K., *J. Indian Chem. Soc.* **45**, 984–986 (1968).
115. Potterton, S. S., and Shults, W. D., *Anal. Lett.* **1**, 11–22 (1967).
116. Pungor, E., *Hung. Sci. Instrum.* **10**, 1–9 (1967).
117. Pungor, E., and Toth, K., *Anal. Chim. Acta* **47**, 291–297 (1969).
118. Rechnitz, G. A., and Hseu, T. M., *Anal. Chem.* **41**, 111–115 (1969).
119. Rechnitz, G. A., and Kenny, N. C., *Anal. Lett.* **2**, 395–402 (1969).
120. Rechnitz, G. A., and Kugler, G. S., *Anal. Chem.* **39**, 1682–1688 (1967).
121. Rechnitz, G. A., and Lin, Z. F., *Anal. Chem.* **40**, 696–699 (1968).
122. Riseman, J. H., Ross, J. W., and Eisemann, G., U.S. Patent 3,357,909 (12 December 1967).
123. Riseman, J. H., Ross, J. W., and Wall, R. A., U.S. Patent 3,282,817 (1 November 1966).
124. Rispens, P., and Hoek, W., *Clin. Chim. Acta* **22**, 291–293 (1968).
125. Robbins, G. D., Foerland, T., and Oestvold, T., *Acta Chem. Scand.* **22**, 3002–3012 (1968).
126. Rogers, W. I., and Wilson, J. A., *Anal. Biochem.* **32**, 31–37 (1969).
127. Ross, J. W., U.S. Patent 3,438,886 (15 April 1969).
128. Ross, J. W., and Frant, M. S., *Anal. Chem.* **41**, 967–969 (1969).
129. Schanne, O. F., Lavallee, M., Laprade, R., and Gagne, S., *Proc. IEEE* **56**, 1072–1082 (1968).
130. Schindler, P., and Waelti, E., *Helv. Chim. Acta* **51**, 539–542 (1968).
131. Schwabe, K., and Dwojak, J., *Z. Phys. Chem.* **64**, 1–11 (1969).
132. Settzo, R. J., and Wise, W. M., Fr. Patent 1,530,500 (28 June 1968).
133. Shainberg, I., and Kaiserman, A., *Soil Sci.* **104**, 410–415 (1967).
134. Shatkay, A., *Anal. Chem.* **40**, 457–458 (1968).
135. Shatkay, A., *Anal. Chem.* **40**, 458 (1968).
136. Shatkay, A., *Anal. Chem.* **41**, 514–517 (1969).
137. Shatkay, A., *Biophys. J.* **8**, 912–919 (1968).
138. Shiller, S. L., and Frant, M. S., U.S. Patent 3,442,782 (6 May 1969).

139. Shiurba, R. A., and Jolly, W. L., *J. Am. Chem. Soc.* **90**, 5289–5291 (1968).
140. Siddiqi, F. A., and Pratap, S., *J. Electroanal. Chem. Interfacial Electrochem.* **23**, 137–146 (1969).
141. Siddiqi, F. A., and Pratap, S., *J. Electroanal. Chem. Interfacial Electrochem.* **23**, 147–156 (1969).
142. Sidorova, M. P., and Fridrikhsberg, D. A., *Vestn. Leningrad Univ.* **22**, 146–150 (1967).
143. Srinivasan, K., and Rechnitz, G. A., *Anal. Chem.* **40**, 509–512 (1968).
144. Srinivasan, K., and Rechnitz, G. A., *Anal. Chem.* **41**, 1203–1208 (1969).
145. Stefanac, Z., and Simon, W., *Anal. Lett.* **1**, 1–9 (1967).
146. Stepinac, M., and Karsulin, M., *Z. Anorg. Allg. Chem.* **355**, 219–224 (1967).
147. Syakais, J. E., and Cramer, A., *Clin. Pathol. Serum. Electrolytes* **1966**, 143–156.
148. Synnott, J. C., and Butler, J. N., *J. Phys. Chem.* **72**, 2474–2477 (1968).
149. Tadros, T. F., and Lyklema, J., *J. Electroanal. Chem. Interfacial Electrochem.* **17**, 267–275 (1968).
150. Tadros, T. F., and Lyklema, J., *J. Electroanal. Chem. Interfacial Electrochem.* **22**, 9–17 (1969).
151. Vanderborgh, N. E., *Talanta* **15**, 1009–1013 (1968).
152. Vanleugenhaghe, C., *Rev. Energ. Primaire* **2**, 8–18 (1966).
153. Van Loon, J. C., *Anal. Lett.* **1**, 393–398 (1968).
154. Vartires, I., *Stud. Cercet. Chim.* **15**, 893–907 (1967).
155. Veksler, M. A., Korolev, A. Y., and Golant, V. A., *Khim.-Farm. Zh.* **2**, 56–60 (1968).
156. Vorob'ev, L. N., *Nature (London)* **217**, 450–451 (1968).
157. Wahid, M. U., Hasan, A., and Siddiqi, F. A., *Bull. Chem. Soc. Jpn.* **40**, 1746–1753 (1967).
158. Warner, T. B., *Anal. Chem.* **41**, 527–529 (1969).
159. Warner, T. B., *Science* **165**, 178–180 (1969).
160. Wasgestian, H. F., *Fresenius' Z. Anal. Chem.* **246**, 237–238 (1969).
161. Webber, H. M., and Wilson, A. L., *Analyst (London)* **94**, 209–220 (1969).
162. Wheeler, A. F., Proceedings of I.E.R.E. Conference on Electronic Engineering Oceanography, University of Southampton, 1966.
163. Whitfield, M., and Leyendekkers, J. V., *Anal. Chim. Acta* **45**, 383–398 (1969).
164. Whitfield, M., and Leyendekkers, J. V., *Anal. Chim. Acta* **46**, 63–70 (1969).
165. Whitfield, M., Leyendekkers, J. V., and Kerr, J. D., *Anal. Chim. Acta* **45**, 399–410 (1969).
166. Wikby, A., and Johansson, G., *J. Electroanal. Chem. Interfacial Electrochem.* **23**, 23–40 (1969).
167. Zinzer, E. J., *Chem. Can.* **21**, 31–35 (1969).

"B" REFERENCES (1970–1971)

1. Abbott, W., *J. Miss. Acad. Sci.* **14**, 11–16 (1968) (Publ. 1969).
2. Aleksandrov, V. V., Berezhnaya, T. A., and Osipenko, L. K., *Electrokhimiya* **6**, 1462–1466 (1970).
3. Andelman, J. B., *Ind. Water. Eng.* **6**(12), 57–59 (1969).
4. Anfalt, T., and Jagner, D., *Anal. Chim. Acta* **50**, 23–30 (1970).
5. Anfalt, T., and Jagner, D., *Anal. Chim. Acta* **53**, 13–22 (1971).
6. Bach, H., and Baucke, F. G. K., *Electrochim. Acta* **16**, 1311–1319 (1971).
7. Badonnel, Y., Crance, J. P., Bertrand, J. M., and Panek, E., *Pharm. Biol.* **6**(61), 149–154 (1969).
8. Bagg, J., *Aust. J. Chem.* **22**, 2467–2470 (1969).

9. Bagg, J., and Chaung, W. P., *Aust. J. Chem.* **24**, 1963–1966 (1971).
10. Bagg, J., Nicholson, O., and Vinen, R., *J. Phys. Chem.* **75**, 2138–2143 (1971).
11. Baker, A. S., and Smith, R., *J. Agr. Food Chem.* **17**, 1284–1287 (1969).
12. Banin, A., and Shaked, D., *Agrochimica* **15**, 238–248 (1971).
13. Bates, R. G., *Glass Microelectrodes* (M. Lavallee, ed.), pp. 1–24, Wiley & Sons, New York (1969).
14. Battistini, A., and Waring, W. W., *Am. Rev. Respir. Dis.* **100**, 237–239 (1969).
15. Baum, G., *Anal. Biochem.* **39**, 65–72 (1971).
16. Baum, G., *Anal. Lett.* **3**, 105–111 (1970).
17. Baum, G., and Wise, W. M., Ger. Offen. 2,024,636 (Cl. G 01n) (10 December 1970).
18. Baum, G., and Ward, F. B., *Anal. Biochem.* **42**, 487–493 (1971).
19. Baum, G., and Ward, F. B., *Anal. Chem.* **43**, 947–948 (1971).
20. Baumann, E. W., *Anal. Chim. Acta* **54**, 189–197 (1971).
21. Baumann, E. W., and Wallace, R. M., *Anal. Chem.* **41**, 2072–2074 (1969).
22. Baumann, E. W., *J. Inorg. Nucl. Chem.* **32**, 3823–3830 (1970).
23. Baumann, E. W., *J. Inorg. Nucl. Chem.* **31**, 3155–3162 (1969).
24. Baumann, E. W., *Anal. Chem.* **42**, 110–111 (1970).
25. Baumann, W. M., and Simon, W., *Helv. Chim. Acta* **52**, 2054–2059 (1969).
26. Bazelle, W. E., *Anal. Chim. Acta* **54**, 29–39 (1971).
27. Beals, R. L., *Anal. Instrum.* **7**, 74–79 (1969).
28. Beebe, C. H., and Strickler, A., Ger. Offen. 2,009,937 (Cl. G 01n) (10 September 1970).
29. Belinskaya, F. A., Materova, E. A., Karmanova, L. A., and Kozyreva, V. I., *Elektrokhimiya* **7**, 214–217 (1971).
30. Berkenbosch, A., *Pfluegers Arch.* **318**, 217–224 (1970).
31. Biheller, J., and Resch, W., *Staub* **31**, 9–11 (1971).
32. Birkeland, J. M., *Caries Res.* **4**, 243–255 (1970).
33. Blaedel, W. J., Easty, D. B., Anderson, L., and Farrel, T. R., *Anal. Chem.* **43**, 890–894 (1971).
34. Bloch, R., Furmansky, M., and Gassner, S., Ger. Offen. 2,109,664 (Cl. G. 01n) (9 September 1971).
35. Bloch, R., Vofsi, D., and Kedem, O., Israeli Patent 26,233 (Cl. B 01k) (29 January 1970).
36. Boksay, Z., Csakvari, B., Havas, J., and Patko, M., *Hung. Sci. Instrum.* **19**, 31–37 (1970).
37. Bond, A. M., and O'Donnell, T. A., *J. Electroanal. Chem. Interfacial Electrochem.* **26**, 137–147 (1970).
38. Bond, A. M., and Hefter, G., *J. Inorg. Nucl. Chem.* **33**, 429–434 (1971).
39. Bond, A. M., and Hefter, G., *Inorg. Chem.* **9**, 1021–1023 (1970).
40. Bottazzini, N., and Crespi, V., *Chim. Ind.* (*Milan*) **52**, 866–869 (1970).
41. Bourdeau, A. M., Sachs, C., Presle, V., and Dromini, M., *Pharm. Biol.* **67**, 527–540 (1970).
42. Brand, M. J. D., Rechnitz, G. A., and Militello, J. J., *Anal. Lett.* **2**, 523–535 (1969).
43. Bruton, L. G., *Anal. Chem.* **43**, 579–581 (1971).
44. Buck, M., and Reussmann, G., *Fluoride* **4**, 5–15 (1971).
45. Buck, R. P., and Nolan, R. W., U.S. Patent 3,558,528 (Cl. 252–520; C 03c, H01b) (26 January 1971).
46. Buck, R. P., *Physical Methods of Chemistry* (A. Weissberger and B. Rossiter, eds.), Vol. I, Part IIA, Chap 2., pp. 61–162, Wiley and Sons, New York (1971).
47. Bunton, N. G., and Crosby, N. T., *Water Treat. Exam.* **18**, 338–342 (1969).
48. Bureau, J., and Collombel, C., *J. Eur. Toxicol.* **4**, 19–35 (1971).
49. Butler, J. N., and Huston, R., *Anal. Chem.* **42**, 1308–1311 (1970).
49a. Bykova, L. N., *Zh. Anal. Khim.* **24**, 1781–1789 (1969).
50. Caille, J. P., and Gagne, S., *Can. J. Physiol. Pharmacol.* **49**, 783–786 (1971).

51. Cammann, K., *Naturwissenschaften* **57**, 298–304 (1970).
52. Cammann, K., *Messtechnik (Brunswick)* **79**(3), 79–83 (1971).
53. Cammann, K., *Beckman Rep.* **1**, 21–25 (1970).
54. Cantliffe, D. J., MacDonald, G. E., and Peck, N. H., *N.Y. Food Life Sci. Bull.* **3**, 1970.
55. Carlson, R. M., and Paul, J. L., *Soil Sci.* **108**, 266–272 (1969).
56. Carlson, R. M., and Kenney, D. R., in *Proceedings of the Symposium on Instrumentation Methods for Analysis of Soils and Plant Tissue, 1970*, pp. 39–65 (1971).
57. Cernik, A. A., Cooke, J. A., and Hall, R. J., *Nature (London)* **227**(5264), 1260–1261 (1970).
58. Chernov'yants, M. S., Zolotov, A. I., Bagdasarov, K. N., Kovalenko, P. N., and Rotinyan, A. L., *Zh. Prikl. Khim. (Leningrad)* **43**, 698–700 (1970).
59. Chowdhury, T. K., *J. Sci. Instrum.* **2**(2), 1087–1090 (1969).
60. Christian, G. D., and Stucky, G. L., *Am. Lab.* , 18–25 (October 1970).
61. Clem, R. G., Jakob, F., and Anderberg, D., *Anal. Chem.* **43**, 292–293 (1971).
62. Clements, R. L., Sergeant, G. A., and Webb, P. J., *Analyst (London)* **96**(1138), 51–54 (1971).
63. Collier, H. B., *Anal. Chem.* **42**, 1443 (1970).
64. Collis, D. E., and Diggens, A. A., *Water Treat. Exam.* **18** (Part 3), 192–202 (1969).
65. Collombel, C., Durand, J. P., Bureau, J., and Cotte, J., *J. Eur. Toxicol.* **3**, 291–299 (1970).
66. Cosgrove, R. E., Mask, C. A., and Krull, I. H., *Anal. Lett.* **3**, 457–464 (1970).
67. Covington, A. K., *Chem. Brit.* **5**, 388 (1969).
68. Csakvari, B., Gomory, P., and Torkos, K., *Proc. IMEKO (Int. Meas. Confed.) Symp. Electrochem. Sens.* **1968**, 51–61.
69. Dahnke, W. C., *Commun. Soil Sci. Plant Anal.* **2**, 73–84 (1971).
70. Dawson, D. G., and Meares, P., *J. Colloid. Interface Sci.* **33**, 117–123 (1970).
71. Deon, G., *Cent. Tech. For. Trop. Nogent-sur-Marne Fr. Note Tech.* **11**, 19 (1970).
72. DiGregorio, J. S., and Morris, M. D., *Anal. Chem.* **42**, 94–97 (1970).
73. Di Martini, R., *Anal. Chem.* **42**, 1102–1105 (1970).
74. Dobbelstein, T. N., and Diehl, H., *Talanta* **16**, 1341–1352 (1969).
75. Dobos, S., *Acta Chim. (Budapest)* **69**, 43–48 (1971).
76. Dobos, S., *Acta Chim. (Budapest)* **69**, 49–57 (1971).
77. Domoryad, I. A., Koltunov, Yu. B., and Koltunov, E. B., *Radiats. Stimulirovannye Protessy Tverd. Talakh*, 139–142 (1969).
78. Duff, E. J., and Stuart, J. L., *Anal. Chim. Acta* **52**, 155–157 (1970).
79. du Pont de Nemours, E. I., and Co., Brit. Patent 1,213,067 (Cl. G 01n) (18 November 1970).
80. Durst, R. A., *Anal. Chem.* **41**, 2089–2090 (1969).
81. Durst, R. A., *Natl. Bur. Stand. (U.S.) Spec. Publ.* **314** (1969).
82. Durst, R. A., *Am. Sci.* **59**, 353–361 (1971).
83. Durst, R. A., and Duhart, B. T., *Anal. Chem.* **42**, 1002–1004 (1970).
84. Eckfeldt, E. L., and Proctor, W. E., Jr., *Anal. Chem.* **43**, 332–337 (1971).
85. Elgquist, B., *J. Inorg. Nucl. Chem.* **32**, 937–944 (1970).
86. El-Swaify, S. A., and Gazdar, M. N., *Soil Sci. Soc. Am. Proc.* **33**, 665–667 (1969).
87. Eriksson, T., and Johansson, G., *Anal. Chim. Acta* **52**, 465–473 (1970).
88. Erlanger, B. F., and Sack, R. A., *Anal. Biochem.* **33**, 318–322 (1970).
89. Eyal, E., and Rechnitz, G. A., *Anal. Chem.* **43**, 1090–1093 (1971).
90. Farren, G. M., Ger. Offen. 2,101,339 (Cl. G 01n) (16 September 1971).
91. Farren, G. M., and Staunton, J. J. J., Ger. Offen. 1,940,353 (Cl. G 01n) (19 February 1970).
92. Farren, G. M., and Staunton, J. J. J., Ger. Offen. 2,006,051 (Cl. G 01n) (3 September 1970).

93. Farren, G. M., and Staunton, J. J. J., Ger. Offen. 2,002,676 (Cl. G 01n) (5 November 1970).
94. Ferren, W. P., and Shane, N. A., *J. Food Sci.* **34**, 317–319 (1969).
95. Ficklin, W. H., *U.S. Geol. Surv. Prof. Pap.* **700-C**, 186–188 (1970).
96. Fioravanti, P., Zuber, R., and Halmi, M., *Mitt. Geb. Lebensmittelunters. Hyg.* **61**, 214–220 (1970).
97. Fishman, M. J., and Feist, O. J., Jr., *U.S. Geol. Surv. Prof. Pap.* **700-C**, 226–228 (1970).
98. Fiskell, J. G. A., and Breland, H. L., *Soil Crop Sci. Soc. Fla. Proc.* **29**, 63–69 (1970).
99. Fleet, B., and Von Storp, H., *Anal. Lett.* **4**, 425–435 (1971).
100. Florence, T. M., *J. Electroanal. Chem. Interfacial Electrochem.* **31**, 77–86 (1971).
101. Florence, T. M., *Proc. R. Aust. Chem. Inst.* **37**(10), 261–270 (1970).
102. Francis, H. J., Jr., Deonarine, J. H., and Persing, D. D., *Microchem. J.* **15**, 580–592 (1969).
103. Frant, M. S., *Plating* **58**, 686–693 (1971).
104. Frant, M. S., and Ross, J. W., Ger. Offen. 1,942,397 (Cl. G 01n) (12 March 1970).
105. Frant, M. S., and Ross, J. W., Ger. Offen. 1,598,453 (Cl. G 01n) (5 August 1971).
106. Frant, M. S., and Ross, J. W., Jr., *Science* **167**(3920), 987–988 (1970).
107. Friconneau, C., and Leboutet, A., Fr. Patent 1,566,161 (Cl. B 01k; G 01n) (9 May 1969).
108. Fry, B. W., and Taves, D. R., *J. Lab. Clin. Med.* **75**, 1020–1025 (1970).
109. Gamsjaeger, H., Schindler, P., and Kleinert, B., *Chimia* **23**(6), 229–230 (1969).
110. Gavach, C., and Bertrand, C., *Anal. Chim. Acta* **55**, 385–393 (1971).
111. Gavach, C., and Guastalla, J., *Membranes Permeabilite Selectivity, Colloquium*, 1967, pp. 165–167 Editions du Centre National de la Recherche Scientifique, Paris (1969).
112. Gavach, C., and Seta, P., *Anal. Chim. Acta* **50**, 407–412 (1970).
113. Gehring, D. G., Dippel, W. A., and Boucher, R. S., *Anal. Chem.* **42**, 1686–1689 (1970).
114. Gelder, R. L., and Neville, J. F., Jr., *Am. J. Clin. Pathol.* **55**, 325–329 (1971).
115. Gibson, E. J., Shiller, S. L., and Riseman, J. H., U.S. Patent 3,467,590 (Cl. 205–195; B 01k) (16 September 1969).
116. Gillingham, J. T., Shirer, M. M., and Page, N. R., *Agron. J.* **61**, 717–718 (1969).
117. Gnusin, N. P., *Visn. Akad. Nauk Ukr. RSR* **1971**(4), 17–27.
118. Goldstein, G., Wolff, C. M., and Schwing, J. P., *Bull. Soc. Chim. Fr.* **1971**, 1195–1201.
119. Grassino, S. L., and Hume, D. N., *J. Inorg. Nucl. Chem.* **33**, 421–428 (1971).
120. Grekovich, A. L., Materova, E. A., and Belinskaya, F. A., *Elektrokhimiya* **6**, 1036–1039 (1970).
121. Grekovich, A. L., Materova, E. A., and Pron'kina, T. I., *Elektrokhimiya* **7**, 436–438 (1971).
122. Grinberg, G. P., Materova, E. A., *Vestn. Leningrad Univ. Fiz. Khim.* **1970**(2), 75–84.
123. Gruen, L. C., and Harrap, B. S., *Anal. Biochem.* **42**, 377–381 (1971).
124. Gruen, L. C., and Harrap, B. S., *J. Soc. Leather Trades Chem.* **55**(4), 131–138 (1971).
125. Guastalla, J., *Nature* (*London*) **227**(5257), 485–486 (1970).
126. Guha, S. K., Bhattacharyya, A. P., and Sen, S., *Trans. Indian Ceram. Soc.* **28**(5), 129–136 (1969).
127. Guignard, J. P., and Friedman, S. M., *J. Appl. Physiol.* **29**, 254–257 (1970).
128. Guilbault, G. G., *Pure Appl. Chem.* **25**, 727–740 (1971).
129. Guilbault, G. G., and Brignac, P. J., Jr., *Anal. Chim. Acta* **56**, 139–142 (1971).
130. Guilbault, G. G., and Hrabankova, E., *Anal. Chem.* **42**, 1779–1783 (1970).
131. Guilbault, G. G., and Hrabankova, E., *Anal. Chim. Acta* **52**, 287–294 (1970).
132. Guilbault, G. G., and Hrabankova, E., *Anal. Chim. Acta* **56**, 285–290 (1971).
133. Guilbault, G. G., and Hrabankova, E., *Anal. Lett.* **3**, 53–57 (1970).
134. Guillou, M., Guillou, D., and Buvet, R., *Membranes Permeabilite Selectivity Colloquium*, 1967, pp. 131–152 Editions du Centre National de la Recherche Scientifique, Paris (1969).

135. Gulanyan, S. A., Andrianov, V. K., Kurella, G. A., and Litvin, F. F., *Biol. Nauki* **1970**(9), 106–111.
136. Gunn, R. B., and Curran, P. F., *Biophys. J.* **11**, 559–571 (1971).
137. Hadjiioannou, T. P., and Papastathopoulos, D. S., *Talanta* **17**, 399–406 (1970).
138. Hahn, C. E. W., *Rev. Sci. Instrum.* **42**, 1164–1168 (1971).
139. Hakoila, E., *Anal. Lett.* **3**, 273–277 (1970).
140. Hale, D. K., and Grovindan, K. P., *J. Electrochem. Soc.* **116**, 1373–1381 (1969).
141. Hansen, S. O., Theodorsen, L., *Clin. Chim. Acta* **31**, 119–122 (1971).
142. Harnoncourt, K., Ger. Offen. 1,963,525 (Cl. G 01n) (27 August 1970).
143. Harnoncourt, K., and Zeiringer, R., Ger. Offen. 2,057,965 (Cl. G 01n) (16 June 1971).
144. Hattner, R. S., Johnson, J. W., Bernstein, D. S., Wachman, A., and Brackman, J., *Clin. Chim. Acta* **28**, 67–75 (1970).
145. Haynes, W. M., and Wagenknecht, J. H., *Anal. Lett.* **4**, 491–495 (1971).
146. Hedwig, G. R., and Powell, H. K. J., *Anal. Chem.* **43**, 1206–1212 (1971).
147. Higaki, K., *Nippon Seirigaku Zasshi* **31**, 617–626 (1969).
148. Higuchi, T., Illian, C. R., and Tossounian, J. L., *Anal. Chem.* **42**, 1674–1676 (1970).
149. Hillier, D. R., Faigh, H. C., and Coon, B. W., Ger. Patent 1,498,612 (Cl. G 01n) (3 December 1970).
150. Hipp, B. W., and Langdale, G. W. *Commun. Soil Sci. Plant Anal.* **2**, 237–240 (1971).
151. Hirata, H., and Higashiyama, K., *Anal. Chim. Acta* **54**, 415–422 (1971).
152. Hirata, H., Higashiyama, K., and Date, K., *Anal. Chim. Acta* **51**, 209–212 (1970).
153. Holmes, P. L., Green, H. E., and Lopez-Majano, V., *Am. J. Clin. Pathol.* **54**, 566–569 (1970).
154. Honig, E. P., and Hengst, J. H. T., *Electrochim. Acta* **15**, 491–499 (1970).
155. Hoole, D. W., and Klein, G. L., Ger. Offen. 1,918,590 (Cl. G 01n) (30 October 1969).
156. Hoole, D. W., Klein, G. L., and Vivian, T. R., U.S. Patent 3,575,834 (Cl. 204–195; G 01n) (20 April 1971).
157. Hozumi, K., and Akimoto, N., *Bunseki Kagaku* **20**, 467–473 (1971).
158. Huston, R., and Butler, J. N., *Anal. Chem.* **41**, 1695–1698 (1969).
159. Ingram, B. L., *Anal. Chem.* **42**, 1825–1827 (1970).
160. Ingram, B. L., and May, I., *U.S. Geol. Surv. Prof. Pap.* **750-B**, 180–184 (1971).
161. Isaev, N. I., Zolotareva, R. I., and Mostovaya, S. A., *Ionoobmen. Membrany Elektrodialize* **1970**, 89–98.
162. Ishibashi, N., *Bunseki Kagaku* **20**, 749–758 (1971).
163. Ivanovskaya, I. S., and Shul'ts, M. M., *Tr. Konf. Anal. Khim. Nevodn. Rastvorov Ikh Fiz.-Khim. Svoistvam 1st* **2**, 124–130 (1968).
164. Ivanovskaya, I. S., Gavrilova, V. I., and Shul'ts, M. M., *Elektrokhimiya* **6**, 1006–1010 (1970).
165. Jacin, H., *Tobacco Sci.* **14**, 28–30 (1970).
166. Jacobs, J. S., Hattner, R. S., and Bernstein, D. S., *Clin. Chim. Acta* **31**, 467–472 (1971).
167. Jacobson, J. S., and Heller, L. I., *Environ. Lett.* **1**, 43–47 (1971).
168. James, R. P., and Nolan, R. W., Ger. Patent 1,448,079 (Cl. G 01n) (11 June 1970).
169. Janaer Glaswerk Schott und Gen., Fr. Patent 1,547,915 (Cl. C 03c; B 01k) (29 November 1968).
170. Jolly, W. L., and Boyle, E. A., *Anal. Chem.* **43**, 514–518 (1971).
171. Jordan, D. E., *J. Ass. Offic. Anal. Chem.* **53**, 447–450 (1970).
172. Juillard, J., *Bull. Soc. Chim. Fr.* **1970**, 2040–2046.
173. Kaplan, A., NASA Contract Rep. NASA-CR-73364 (1969), 54 pp.
174. Karlberg, B., and Johansson, G., *Talanta* **16**, 1545–1551 (1969).
175. Kater, J. A. R., and Leonard, J. E., U.S. Patent 3,498,899 (Cl. 204–195; B 01k) (3 March 1970).

176. Kawamura, M., and Kashime, T., *Kyoritsu Yakka Daigaku Daigaku Kenkyu Nempo* **1970**(15), 11–20.
177. Ke, P. J., and Regier, L. W., *Anal. Chim. Acta* **53**, 23–29 (1971).
178. Ke, P. J., and Regier, L. W., *J. Fish Res. Bd. Can.* **28**, 1055–1056 (1971).
179. Kedem, O., Loebel, E., and Furmansky, M., Ger. Offen. 2,027,128 (Cl. G 01n) (23 December 1970).
180. Keeney, D. R., Byrnes, B. H., and Genson, J. J., *Analyst* (*London*) **95**(1129), 363–386 (1970).
181. Khuri, R. N., Agulian, S. K., and Wise, W. M., *Pfleugers Arch.* **322**, 39–46 (1971).
182. Kleboth, K., *Monatsh. Chem.* **101**, 767–775 (1970).
183. Knupp, R. C., *Am. Ceram. Soc. Bull.* **49**, 773–776 (1970).
184. Krijgsman, W., Mansveld, J. F., and Griepink, B. F. A., *Fresenius' Z. Anal. Chem.* **249**, 368–370 (1970).
185. Krull, I. H., Mask, C. A., and Cosgrove, R. E., *Anal. Lett.* **3**, 43–51 (1970).
186. Kryukov, P. A., and Starostina, L. I., *Izv. Sib. Otd. Akad. Nauk SSSR Ser. Khim. Nauk* **1970**(3), 27–36.
187. Kubota, H., *Anal. Chem.* **42**, 1593–1596 (1970).
188. La Croix, R. L., Kenney, D. R., and Walsh, L. M., *Commun. Soil Sci. Plant Anal.* **1**, 1–6 (1970).
189. Lal, S., *Fresenius' Z. Anal. Chem.* **255**, 209–210 (1971).
190. Lal, S., and Christian, G. D., *Naturwissenschaften* **58**, 362 (1971).
191. Lal, S., and Christian, G. D., *Anal. Chim. Acta* **52**, 41–46 (1970).
192. Lal, S., and Christian, G. D., *Anal. Chem.* **43**, 410–421 (1971).
193. Lal, S., and Christian, G. D., *Anal. Lett.* **3**, 11–15 (1970).
194. Langmuir, D., and Jacobson, R. L., *Environ. Sci. Technol.* **4**, 834–838 (1970).
195. Lavallee, M., Schanne, O. F., and Hebert, N. C., eds., *Glass Microelectrodes*, Wiley and Sons, New York (1969).
196. Levaggi, D. A., Oyung, W., and Feldstein, M., *J. Air Pollut. Contr. Assoc.* **21**, 277–279 (1971).
197. Levchenko, V. M., Minkin, M. B., Dukhnina, T. P., and Mikhalenko, B. S., *Gidrokhim. Mater.* **52**, 146–151 (1969).
198. Levins, R. J., *Anal. Chem.* **43**, 1045–1047 (1971).
199. Leyendekkers, J. V., and Whitfield, M., *Anal. Chem.* **43**, 322–327 (1971).
200. Leyendekkers, J. V., and Whitfield, M., *J. Phys. Chem.* **75**, 957–976 (1971).
201. Li, Ting-Kai, and Piechocki, J. T., *Clin. Chem.* **17**, 411–416 (1971).
202. Liberti, A., and Mascini, M., *Fluoride* **4**(2), 49–56 (1971).
203. Licis, J., *Lalv. PSR Zinat. Akad. Vestis Kim Ser.* **1970**(1), 122–123.
204. Light, T. S., *Natl. Bur. Stand.* (*U.S.*) *Spec. Publ.* **314**, 349–374 (1969).
205. Light, T. S., *Ind. Water Eng.* **6**(9), 33–37 (1969).
206. Liteanu, C., and Ghergariu-Mirza, L., *Rev. Roum. Chim.* **15**, 1871–1882 (1970).
207. Liteanu, C., Hopirtean, E., and Mioscu, M., *Stud. Cercet. Chem.* **18**, 241–272 (1970).
208. Liteanu, C., and Hopirtean, E., *Talanta* **17**, 1067–1074 (1970).
209. Liteanu, C., and Hopirtean, E., *Rev. Roum. Chim.* **15**, 1331–1336 (1970).
210. Liteanu, C., and Hopirtean, E., *Stud. Univ. Babes-Bolyai Ser. Chem.* **15**(2), 41–44 (1970).
211. Liteanu, C., and Hopirtean, E., *Rev. Roum. Chim.* **16**, 55–62 (1971).
212. Liteanu, C., and Hopirtean, E., *Rev. Roum. Chim.* **16**, 559–563 (1971).
213. Maas, A. H. J., *J. Appl. Physiol.* **30**, 248–250 (1971).
214. Maas, A. H. J., *Clin. Chim. Acta* **28**, 373–390 (1970).
215. Mack, A. R., and Sanderson, R. B., *Can. J. Soil Sci.* **51**, 95–104 (1971).
216. Makabe, H., Ger. Offen. 1,933,642 (Cl. G 01n) (29 January 1970).
217. Manahan, S. E., U.S. Clearinghouse Fed. Sci. Tech. Inform., PB Rep., No. 195167 (1970).

218. Manahan, S. E., *Anal. Chem.* **42**, 128–129 (1970).
219. Manahan, S. E., *Trace Substance Environmental Health —3, Proc. Univ. Mo. Ann. Conf., 3rd.*, 1969, pp. 353–357, University of Missouri, Columbia, Missouri (1970).
220. Manahan, S. E., Smith, M., Alexander, D., and Hamilton, R., U.S. Clearinghouse Fed. Sci. Tech. Inform., PB Rep., No. 192809, 18 pp. (1970).
221. Marton, A., and Pungor, E., *Anal. Chim. Acta* **54**, 209–219 (1971).
222. Marton, A., and Pungor, E., *Magy. Kem. Foly.* **77**, 390–396 (1971).
223. Mascini, M., *Inquinamento* **13**(2), 21–29 (1971).
224. Mask, C. A., and Krull, I. H., Ger. Offen. 2,062,061 (Cl. G 01n) (24 June 1971).
225. Matsuyama, G., Carlsen, E. N., and Jolley, W. B., 20th Annual Meeting of the American Physiological Society, University of California—Davis (August 1969).
226. Matsui, M., and Freiser, H., *Anal. Lett.* **3**, 161–167 (1970).
227. Matsushita, H., and Furuta, S., *Kogyo Kagaku Zasshi* **73**, 1817–1821 (1970).
228. Matsushita, H., and Furuta, S., *Kogyo Kagaku Zasshi* **73**, 2051–2053 (1970).
229. Matsushita, H., and Furuta, S., *Kogyo Kagaku Zasshi* **73**, 2119–2123 (1970).
230. Maurer, O., and Thieme, R., Ger. Offen. 1,944,200 (Cl. G 01n) (25 March 1971).
231. Mellon, E. F., and Gruber, H. A., *J. Am. Leather Chem. Assoc.* **65**, 154–163 (1970).
232. Metz, L. T., Van Houwelingen, J., and De Jong, H., U.S. Patent 3,498,901 (Cl. 204-195; G 01n) (3 March 1970).
233. Michel, E., and Papst., K. H., Ger. Offen. 1,804,962 (Cl. G 01n) (11 June 1970).
234. Milham, P. J., *Analyst* (*London*) **95**(1133), 758–759 (1970).
235. Milham, P. J., Awad, A. S., Paull, R. E., and Bull, J. H., *Analyst* (*London*) **95**(1133), 751–757 (1970).
236. Mirna, A., *Fresenius' Z. Anal. Chem.* **254**, 114–116 (1971).
237. Miyada, D. S., Inami, K., and Matsuyama, G., *Clin. Chem.* **17**, 27–30 (1971).
238. Montalvo, J. G., Jr., *Anal. Chem.* **41**, 2093–2094 (1969).
239. Montalvo, J. G., Jr., *Anal. Biochem.* **38**, 357–363 (1970).
240. Montalvo, J. G., Jr., and Guilbault, G. G., *Anal. Chem.* **41**, 1697–1699 (1969).
241. Moody, G. J., Oke, R. B., and Thomas, J. D. R., *Analyst* (*London*) **95**(1136), 910–918 (1970).
242. Moody, G. J., Oke, R. B., and Thomas, J. D. R., *Lab. Pract.* **18**, 941–945 (1969).
243. Moody, G. J., and Thomas, J. D. R., *Lab. Pract.* **20**, 307–311 (1971).
244. Moody, G. J., and Thomas, J. D. R., *Selective Ion Sensitive Electrodes*, Merrow Pub. Co. Ltd., Watford, Hertfordshite, England (1971).
245. Moody, G. J., Oke, R. B., and Thomas, J. D. R., *Lab. Pract.* **18**, 1056–1062 (1969).
246. Moore, E. W., *J. Clin. Invest.* **49**, 318–334 (1970).
247. Moriguchi, Y., and Hosokawa, I., *Nippon Kagaku Zasshi* **92**, 56–60 (1971).
248. Mowbray, J. H., *Biomed. Eng.* **4**, 360–361 (1969).
249. Mueller, D. C., West, P. W., and Mueller, R. H., *Anal. Chem.* **41**, 2038–2040 (1969).
250. Mueller, R. H., Anal. Chem. **41**(12), 113A–114A, 116A (1969).
251. Muldoon, P. J., and Liska, B. J., *J. Dairy Sci.* **54**, 117–119 (1971).
252. Mulyanov, P. V., and Rodionov, Y. A., U.S.S.R. Patent 259,462 (Cl. G 01n) (12 December 1969).
253. Nagelberg, I., Braddock, L. I., and Barbero, G. J., *Science* **166**(3911), 1403–1404 (1969).
254. Nakayama, F. S., *J. Chem. Eng. Data* **16**, 178–181 (1971).
255. Natarajan, R., and Rajawat, M. S., *Indian J. Technol.* **8**(2), 76–78 (1970).
256. Naumann, R., and Weber, C., *Fresenius' Z. Anal. Chem.* **253**, 111–113 (1971).
257. Neefus, J. D., Cholak, J., and Saltzman, B. E., *Am. Ind. Hyg. Assoc. J.* **31**, 96–99 (1970).
258. Neff, G. W., *Clin. Chem.* **16**, 781–785 (1970).
259. Neti, R. M., and Bing, C. C., Ger. Offen. 2,019,523 (Cl. G 01n) (4 March 1971).
260. Neumann, G., *Ark. Kemi* **32**(20), 229–247 (1970).
261. Newman, L., Klotz, P., Mukherji, A., and Feldberg, S., *Inorg. Chem.* **10**, 740–743 (1971).

262. Nishimoto, K., Japan. Patent 6,924,424 (Cl. 21 A 295) (16 October 1969).
263. Nizhel'skaya, L. V., and Vrevskii, B. M., *Ionnyo Obmen Ionity* **1970**, 203–205.
264. Nizhel'skaya, L. V., and Vrevskii, B. M., *Sb. Nauch. Tr. Leningrad Inst. Tekst. Legk. Prom.* **1969**(10), 118–119.
265. Nutbourne, D. M., *Analyst (London)* **95**(1131), 609–613 (1970).
266. Oehme, F., *Gavanotechnik* **61**(2), 133–141 (1970).
267. Oehme, F., Ertl, S., and Dolezalova, L., *Oberflaeche-Surface* **10**, 597–610 (1969).
268. Oehme, F., Ger. Offen. 2,006,194 (Cl. G 01n) (11 February 1971).
269. Oeien, A., and Selmer-Olsen, A. R., *Analyst (London)* **94**(1123), 888–894 (1969).
270. Okada, M., and Matsushita, H., *Kogyo Kagaku Zasshi* **72**, 1407–1409 (1969).
271. Oliver, R. T., and Clayton, A. G., *Anal. Chim. Acta* **51**, 409–415 (1970).
272. Oliver, R. T., Lenz, G. F., and Frederick, W. P., *Adv. Automat. Anal. Technicon Int. Congr. 1969*, **2**, 309–314 (1970).
273. Onishchenko, N. A., Belyustin, A. A., Kantere, V., Kolelishvili, R. I., Akeliene, D., and Andriadze, N. A., *Biofizika* **15**, 453–458 (1970).
274. Onken, A. R., and Sunderman, H. D., *Commun. Soil Sci Plant Anal.* **1**, 155–161 (1970).
275. Overman, R. F., *Anal. Chem.* **43**, 616–617 (1971).
276. Padova, J., *J. Phys. Chem.* **74**, 4587–4590 (1970).
277. Paglia-Dubini, E., Mussini, T., and Galli, R., *Z. Naturforsch. A* **26**, 154–158 (1971).
278. Pain, B. K., and Mukherjee, S. K., *J. Indian Soc. Soil Sci.* **17**, 407–410 (1969).
279. Pain, B. K., and Mukherjee, S. K., *J. Indian Soc. Soil Sci.* **17**, 209–215 (1969).
280. Pajdowski, L., and Joh Ewa, *Chem. Anal.* **15**, 527–532 (1970).
281. Paletta, B., *Mikrochim. Acta* **1969**, 1210–1214.
282. Paletta, B., and Panzenbeck, K., *Clin. Chim. Acta* **26**, 11–14 (1969).
283. Patel, P. R., Moreno, E. C., and Patel, J. M., *J. Res. Natl. Bur. Stand. Sect. A* **75**, 205–211 (1971).
284. Patko, M., and Doktor, E., *Proc. IMEKO (Int. Meas. Confed.) Symp. Electrochem. Sens.* **1968**, 63–68.
285. Patterson, S. J., Bunton, N. G., and Crosby, N. T., *Water Treat. Exam.* **18** (Part 3), 182–191 (1969).
286. Pavel, J., Kuebler, R., and Wagner, H., *Microchem. J.* **15**, 192–198 (1970).
287. Pearson, J. T., *Chem. Drug.* **195**(4745–4746), 218–219 (1971).
288. Pearson, J. T., and Humphreys, K. J., *Chem. Drug. (Suppl.)* **195**, 126S–130S (1971).
289. Peters, M. A., and Ladd, D. M., *Talanta* **18**, 655–664 (1971).
290. Peterson, A. J., U.S. Patent 3,486,997 (Cl. 204-195; B 01k) (30 December 1966).
291. Peterson, A. J., U.S. Patent 3,598,712 (Cl. 204-195; G 01n) (10 August 1971).
292. Pillai, G. K., and Pandit, D. R., *J. Indian Chem. Soc.* **47**, 669–672 (1970).
293. Pioda, L. A. R., Simon, W., Boshard, H. R., and Curtius, H. C., *Clin. Chim. Acta* **29**, 289–293 (1970).
294. Pioda, L. A. R., Stankova, V., and Simon, W., *Anal. Lett.* **4**, 665–674 (1969).
295. Pouget, R., *Chim. Anal. (Paris)* **53**, 479–483 (1971).
296. Pungor, E., and Toth, K., *Analyst (London)* **95**(1132), 625–648 (1970).
297. Pungor, E., and Toth, K., *Hung. Sci. Instrum.* **1968**(14), 15–20.
298. Pungor, E., and Toth, K., *Hung. Sci. Instrum.* **1970**(18), 1–8.
299. Pungor, E., and Toth, K. *Cron. Chim.* **22**, 12–19 (1968).
300. Rechnitz, G. A., *Acc. Chem. Res.* **3**, 69–74 (1970).
301. Rechnitz, G. A., and Kenny, N. C., *Anal. Lett.* **3**, 509–514 (1970).
302. Rechnitz, G. A., *Anal. Chem.* **41**(12), 109A–113A (1969).
303. Rechnitz, G. A., *Science* **166**(3904), 532 (1969).
304. Rechnitz, G. A., and Kenny, N. C., *Anal. Lett.* **3**, 259–271 (1970).
305. Rechnitz, G. A., and Llenado, R., *Anal. Chem.* **43**, 283 (1971).

306. Rechnitz, G. A., and Mohan, M. S., *Science* **168**(3938), 1460 (1970).
307. Riande, E., Guzman, G. M., and Dominguez, M., *An. Fis.* **66**, 57–61 (1970).
308. Ridden, J. M. C., Barefoot, R. R., and Roy, J. G., *Anal. Chem.* **43**, 1109–1110 (1971).
309. Riseman, J. M., *Am. Lab.* **1969**(July), 32–39.
310. Riseman, J. M., *Ind. Wastes* **1970** (September–October), 12, 14, 16, 18.
311. Rodkey, F. L., Collision, H. A., O'Neal, J. D., and Sendroy, J., Jr., *J. Appl. Physiol.* **30**, 178–185 (1971).
312. Rogers, W. I., and Wilson, J. A., *Anal. Biochem.* **32**, 31–37 (1969).
313. Rolia, E., and Ingles, J. C., *Can. Mining J.* **92**(6), 94–95, 97–100, 102 (1971).
314. Ross, J. W., Jr., and Frant, M. S., *Anal. Chem.* **41**, 1900–1902 (1969).
315. Ross, J. W., Jr., U.S. Patent 3,497,424 (Cl. 204-1; B 01k, G 01n) (24 February 1970).
316. Rossotti, F. J. C., Henry, R. P., Prue, J. E., and Whewell, R. J., *J. Chem. Soc. D* **1971**, 868–869.
317. Ruzicka, J., and Tjell, J. C., *Anal. Chim. Acta* **49**, 346–348 (1970).
318. Ruzicka, J., Lamm, C. G., and Tjell, J. C., Ger. Offen. 2,034,686 (Cl. G 01n) (11 February 1971).
319. Ruzicka, J., and Lamm, C. G., *Anal. Chim. Acta* **53**, 206–208 (1971).
320. Ruzicka, J., and Lamm, C. G., *Anal. Chim. Acta* **54**, 1–12 (1971).
321. Sachs, C. E., and Bourdeau, A. M., *J. Physiol. (Paris) Suppl.* **62**, 313–314 (1970).
322. Sachs, C., Bourdeau, A. M., Balsan, S., and Presle, V., *Ann. Biol. Clin. (Paris)* **27**, 487–510 (1969).
323. Sanz, M. C., and Staunton, J. J. J., Ger. Offen. 1,817,313 (Cl. G 01n) (16 October 1969).
324. Savvin, N. I., Syrchenkov, A. Y., Shterman, V. S., and Gordievskii, A. V., *Tr. Mosk. Khim.-Tekhnol. Inst.* **1969**(62), 213–216.
325. Schneeweiss, F., and L'Orange, R., *Z. Naturforsch. B* **26**, 624–625 (1971).
326. Schoeller, F., and Kasper, W., *Gas, Wasser, und Waerme* **24**(6), 115–119 (1970).
327. Scholer, R. P., and Simon, W., *Chimia* **24**(10), 372–374 (1970).
328. Schwartz, H. D., McConville, B. C., and Christopherson, E. F., *Clin. Chim. Acta* **31**, 97–107 (1971).
329. Scibona, G., Danesi, P. R., Conte, A., and Scuppa, B., *J. Colloid Interface Sci.* **35**, 631–635 (1971).
330. Scibona, G., Danesi, P. R., Salvemini, F., and Scuppa, B., *J. Phys. Chem.* **75**(4), 54–61 (1971).
331. Scibona, G., Mantella, L., and Danesi, P. R., *Anal. Chem.* **42**, 844–848 (1970).
332. Scibona, G., and Scuppa, B., *Corsi Semin. Chim.* **1968**(12), 23–25.
333. Selig, W., *Mikrochim. Acta* **1970**, 229–234.
334. Selif, W., *Mikrochim. Acta* **1971**(1), 46–53.
335. Selig, W., *Fresenius' Z. Anal. Chem.* **249**, 30–34 (1970).
336. Selig, W., *Mikrochim. Acta* **1970**, 564–571.
337. Selig, W., *Microchem. J.* **15**, 452–458 (1970).
338. Semere, E., Fr. Patent 1,566,637 (Cl. G 01n) (9 May 1969).
339. Shainberg, I., *Trans. Int. Congr. Soil Sci., 9th, 1968*, **1**, 577–586.
340. Sharp, M., and Johansson, G., *Anal. Chim. Acta* **54**, 13–21 (1971).
341. Shatkay, A., *Electrochim. Acta* **15**, 1759–1767 (1970).
342. Shearer, D. A., and Morris, G. F., *Microchem. J.* **15**, 199–204 (1970).
343. Shterman, V. S., Rozenkevich, N. A., Gordievskii, A. V., and Filippov, E. L., *Zh. Fiz. Khim.* **43**, 1552–1555 (1969).
344. Shterman, V. S., Gordievskii, A. V., Filippov, E. L., and Bruk, S. V., *Zh. Fiz. Khim.* **44**, 2059–2060 (1970).
345. Shul'ts, M. M., *Dokl. Akad. Nauk SSSR (Phys. Chem)* **194**, 377–380 (1970).
346. Shul'ts, M. M., and Stefanova, O. K., *Vest. Leningrad Univ. Fiz. Khim.* **1971**(1), 22–29.

347. Simon, W., Wuhrmann, H. R., Vasak, M., Pioda, L. A. R., Dohner, R., and Stefanac, Z., *Angew Chem. Int. Ed. Engl.* **9**, 445–455 (1970).
348. Simon, W., and Wegmann, D., Swiss Patent 489,017 (Cl. G 01n) (29 May 1970).
349. Simon, W., Moeller, W., and Dohner, R., Ger. Offen. 2,021,318 (Cl. G 01n) (7 January 1971).
350. Simon, W., Swiss Patent 479,870 (Cl. G 01n) (28 November 1969).
351. Simon, W., *Pure Appl. Chem.* **25**, 811–823 (1971).
352. Simpson, R. J., *Effluent Water Treat. J.* **11**(2), 96–98 (1971).
353. Singer, L., and Armstrong, W. D., *Arch. Oral Biol.* **14**, 1343–1347 (1969).
354. Singer, L., Jarvey, B. A., Venkateswarly, P., and Armstrong, W. D., *J. Dent. Res.* **49**, 455 (1970).
355. Sipos, J. H., Kalman, I., and Mikes, J., *Acta Chim. (Budapest)* **62**, 141–149 (1969).
356. Slanina, J., Buysman, E., Agterdenbos, J., and Griepink, B., *Mikrochim. Acta* **1971**, 657–661.
357. Smith, W. C., and Souza, R. L., U.S. Patent 3,485,740 (Cl. 204–195; B 01k) (23 December 1969).
358. Stefenac, Z., and Simon, W., *Microchem. J.* **12**, 125 (1967).
359. Stock, J. T., *J. Chem. Educ.* **47**, 593 (1970).
360. Strafelda, F., *Sb. Vys. Sk. Chem.-Technol. Praze Anal. Chem.* **1969**(4), 209–221.
361. Stucky, G. L., Ger. Offen. 2,057,114 (Cl. G 01n) (3 June 1971).
362. Sucha, L., and Suchanek, M., *Anal. Lett.* **3**, 613–621 (1970).
363. Sudrabin, L. P., U.S. Patent 3,471,394 (Cl. 204–194; B 01k) (7 October 1969).
364. Swartz, J. L., and Light, T. S., *Tappi* **53**, 90–95 (1970).
365. Tackett, S. L., *Anal. Chem.* **41**, 1703–1705 (1969).
366. Tacussel, J. R. J. A., Fr. Patent 1,566,507 (Cl. B 01k, G 01n) (9 May 1969).
367. Tamamushi, R., and Goto, S., *Bull. Chem. Soc. Jpn.* **43**, 3420–3424 (1970).
368. Tanaka, T., Hiiro, K., and Kinoyama, T., *Osaka Kogyo Gijutsu Shikensho Kiho* **21**(2), 93–96 (1970).
369. Tateda, A., Fritz, J. S., and Itani, S., *Mem. Fac. Sci. Kyushu Univ. Ser. C* **7**(2), 147–154 (1970).
370. Taubinger, R. P., *Prod. Finish (London)* **24**(3), 32–35 (1971).
371. Taylor, J. T., U.S. Patent 3,528,903 (Cl. 204–195; G 01n) (15 September 1970).
372. Teraoka, I., Hatanaka, S., Fukushima, Y., and Ishihara, T., *Sekko To Sekkai* **103**, 312–317 (1969).
373. Terry, M. B., and Kasler, F., *Mikrochim. Acta* **1971**, 569–572.
374. Toth, K., and Pungor, E., *Proc. IMEKO (Int. Meas. Confed.) Symp. Electrochem. Sens.* **1968**, 35–50.
375. Toyoshima, Y., and Nozaki, H., *J. Phys. Chem.* **74**, 2704–2710 (1970).
376. Trutnovsky, H., *Z. Klin. Chem. Klin. Biochem.* **9**, 341–345 (1971).
377. Tusl, J., *Clin. Chim. Acta* **27**, 216–218 (1970).
378. Tusl, J., *Chem. Listy* **64**, 322–324 (1970).
379. Verloo, M., and Cottenie, A., *Meded. Fac. Landbouwwetensch. Univ. Gent* **34**, 137–152 (1969).
380. Verloo, M., and Cottenie, A., *Meded. Fac. Landbouwwetensch. Univ. Gent* **35**, 291–299 (1970).
381. Vermeer, W., Ger. Patent 1,598,765 (Cl. G 01n) (14 January 1971).
382. Vesely, J., *Chem. Listy* **65**, 86–90 (1971).
383. Vihavainen, T., *Valtion Tek. Tutkimuslaitos Tiedotus Sar. 1*, **1970**(53).
384. Visser, B. F., Neth. Appl. 69 13,750 (Cl. G 01n) (12 March 1971).
385. Vorob'ev, L. N., and Khitrov, Y. A., *Stud. Biophys.* **26**, 49–56 (1971).
386. Walker, J. L., Jr., *Anal. Chem.* **43**(3), 89A–91A (1971).

387. Wallace, R. A., *J. Appl. Phys.* **42**, 3121–3124 (1971).
388. Warner, T. B., U.S. Clearinghouse Fed. Sci. Tech. Inform., AD 1969, AD-690146.
389. Warner, T. B., *Water Res.* **5**, 459–465 (1971).
390. Warner, T. B., U.S. Clearinghouse Fed. Sci. Tech. Inform., AD 1970, No. 717595.
391. Watanabe, H., and Gafford, R. D., U.S. Patent 3,530,849 (Cl. 128/2.1; A 61b) (29 September 1970).
392. Watanabe, H., and Gafford, R. D., U.S. Patent 2,009,938 (Cl. G 01n) (22 October 1970).
393. Watters, J. I., Kalliney, S., and Machen, R., *J. Inorg. Nucl. Chem.* **31**, 3823–3829 (1969).
394. Weiss, D., *Chem. Listy* **63**, 1152–1156 (1969).
395. Weiss, D., *Chem. Listy* **65**, 305–312 (1971).
396. Westcott, C. C., *Food Technol.* (*Chicago*) **25**, 709–710 (1971).
397. Whitfield, M., and Leyendekkers, J. V., *Anal. Chem.* **42**, 444–448 (1970).
398. Wise, W. M., U.S. Patent 3,502,560 (Cl. 204–195; B 01k) (24 March 1970).
399. Wise, W. M., Jurey, M. J., and Baum, G., *Clin. Chem.* **16**, 103–106 (1970).
400. Woodson, J. H., and Liebhafsky, H. A., *Anal. Chem.* **41**, 1894–1897 (1969).
401. Woodward, B., Taylor, N. F., and Brunt, R. V., *Anal. Biochem.* **36**, 303–309 (1970).
402. Yamazoe, F., *Nippon Dojo-Hiryogaku Zasshi* **42**, 44 (1971).
403. Yoshimori, T., *Denki Kagaku* **38**, 869–875 (1970).
404. Yumatov, E. A., *Fiziol. Zh. SSSR im-I.M. Sechenova* **56**, 1657–1660 (1970).
405. Zavgorodnii, S. F., and Kamyshnikov, I. F., *Sovrem. Metody Khim. Tekhnol. Kontr. Proizvod.* **1968**, 51–52.
406. Zeuthen, T., *Acta Physiol. Scand.* **81**, 141–142 (1971).

"C" REFERENCES (1972–1973)

1. Abramov, A. A., Mashevskii, G. N., Roi, N. I., and Fedoricheva, T. V., *Obogashch. Rud.* **16**(4), 27–31 (1971) (Russ.).
2. Abramov, A. A., Alekseev, V. P., Gordeev, A. A., Mashevskaya, G. V., Mashevski,, G. N., Serov, G. V., and Fedoricheva, T. V., *Otkrytiya Izobret. Prom. Obraztsy Tovarnyne Znaki* **49**(13), 181 (1972).
3. Adametzova, H., and Gregr, J., *Chem. Prum.* **21**(10), 506–509 (1971) (Czech.).
4. Adhikari, M., and Ghosh, D., *J. Inst. Chem. Calcutta* **44**(Part 6), 194–196 (1973) (Eng.).
5. Adhikari, M., and Biswas, G. G., *J. Inst. Chem. Calcutta* **45**(Part 1), 14–16 (1973) (Eng.).
6. Agarwal, R. P., and Moreno, E. C., *Talanta* **18**(9), 873–890 (1971).
7. Ahrland, S., and Kullberg, L., *Acta Chem. Scand.* **25**(9), 3457–3470 (1971).
8. Akimoto, N., and Hozumi, K., *Bunseki Kagaku* **20**(9), 1186–1191 (1971) (Japan).
9. Akimoto, N., and Hozumi, K., *Bunseki Kagaku* **21**(11), 1490–1497 (1972) (Japan.).
10. Alimarin, I. P., Petrikova, M. M., and Kokina, T. A., *Ref. Zh. Khim.* Abstr. No. 4G25 (1973).
11. Allam, A. I., and Hollis, J. P., *Soil Sci.* **114**(6), 456–467 (1972).
12. Ammann, D., Pretsch, E., and Simon, W., *Anal. Lett.* **5**(11), 843–850 (1972).
13. Ammann, D., Pretsch, E., and Simon, W., *Tetrahedron Lett.* **1972**(24), 2473–2476.
14. Ammann, D., Pretsch, E., and Simon, W., *Helv. Chim. Acta* **56**(5), 1780–1787 (1973) (Ger.).
15. Andreeva, O. S., and Danilkin, V. I., *Elektrokhimiya* **8**(1), 56–58 (1972) (Russ.).
16. Anfalt, T., and Jagner, D., *Anal. Chim. Acta* **56**(3), 477–481 (1971).
17. Anfalt, T., and Jagner, D., *Anal. Chim. Acta* **66**(1), 152–155 (1973).
18. Anon., *Chem. Engl. News* **50**(6), 29 (1972).
19. Aoki, M., *Engei Shikenjo Hokoku Ser. B* **1972**(12), 163–170 (Japan.).

20. Armstrong, W. McD., and Lee, C. O., *Science* **171**, 413 (1971).
21. Back, S. G., *Anal. Lett.* **4**(11), 793–798 (1971).
22. Back, S. G., *Anal. Chem.* **44**(9), 1696–1698 (1972).
23. Back, S., and Sandblom, J., *Anal. Chem.* **45**(9), 1680–1684 (1973).
24. Bagg, J., and Rechnitz, G. A., *Anal. Chem.* **45**(2), 271–276 (1973).
25. Baker, R. L., *Anal. Chem.* **44**(7), 1326–1327 (1972).
26. Bamburov, V. G., Bausova, N. V., Manokova, L. I., and Sivoplyas, A. P., *Otkrytiya Izobret. Prom. Obraztsy Tovarnye Znaki* **49**(20), 154 (1972).
27. Banin, A., and Shaked, D., in *Role Fertilization Intensification for Agricultural Products, Proceedings of the Ninth Congress Inst. Potash Inst., 1970*, pp. 181–191, Inst. Int. Potasse, Bern (1971).
28. Banwart, W. L., Tabatabai, M. A., and Bremner, J. M., *Commun. Soil Sci. Plant. Anal.* **3**(6), 449–458 (1972).
29. Bartusek, M., Senkyr, J., Janosova, J., and Polasek, M., in *Ion-Selective Electrodes Symposium* 1972, pp. 173–182 (E. Pungor, ed.), Akad. Kiado, Budapest (1973).
30. Bast, J. C., *Chem.-Ztg.* **96**(2), 108–111 (1972) (Ger.).
31. Baucke, F. G. K., Ger. Offen., Patent 2,040,200 (17 February 1972).
32. Baucke, F. G. K., *Z. Naturforsch. A* **26**(10), 1778 (1971).
33. Baum, G., Ger. Offen. Patent 2,129,395 (23 December 1971).
34. Baum, G., U.S. Patent 3,632,483 (4 January 1972).
35. Baum, G., and Ward, F. B., *Anal. Chem.* **43**(7), 947–948 (1971).
36. Baum, G., *J. Phys. Chem.* **76**(13), 1872–1875 (1972).
37. Baum, G., Lynn, M., and Ward, F. B., *Anal. Chim. Acta* **65**(2), 385–391 (1973).
38. Baum, G., and Lynn, M., *Anal. Chim. Acta* **65**(2), 393–403 (1973).
39. Baumann, E. W., *Anal. Chim. Acta* **64**(2), 284–288 (1973).
40. Baumung, H., *Muenchner Beitr. Abwasser. Fisch.-Flussbiol.* **19**, 271–283 (1971) (Ger.).
41. Beg, M. A., and Pratap, S., *Aust. J. Chem.* **25**(9), 1837–1842 (1972).
42. Beg, M. A., and Pratap, S., *J. Electroanal. Chem. Interfacial Electrochem.* **36**(2), 349–353 (1972).
43. Belinskaya, F. A., Krunchak, V. G., Rodichev, A. G., and Sosnovskii, R. I., *Otkrytiya Izobret. Prom. Obraztsy Tovarnye Znaki* **49**(12), 168 (1972).
44. Belinskaya, F. A., Krunchak, V. G., Rodichev, A. G., and Sosnovskii, R. I., *Otkrytiya Izobret. Prom. Obraztsy. Tovarnye Znaki* **49**(14), 146 (1972).
45. Belack, E., *J. Am. Water Works Ass.* **64**(1), 62–66 (1972).
46. Belokurov, O. G., *Lab. Delo* **1972**(2), 122 (Russ.).
47. Bergmann, M., Goischke, E. M., and Goischke, H. K., *Deut. Gesundheitsw.* **28**(22), 1022–1024 (1973) (Ger.).
48. Beukers, H., *Tijdschr. Chem. Instrum.* **5**(4), 7708, 80–83 (1972); **5**(5), 101–102, 104–109 (1972) (Neth.).
49. Blay, J. A., and Ryland, J. H., *Anal. Lett.* **4**(10), 653–663 (1971).
50. Bobrov, V. S., Shul'ts, M. M., and Evstaf'eva, R. I., *Zh. Prikl. Khim.* (*Leningrad*) **45**(3), 529–534 (1972) (Russ.).
51. Bobrov, V. S., Kochetova, T. I., and Shul'ts, M. M., *Elektrokhimiya* **8**(8), 1224–1226 (1972) (Russ.).
52. Boesch, H., and Weingerl, H., *Fresenius' Z. Anal. Chem.* **262**(2), 104–109 (1972) (Ger.).
53. Bond, A. M., and Hefter, G., *J. Inorg. Nucl. Chem.* **34**(2), 603–607 (1972).
54. Bonnecaze, G., Lichanot, A., and Gromb, S., *C.R. Acad. Sci. Ser. B* **274**(17), 1032–1035 (1972) (Fr.).
55. Borisov, B., *Ref. Zh. Met.* Abstr. No. 2G147 (1972).
56. Borisov, B. M., and Mashevskii, G. N., *Obogashch. Rud* **17**(4), 46–48 (1972) (Russ.).
57. Botre, C., Memoli, A., Borghi, S., and Benignetti, M. T., *Experientia Suppl.* **18**, 171–181 (1971).

58. Botre, C., Mascini, M., Bencivenga, B., and Pallotti, G., *Farmaco Ed. Prat.* **28**(4), 218–225 (1973) (Ital.).
59. Bound, G. P., Fleet, B., Von Storp, H., and Evans, D. H., *Anal. Chem.* **45**(4), 788–789 (1973).
60. Bourdeau, A. M., Sachs, Ch., Presle, V., and Dromini, M., *Pharm. Biol.* **6**(67), 527–540 (1970) (Fr.).
61. Braud, M., and Dubernard, J., *Cotton Fibres Trop.* **27**(4), 411–413 (1972) (Fr.).
62. Bronstein, H. R., and Manning, D. L., *J. Electrochem. Soc.* **119**(2), 125–128 (1972).
63. Brown, J. F., Jr., Slusarczuk, G. M. J., and Lablanc, O. H., Jr. Ger. Offen. Patent 2,250,623 (26 April 1973).
64. Brown, J. F., Jr., Leblanc, O. H., Jr., Grubb, W. T., Ger. Offen. Patent 2,250,714 (26 April 1973).
65. Brudevold, F., Moreno, E., and Bakhos, Y., *Arch. Oral Biol.* **17**(8), 1155–1163 (1972).
66. Buck, R. P., *Anal. Chem.* **45**(4), 654–665 (1973).
67. Buck, R. P., and Sandifer, J. R., *J. Phys. Chem.* **77**(17), 2122–2128 (1973).
68. Buhl, F., and Goryl, W., *Pr. Nauk. Uniw. Slask. Katowicack* **27**, 25–32 (1972) (Pol.).
69. Burdin, A., Mesplede, J., and Porthault, M., *C.R. Acad. Sci. Ser. C* **276**(2), 173–176 (1973) (Fr.).
70. Burr, R. G., *Clin. Chim. Acta* **43**(3), 311–316 (1973).
71. Busenberg, E., and Clemency, C. V., *Clays Clay Miner* **21**(4), 213–217 (1973).
72. Busev, V. M., *Nauch. Soobshch. Inst. Biol. Morya Dal'nevost Nauch. Tsentr. Akad. Nauk SSSR* **2**, 38–41 (1971) (Russ.).
73. Cammann, K., *Das Arbeiten mit lonenselecktioen Electroden*, pp. 226, Springer-Verlag, Berlin (1973).
74. Carr, C. W., *Meth. Enzymol.* **26**(Pt. C), 182–193 (1972).
75. Carruth, W. L., and Negus, R. W., U.S. Patent 3,615,321 (26 October 1971).
76. Chamberland, E., and Doiron, E. B., *Can. J. Plant Sci.* **53**(1), 233–235 (1973) (Fr.).
77. Chernoberezhskii, Yu. M., and Omarova, K. I., *Vestn. Leningrad Univ. Fiz. Khim.* **1972**(3), 125–131 (Russ.).
78. Ching, W., and McCartney, E. R., *J. Appl. Chem. Biotechnol.* **23**(6), 441–450 (1973).
79. Ching, W., and McCartney, E. R., *J. Appl. Chem. Biotechnol.* **23**(6), 451–455 (1973).
80. Christian, G. D., U.S. 3,655,526 (11 April 1972).
81. Christopher, A. J., and Fennell, T. R. F. W., *Chem. Anal.* (*Warsaw*) **17**(3), 663–668 (1972).
82. Clerc, J. T., and Pretsch, E., *Chimia* **26**(1), 29–31 (1972) (Ger.).
83. Clerc, J. T., Kahr, G., Pretsch, E., Scholer, R. P., and Wuhrmann, H. R., *Chimia* **26**(6), 287–298 (1972) (Ger.).
84. Coetzee, C. J., and Basson, A. J., *Anal. Chim. Acta* **56**(2), 321–324 (1971).
85. Coetzee, C. J., and Basson, A. J., *Anal. Chim. Acta* **57**(2), 478–480 (1971).
86. Coetzee, C. J., and Basson, A. J., *J. S. Afr. Chem. Inst.* **26**(2), 39–44 (1973) (Afrik.).
87. Coetzee, C. J., and Basson, A. J., *Anal. Chim. Acta* **64**(2), 300–304 (1973).
88. Collombel, C., Bureau, J., and Cotte, J., *Ann. Pharm. Fr.* **29**(11), 541–552 (1971) (Fr.).
89. Conrad, F. J., *Talanta* **18**(9), 952–995 (1971).
90. Conti, U., and Wilde, P., *J. Mar. Technol. Soc.* **6**(2), 17–23 (1972).
91. Cornish, D. C., and Simpson, R. J., *Meas. Contr.* **4**(11), 308–311 (1971).
92. Cosgrove, R. E., Krull, I. H., and Mask, C. A., Ger. Offen. Patent 2,136,023 (27 January 1972).
93. Cosgrove, R. E., Krull, I. H., and Mask, C. A., Ger. Offen. Patent 2,132,333 (13 January 1972).
94. Cosgrove, R. E., II, and Mask, C. A., U.S. Patent 3,729,401 (24 April 1973).
95. Covington, A. K., and Thain, J. M., *J. Chem. Educ.* **49**(8), 554–556 (1972).
96. Crochet, K. L., and Montalvo, J. G., Jr., *Anal. Chim. Acta* **66**(2), 259–269 (1973).

97. Csvakvari, B., Boksay, Z., Bouquet, G., and Ivanovskaya, I., *Stekloabraznoe Sostoyanie Tr. Vses. Soveshch.* 5th 1969 (E. A. Poraikoshits, ed.), pp. 310–313, Nauka, Leningrad (1971).
98. Csakvari, B., *Hung. Sci. Instrum.* **25**(November), 11–13 (1972).
99. Czaban, J. D., and Rechnitz, G. A., *Anal. Chem.* **45**(3), 471–447 (1973).
100. Danesi, P. R., Scibona, G., and Scuppa, B., *Anal. Chem.* **43**(13), 1892–1895 (1971).
101. Davies, J. E. W., Moody, G. J., and Thomas, J. D. R., *Analyst (London)* **97**(1151), 87–94 (1972).
102. Davies, J. E. W., Moody, G. J., Price, W. M., and Thomas, J. D. R., *Lab. Pract.* **22**(1), 20–25 (1973).
103. De Baenst, G., Mertens, J., Van den Winkel, P., and Massart, D. L., *J. Pharm. Belg.* **28**(2), 188–194 (1973) (Fr.).
104. Degenhart, H. J., Abein, G., Bevaart, B., and Baks, J., *Clin. Chim. Acta* **38**(1), 217–220 (1972).
105. Deschamps, P., and Deschamps, P., *C.R. Acad. Sci. Ser. D* **274**(8), 1215–1217 (1972) (Fr.).
106. Deschreider, A. R., and Meaux, R., *Analusis* **2**(6), 442–445 (1973) (Fr.).
107. Dhaneshwar, M. R., and Dhaneshwar, R. G., *Trans. Soc. Adv. Electrochem. Sci. Technol.* **8**(2), 61–65 (1973).
108. Diggens, A. A., Parker, K., and Webber, H. M., *Analyst (London)* **97**(1152), 198–203 (1972).
109. Doyle, N. E., Jr., U.S. Patent 3,717,565 (20 February 1973).
110. Dubini-Paglia, E., Galli, R., and Mussini, T., *Experientia Suppl.* **18**, 259–268 (1971) (Eng.).
111. Ducksbury, A. N., and Telford, B. C., *Proc. R. Aust. Chem. Inst.* **39**(11), 323–327 (1972).
112. Duff, E. J., and Stuart, J. L., *Talanta* **19**(1), 76–80 (1972).
113. Dunsmore, H. S., and Midgley, D., *J. Chem. Soc. A* **1971**(20), 3238–3240.
114. Eisenman, G., *Membranes—A Series of Advances*, Marcel Dekker, New York (1972) (in two volumes).
115. Electrofact, N.V., Neth. Appl. Patent 71 11,344 (20 February 1973).
116. Entwistle, J. R., Weedon, C. J., and Hayes, T. J., *Chem. Ind. (London)* **1973**(9), 433–434.
117. Eriksson, T., *Anal. Chim. Acta* **58**(2), 437–444 (1972).
118. Eriksson, T., *Anal. Chim. Acta* **65**(2), 417–424 (1973).
119. Eriksson, T., and Johansson, G., *Anal. Chim. Acta* **63**(2), 445–453 (1973).
120. Evans, D. H., *Anal. Chem.* **44**(4), 875–887 (1972).
121. Evans, P. A., Moody, G. J., and Thomas, J. D. R., *Lab. Pract.* **20**(8), 644–650, 652 (1971).
122. Farren, G. M., U.S. Patent 3,657,093 (18 April 1972).
123. Farren, G. M., Ger. Offen. Patent 2,139,963 (25 May 1972).
124. Ficklin, W. H., and Gotschall, W. C., *Anal. Lett.* **6**(3), 317–324 (1973).
125a. Fiori, G., and Formaro, L., *Chim. Ind. (Milan)* **54**(10), 883–894 (1972) (Ital.).
125b. Fishman, G. I., and Pevzner, I. D., U.S.S.R. Patent 327,406 (26 January 1972).
125c. Flaschka, H., and Paschal, D., *Anal. Lett.* **6**(1), 101–113 (1973).
126. Fleet, B., and Von Storp, H., *Anal. Chem.* **43**(12), 1575–1581 (1971).
127. Fleet, B., and Ho, A. Y. W., *Ion-Selective Electrodes Symposium* 1972 (E. Pungor, ed.), pp. 17–35, Akad. Kiado, Budapest (1973).
128. Frant, M. S., *Galvanotecknik* **63**(8), 745–748 (1972) (Ger.).
129. Frant, M. S., Ross, J. W., Jr., and Riseman, J. H., *Anal. Chem.* **44**(13), 2227–2230 (1972).
130. Fuchs, C., and Paschen, K., *Deut. Med. Wochenschr.* **97**(1), 23–24 (1972) (Ger.).
131. Fuchs, C., Paschen, K., Spieckermann, P. G., and Von Westberg, C., *Klin. Wochenschr.* **50**(17), 824–832 (1972) (Ger.).
132. Gavach, C., *C.R. Acad. Sci. Ser. C* **273**(7), 489–492 (1971) (Fr.).
133. Gebert, G., *Aerztl. Forsch.* **26**(11), 379–385 (1972) (Ger.).

134. Geddes, L. A., *Electrodes and the Measurement of Bioelectric Events*, pp. 382, Interscience, Chichester, England (1972).
135. Gershov, V. M., and Purins, B., *Latv. PSR Zinat. Akad. Vestis Kim. Ser.* **1971**(4), 494 (Russ.).
136. Gershov, V. M., and Purin, B., *Latv. PSR Zinat. Akad. Vestis Khim. Ser.* **1971**(5), 528–532 (Russ.).
137. Gershov, V. M., Vossel, I. K., and Purins, B., *Latv. PSR Zinat. Akad. Vestis Kim. Ser.* **1971**(5), 619–620 (Russ.).
138. Gershov, V. M., Purins, B., and Ozols-Kalnins, G., *Elektrokhimiya* **8**(5), 673–675 (1972) (Russ.).
139. Gilbert, T. R., and Clay, A. M., *Anal. Chem.* **45**(9), 1757–1759 (1973).
140. Giulimondi, G., and Piloto, A., *Cellul. Carta* **23**(5), 11–16 (1972) (Ital.).
141. Glascoe, P. K., and Bush, C. N., *Anal. Chem.* **44**(4), 833–834 (1972).
142. Goodfellow, G., and Webber, H. M., *Analyst (London)* **97**(1151), 95–103 (1972).
143. Gordievskii, A. V., Syrchenkov, A. Ya., Sergievskii, V. V., and Savvin, N. I., *Elektrokhimiya* **8**(4), 520–521 (1972) (Russ.).
144. Gordievskii, A. V., Shterman, V. S., Syrchenkov, A. Ya., Savvin, N. I., Zhukov, A. F., and Urusov, Yu. I., *Zh. Anal. Khim.* **27**(11), 2170–2174 (1972) (Russ.).
145. Gordievskii, A. V., Shterman, V. S., Syrchenkov, A. Ya., Savvin, N. I., and Zhukov, A. F., *Zh. Anal. Khim.* **27**(4), 772–775 (1972) (Russ.).
146. Gordievskii, A. V., Syrchenkov, A. Ya., Savvin, N. I., Shterman, V. S., and Khozhukhova, G. G., *Zavod. Lab.* **38**(3), 265–268 (1972) (Russ.).
147. Grazan, A. M., Helmy, A. K., and Tschapek, M., *J. Electroanal. Chem. Interfacial Electrochem.* **36**(2), 503–506 (1972).
148. Grekovich, A. L., Materova, E. A., and Belinskaya, F. A., *Elektrokhimiya* **7**(9), 1275–1279 (1971) (Russ.).
149. Grekovich, A. L., Materova, E. A., and Yurinskaya, V. E., *Zh. Anal. Khim.* **27**(6), 1218–1220 (1972) (Russ.).
150. Grekovich, A. L., Materova, E. A., and Garbuzova, N. V., *Zh. Anal. Khim.* **28**(6), 1206–1209 (1973) (Russ.).
151. Griffith, F. D., and Barnes, J. R., *Laboratory Diagnosis of Disease Caused Toxic Agents*, Ed. Proc. Appl. Semin. 1970 (F. W. Sungderman, ed.), pp. 215–217, Warren H. Green, Inc., St. Louis (1970).
152. Griffiths, G. H., Moody, G. J., and Thomas, J. D. R., *Analyst (London)* **97**(1155), 420–427 (1972).
153. Griffiths, G. H., Moody, G. J., and Thomas, J. D. R., *J. Inorg. Nucl. Chem.* **34**(10), 3043–3048 (1972).
154. Grof, O., *Vod. Hospod. B* **22**(2), 41–45 (Czech.).
155. Grubb, W. T., Ger. Offen. Patent 2,229,135 (28 December 1972).
156. Grubb, W. T., Ger. Offen. Patent 2,158,243 (29 June 1972).
157. Grubb, W. T., Ger. Offen. 2,256,710 (24 May 1973).
158. Guilbault, G. G., and Hrbankova, E., *Anal. Chim. Acta* **56**(2), 285–290 (1971).
159. Guilbault, G. G., and Shu, F. R., *Anal. Chim. Acta* **56**(3), 333–338 (1971).
160. Guilbault, G. G., *Biotechnol. Bioeng. Symp.* **3**, 361–376 (1972).
161. Guilbault, G. G., and Shu, F. R., *Anal. Chem.* **44**(13), 2161–2166 (1972).
162. Guilbault, G. G., and Nagy, G., *Anal. Lett.* **64**(4), 301–312 (1973).
163. Guilbault, G. G., and Nagy, G., *Anal. Chem.* **45**(2), 417–419 (1973).
164. Gyenge, R., Liptak, J., and Szava, J., in *Lucr. Conf. Nat. Chim. Anal., 3rd* 1971, Vol. 1, pp. 259–264, Inst. Cent. Cercet. Chim., Bucharest (1971) (Rom.).
165. Gyenge, R., and Liptak, J., in *Proceedings of the 2nd Congress on Applied Physical Chemistry*, 1971 (I. Buzas, ed.), Vol. 1, pp. 639–644, Akad. Kiado, Budapest (1971).
166. Hall, D. A., and Kadish, A. H., Ger. Patent 1,598,079 (29 June 1972).

167. Hall, L. L., Smith, F. A., De Lopez, O. H., and Gardner, D. E., *Clin. Chem.* **18**(12), 1455–1458 (1972).
167a. Hansen, E. H., Lamm, C. H., and Ruzicka, J., *Anal. Chim. Acta* **59**(3), 403–426 (1972).
168. Hanson, W. C., and Lloyd, D. J., *Chem. Ind. (London)* **1972**(1), 41–42.
169. Harzdorf, C., *Fresenius' Z. Anal. Chem.* **262**(3), 167–170 (1972) (Ger.).
169a. Havas, J., Magi, A., Szoke, I., and Varro, A., in *Proceedings of the 2nd Conference on Applied Physical Chemistry*, 1971, (I. Buzas, ed.), Vol. 1, pp. 625–629, Akad. Kiado, Budapest (1971).
170. Havas, J., and Kiszel, J., in *Proceedings of the 2nd Conference on Applied Physical Chemistry*, 1971, (I. Buzas, ed.), Vol. 1, pp. 553–559, Akad. Kiado, Budapest (1971).
171. Havas, J., in *Proceedings of the 2nd Conference on Applied Physical Chemistry*, 1971, (I. Buzas, ed.), Vol. 1, pp. 631–638, Akad. Kiado, Budapest (1971).
172. Havas, J., *Kem. Kozlem* **37**(3), 315–350 (1972) (Hung.).
173. Hawley, J. E., and Pytkowicz, R. M., *Mar. Chem.* **1**(3), 245–250 (1973).
174. Haynes, S. J., and Clark, A. H., *Econ. Geol.* **67**(3), 378–382 (1972).
175. Heckel, E., and Marsh, P. F., *Anal. Chem.* **44**(14), 2347–2351 (1972).
176. Heerman, L. F., and Reschnitz, G. A., *Anal. Chem.* **44**(9), 1655–1658 (1972).
177. Hehir, A. F., Beck., C. G., and Prettejohn, T. H. F., *Aust. J. Dairy Technol.* **26**(3), 110–111 (1971).
178. Helesic, L., *Collect. Czech. Chem. Commun.* **37**(5), 1514–1519 (1972) (Ger.).
179. Henriet, D., *Cent. Doc. Siderung. Circ. Inform. Tech.* **29**(10), 2275–2288 (1972) (Fr.).
180. Henscheid, T., Schoenrock, K., and Berger, P., *J. Am. Soc. Sugar Beet Technol.* **16**(6), 482–495 (1971).
181. Herrmann, G., Ger. Patent 2,064,822 (29 June 1972).
182. Hersh, L. S., and Teter, M. P., *J. Phys. Chem.* **76**(24), 3633–3638 (1972).
183. Higashiyama, K., and Hirata, H., Ger. Offen. Patent 2,210,529 (2 November 1972).
184. Higashiyama, K., and Hirata, H., Ger. Offen. Patent 2,210,528 (2 November 1972).
185. Higashiyama, K., and Hirata, H., Ger. Offen. Patent 2,210,526 (26 October 1972).
186. Higashiyama, K., and Hirata, H., Ger. Offen. Patent 2,210, 531 (26 October 1972).
187. Higashiyama, K., and Hirata, H., Ger. Offen. Patent 2,210,532 (2 November 1972).
188. Higashiyama, K., and Hirata, H., Ger. Offen. Patent 2,210,530 (2 November 1971).
189. Higashiyama, K., and Hirata, H., Ger. Offen. Patent 2,210,525 (2 November 1972).
190. Higashiyama, K., and Hirata, H., Ger. Offen. Patent 2,210,527 (2 November 1972).
191. Higashiyama, K., and Hirata, H., Ger. Offen. Patent 2,210,533 (2 November 1972).
192. Hirata, H., and Higashiyama, K., *Bull. Chem. Soc. Jpn.* **44**(9), 2420–2423 (1971).
193. Hirata, H., and Higashiyama, K., *Anal. Chim. Acta* **57**(2), 476–477 (1971).
194. Hirata, H., and Higashiyama, K., *Fresenius' Z. Anal. Chem.* **257**(2), 104–107 (1971) (Engl.).
195. Hirata, H., and Date, K., *Bull. Chem. Soc. Jpn.* **46**(5), 1468–1471 (1973).
196. Hnik, P., Vyskocil, F., Kriz, N., and Holas, M., *Brain Research* **40**, 559–562 (1972).
197. Hnik, P., Kriz, N., Vyskocil, F., Smiesko, V., Mejsnar, J., Ujec, E., and Holas, M., *Pfluegers Arch.* **338**(2), 177–181 (1973).
198. Hopirtean, E., Tirsar, A., and Liteanu, C., *Stud. Univ. Babes-Bolyai Ser. Chim.* **17**(2), 99–103 (1972) (Rom.).
199. Hozumi, K., *Kagaku No Ryoiki* **26**(11), 881–888 (1972) (Japan.).
200. Hukushima, M., Hukishima, H., and Kuroda, T., *Bunseki Kagaku* **21**(4), 522–526 (1972) (Japan.).
201. Hulanicki, A., *Chem. Anal. (Warsaw)* **17**(2), 217–243 (1972) (Pol.).
202. Hulanicki, A., and Trojamowicz, M., *Talanta* **20**(7), 599–608 (1973).
203. Hussein, W. R., Von Storp, L. H., and Guilbault, G. G., *Anal. Chim. Acta* **61**(1), 89–97 (1972).

204. Ijsseling, F. P., *Chem. Tech. (Amsterdam)* **27**(9), 233–239 (1972) (Neth.).
205. Ikeda, S., Satake, H., Hisano, T., and Terazawa, T., *Talanta* **19**(12), 1650–1654 (1972).
206. Ikeda, S., Hirata, J., and Satake, H., *Nippon Kagaku Kaishi* **1973**(8), 1473–1477 (Japan.).
207. Iofis, B. G., Savvin, N. I., Vishnyakov, A. V., and Gordievskii, A. V., *Zavod. Lab.* **39**(3), 267–268 (1973) (Russ.).
208. Ishibashi, N., and Kohara, H., *Anal. Lett.* **4**(1), 785–792 (1971).
209. Ishibashi, N., and Kohara, H., *Bunseki Kagaku* **21**(1), 100–101 (1972) (Japan.).
210. Ishibashi, N., Kohara, H., and Uemura, N., *Bunseki Kagaku* **21**(8), 1072–1078 (1972) (Japan.).
211. Ishibashi, N., *Kagaku Kogyo* **23**(6), 795–799 (1972) (Japan.).
212. Ishibashi, N., Kina, K., and Maekawa, N., *Chem. Lett.* **1973**(2), 119–120.
213. Ishibashi, N., Jyo, A., and Yonemitsu, M., *Chem. Lett.* **1973**(5), 483–484.
214. Ishibashi, N., and Jyo, A., *Microchem. J.* **18**(3), 220–225 (1973).
215. Ishibashi, N., Kohara, H., and Horinouchi, K., *Talanta* **20**(9), 867–874 (1973).
216. Ishibashi, N., and Kina, K., *Bull. Chem. Soc. Jpn.* **46**(8), 2454–2456 (1973).
217. Ivanov, N. A., *God. Vissh. Khimikoteknol. Inst. Sofia* 1969 **16**(2), 187–198 (1971) (Bulg.).
218. Ixfeld, H., Reusmann, G., and Pallasch, H., Ger. Offen. Patent 2,136,625 (8 February 1973).
219. Jacob, H. E., *Z. Allg. Mikrobiol.* **11**(8), 691–734 (1971) (Ger.).
220. Jacobson, J. S., and Heller, L. I., in *Proceedings of the 2nd International Clean Air Congress*, 1970 (H. M. Englund, ed.), pp. 459–462, Academic Press, New York (1971).
221. Jacobson, J. S., and McCune, D. C., *J. Ass. Offic. Anal. Chem.* **55**(5), 991–998 (1972).
222. Jagner, D., and Pavlova, V., *Anal. Chim. Acta* **60**(1), 153–158 (1972).
223. James, H. J., Carmack, G. P., and Freiser, H., *Anal. Chem.* **44**(4), 853–855 (1972).
224. Jerome, R., and Desreux, V., *J. Chim. Phys. Physiochim. Biol.* **69**(7–8), 1177–1182 (1972) (Fr.).
225. Jerome, R., and Desreux, V., *J. Appl. Polym. Sci.* **16**(11), 2739–2749 (1972).
226. Johansson, G., and Edstrom, K., *Talanta* **19**(12), 1623–1632 (1972).
227. John, E., and Pajdowski, L., *Pr. Nauk Univ. Slask. Katowicach* **9**, 25–31 (1970) (Pol.).
228. Kalman, J., Toth, K., and Kuttle, D., *Acta Pharm. Hung.* **41**(6), 267–272 (1971) (Hung.).
229. Kaludjercic, P., *Kem. Ind.* **19**(9), 449–453 (1970) (Croat.).
230. Karlberg, B., *J. Electroanal. Chem. Interfacial Electrochem.* **42**(1), 115–126 (1973).
231. Karlberg, B., *J. Electroanal. Chem. Interfacial Electrochem.* **45**(1), 127–139 (1973).
232. Karlberg, B., *Anal. Chim. Acta* **66**(1), 93–103 (1973).
233. Karlmark, B., Sohtell, M., and Ulfendahl, H. R., *Acta Soc. Med. Upsal.* **76**(1–2), 58–62 (1971).
234. Kawai, K., Takahashi, T., Machida, S., and Watanuki, T., *Igaku No Ayumi* **85**(5), 298–299 (1973) (Japan.).
235. Kawamura, M., and Kashima, T., *Kyoto Daigaku Kekkaku Kenkyusho Nempo* **1971**(16), 1–8 (Japan.).
236. Kawamura, M., Shimizu, H., and Kashima, T., *Kyoritsu Yakka Daigaku Kenkyu Nempo* **1972**(17), 38–48 (Japan.).
237. Khitrov, Yu. A., and Vorob'ev, L. N., *Fiziol. Rast.* **18**(6), 1169–1174 (1971) (Russ.).
238. Khuri, R. N., Agulian, S. K., and Wise, W. M., *Pflugers Arch.* **322**, 39 (1971).
239. Khuri, R. N., Agulian, S. K., and Kalloghlian, A., *Pflugers Arch.* **335**, 297–308 (1972).
240. Khuri, R. N., Hajjar, J. J., and Agulian, S. K., *J. Appl. Physiol.* **32**(3), 419–422 (1972).
241. Khutsishvili, A. N., and Kakhadze, A. E., *Ref. Zh. Khim.*, Abstr. No. 18B1179 (1970).
242. Knevel, A. M., and Kehr, P. F., *Anal. Chem.* **44**(11), 1863–1865 (1972).
243. Kojima, T., Ichise, M., and Seo, Y., *Talanta* **19**(4), 539–547 (1972).

244. Komar, N. P., *Ocherki Sovrem. Geokhim. Anal. Khim.* **1972**, 604–610 (Russ.).
245. Komiya, H., *Bunseki Kagaku* **21**(7), 911–916 (1972) (Japan.).
246. Kondo, T., *Nippon Eiseigaku Zasshi* **24**(5–6), 491–496 (1970) (Japan.).
247. Koros, E., and Burger, M., in *Ion-Selective Electrodes Symposium* 1972 (E. Pungor, ed.), pp. 191–203, Akad. Kiado, Budapest (1973).
248. Koryta, J., *Anal. Chim. Acta* **61**(3), 329–411 (1972).
249. Kotenko, O. M., *Farm. Zh. (Kiev)* **27**(3), 49–51 (1972) (Ukrain.).
250. Kramarz, W., *Czas. Tech. M* **1971**(5), 36–40 (Pol.).
251. Krebs, W. M., *Anal. Chem.* **44**(1), 187–190 (1971).
252. Krull, I. H., and Mask, C. A., U.S. Patent 3,617,460 (2 November 1971).
253. Krunchak, V. G., Raskina, I. G., Rodichev, A. G., Sosnovskii, R. I., and Shagin, A. V., *Tr. Vses. Nauch.-Issled. Inst. Tsellyul.-Bum. Prom.* **59**, 153–158 (1971) (Russ.).
254. Krupskii, N. K., Boriskova, Zh. L., Aleksandrova, A. M., and Dolidze, V. A., *Agrokhimiya* **1973**(3), 132–138 (Russ.).
255. Kubec, Z., *Chem. Prum* **21**(11), 564–565 (1971) (Czech.).
256. Kubec, Z., and Maicrova, E., *Chem. Prum.* **21**(8), 388–389 (1971) (Czech.).
257. Kublanovskii, V. S., Kladnitskaya, K. B., and Zosimovich, D. P., *Ukr. Khim. Zh.* **38**(3), 283–285 (1972) (Russ.).
258. Kucsera-Payay, M., *Kem. Kozlem* **39**(1), 35–42 (1973) (Hung.).
259. Kuettel, D., *Kem. Kozlem* **38**(2), 235–243 (1972) (Hung.).
260. Kuriya, Y., Maeda, T., and Konishi, S., *Maruzen Sekiya Giho* **17**, 49–53 (1972) (Japan.).
261. Kushal'nikov, V. Y., Kryukova, N. P., Zavalova, O. V., and Labaznikov, M. G., *Steklo Keram.* **1972**(11), 36 (Russ.).
262. Laborit, G., and Garcia, A., *Ann. Anesthesiol. Fr.* **14**(3), 317–318 (1973) (Fr.).
263. Ladenson, J. H., and Bowers, G. N., Jr., *Clin. Chem.* **19**(6), 565–574 (1973).
264. Laemmke, A., *Mitt. Deut. Ges. Holzforsch.* **57**, 60–65 (1971) (Ger.).
265. Lakhitikova, V. N., and Chikunova, M. P., U.S.S.R. Patent 371,496 (22 February 1973).
266. Lakshminarayanaiah, N. K., *Electrochemistry* **2**, 203–286 (1972).
267. Lal, S., and Christian, G. D., *Indian J. Chem.* **10**(1), 123–124 (1972).
267a. Lamm, C. G., Hansen, E. H., and Ruzicka, J., *Anal. Lett.* **5**(7), 451–459 (1972).
268. Land, J. E., and Osborne, C. V., *J. Less-Common Metals* **29**(2), 147–153 (1972).
269. Larsen, M. J., Kold, M., and Von der Fehr, F. R., *Caries Res.* **6**(3), 193–202 (1972).
270. Laskowski, M., Jr., and Finkenstadt, W. R., *Meth. Enzymol.* **26** (Pt. C), 193–227 (1972).
271. LeBlanc, P. J., and Sliwinski, J. F., *Am. Lab.* **5**(7), 51–54 (1973).
272. Lee, C. O., and Armstrong, W. McD., *Science* **175**, 1261 (1972).
273. Lemahieu, G., Lemahieu-Hode, C., and Resibois, B., *Analysis* **1**(2), 110–116 (1972) (Fr.).
274. Lenz, B. L., and Mold, J. R., *Tappi* **54**(12), 2051–2055 (1971).
275. Levins, R. J., Ger. Offen. Patent 2,244,721 (29 March 1973).
276. Leyendekkers, J. V., *Anal. Chem.* **43**(13), 1835–1843 (1971).
277. Liberti, A., and Mascini, M., in *Proceedings of the 2nd International Clean Air Congress* 1970 (H. M. Englund, ed.), pp. 519–522, Academic Press, New York (1971).
278. Lingarde, F., *Clin. Chim. Acta* **40**(2), 477–484 (1972).
279. Liteanu, C., and Hopirtean, E., *Stud. Univ. Babes-Bolyai Ser. Chem.* **16**(1), 83–87 (1971) (Rom.).
280. Liteanu, C., and Hopirtean, E., in *Lucr. Conf. Nat. Chim. Anal., 3rd* 1971, Vol. 1, pp. 235–240, Inst. Cent. Cercet. Chim., Bucharest (1971) (Rom.).
281. Liteanu, C., Popescu, I. C., and Hopirtean, E., *Ion-Selective Electrodes Symposium* 1972 (E. Pungor, ed.), pp. 51–96. Akad. Kiado, Budapest (1973).
282. Liteanu, C., and Popescu, I. C., *Stud. Univ. Babes-Bolyai Ser. Chem.* **17**(2), 45–49 (1972) (Rom.).

283. Liteanu, C., and Hopirtean, E., *Chem. Anal. (Warsaw)* **17**(4), 859–869 (1972).
284. Liteanu, C., and Hopirtean, E., *Talanta* **19**(8), 971–974 (1972).
285. Liteanu, C., Popescu, I. C., and Nascu, H., *Rev. Roum. Chim.* **17**(9), 1615–1619 (1972) (Eng.).
286. Llenado, R. A., and Rechnitz, G. A., *Anal. Chem.* **44**(3), 468–471 (1972).
287. Llenado, R. A., and Rechnitz, G. A., *Anal. Chem.* **44**(8), 1366–1370 (1972).
288. L'Orange, R., and Frommer-Evers, B., *Z. Naturforsch. B.* **27**(8), 996–1000 (1972) (Ger.).
289. Louw, C. W., and Richards, J. F., *Analyst (London)* **97**(1154), 334–339 (1972).
290. Luck, J. R., U.S. Patent 3,629,089 (21 December 1971).
291. Maienthal, E. J., and Taylor, J. K., *Water Water Pollut. Handb.* **4**, 1751–1800 (1973).
292. Mainka, E., and Coerdt, W., *Nucl. Sci. Abst.* **25**(16), 37268 (1971).
293. Majer, V., Vesely, J., and Stulik, K., *Anal. Lett.* **6**(6), 577–584 (1973).
294. Manahan, S. E., Smith, M. J., Alexander, D., and Robinson, P., *Gov. Rep. Announce. (U.S.)* **72**(3), 137 (1972).
295. Mangiuca, M., and Schneider, F., *Fiziol. Norm. Patol.* **19**(1), 25–30 (1973) (Rom.).
296. Marounek, M., and Ulrich, V., *Listy Cukrov.* **89**(7), 155–159 (1973) (Czech.).
297. Marsoner, H. J., and Harnoncourt, K., *Aerztl. Lab.* **18**(11), 397–402 (1972) (Ger.).
298. Martin, C., and Poudou, A. M., *Trav. Soc. Pharm. Montpellier* **31**(4), 371–380 (1971) (Fr.).
299. Martin, C., and Mandrou, B., *Trav. Soc. Pharm. Montpellier* **33**(2), 229–232 (Fr.).
300. Mascini, M., and Liberti, A., *Anal. Chim. Acta* **53**(1), 202–205 (1971).
301. Mascini, M., and Liberti, A., *Anal. Chim. Acta* **60**(2), 405–412 (1972).
302. Mascini, M., *Anal. Chem.* **45**(3), 614–615 (1973).
303. Mascini, M., and Liberti, A., *Anal. Chim. Acta* **64**(1), 63–70 (1973).
304. Mascini, M., *Analyst (London)* **98**(1166), 325–328 (1973).
305. Mason, W. D., Needham, T. E., and Price, J. C., *J. Pharm. Sci.* **60**(11), 1756–1757 (1971).
306. Materova, E. A., and Muzhovikov, V. V., *Elektrokhimiya* **7**(11), 1741–1744 (1971) (Russ.).
307. Materova, E. A., Grekovich, A. L., and Didina, S. E., *Elektrokhimiya* **8**(12), 1829–1832 (1972) (Russ.).
308. Materova, E. A., and Mukhovikov, V. V., *Vestn. Leningrad Univ. Fiz. Khim.* **1973**(2), 148–151 (Russ.).
309. Matsuo, K., Japan Patent 73 01,097 (13 January 1973).
310. Matsushita, H., and Furuta, S., *Chubu Kogyo Daigaku Kyo* **7**, 103–109 (1971).
311. Maurer, O., Ger. Offen. Patent 2,103,901 (10 August 1972).
312. McBryde, W. A. E., *Analyst (London)* **96**(1147), 739–740 (1971).
313. McClelland, N. I., and Mancy, K. H., *J. Am. Water Works Assoc.* **64**(12), 795–803 (1972).
314. Medisch, J. O., *Rev. Quim. Ind. (Rio de Janeiro)* **41**(486), 13–15 (1972) (Port.).
315. Mel'nichenko, V. P., *Patol. Fiziol. Eksp. Ter.* **1972**(5), 83–86 (Russ.).
316. Mertens, J., Van den Winkel, P., and Massart, D. L., *Anal. Lett.* **6**(1), 81–88 (1973).
317. Mertens, J., and Massart, D. L., *Bull. Soc. Chim. Belg.* **82**(3–4), 179–190 (1973).
318. Milicka, L., *Z. Phys. Chem. (Leipzig)* **249**(1–2), 63–72 (1972) (Ger.).
319. Milicka, L., *J. Phys. Chem. (Leipzig)* **249**(3–4), 177–180 (1972) (Ger.).
319a. Moebius, H. H., Ger. Offen. Patent 2,052,422 (27 April 1972).
320. Moiseev, V. V., Permyakova, T. V., and Plotmikova, M. N., in *Stekloobraznoe Sostoyanie Tr. Vses. Soveshch., 5th* 1969 (E. A. Porai-Koshits, ed.), pp. 314–317, Nauka, Leningrad (1971) (Russ.).
321. Montalvo, J. G., Jr., *Anal. Chim. Acta* **65**(1), 189–197 (1973).
322. Moon, S. C., *Daehan Hwahak Hwoejee* **16**(3), 149–156 (1972).
323. Moore, J. H., and Schechter, R. S., *AIChE. J.* **19**(4), 741–747 (1973).
324. Moorhead, E. D., and The, P. W., *J. Appl. Chem. Biotechnol.* **22**(4), 441–454 (1972).

325. Moreno, M. F., De la Torre Boronat, M. C., and Serrat-Vilardell, M., *Circ. Farm.* **29**(233), 315–331 (1971) (Span.).
326. Mort, W. E., and Simon, W., *Helv. Chim. Acta* **54**(8), 2683–2704 (1971) (Ger.).
327. Morie, G. P., Ledford, C. J., and Glover, C. A., *Anal. Chem. Acta* **60**(2), 397–403 (1972).
328. Mukherjee, L. M., and Boden, D. P., *Electrochim. Acta* **17**(5), 965–971 (1972).
329. Murenkov, A. M., and Tsymalaya, A. I., *Khim. Tekhnol. (Kiev.)* **1973**(3), 37–38 (Russ.).
330. Mussini, T., Galli, R., and Dubini-Paglia, E., *J. Chem. Soc. Faraday Trans. 1* **68** (Pt. 7), 1322–1327 (1972).
331. Muto, G., Lee, Y. K., Whang, K. J., and Nozaki, K., *Bunseki Kagaku* **20**(10), 1271–1277 (1971) (Japan.).
332. Nadareishvili, K. Sh., Sanaya, T. V., and Tevdoradze, V. V., *Soobschch. Akad. Nauk. Gruz. SSR* **68**(2), 421–424 (1972) (Russ.).
333. Narasimhan, S. C., and Venkateswarlu, K. S., A.E.C., Bhabha At. Res. Cent. (Rep). 1972, B.A.R.C.-652, 4 pp.
334. Natarajan, R., and Rajawat, M. S., *Indian J. Technol.* **10**(4), 139–141 (1972).
335. Ngo, T. T., and Shargool, P. D., *Anal. Biochem.* **54**(1), 247–261 (1973).
336. Niklas-Salminen, R., *J. Soc. Leather Trades' Chem.* **56**(4), 139–145.
337. Normura, T., and Nakagawa, G., *Bunseki Kagaku* **20**(12), 1570–1575 (1971) (Japan.).
338. Noshiro, M., and Jitsugiri, Y., *Nippon Kagaku Kaishi* **1972**(2), 350–352 (Japan.).
339. Noshiro, M., and Jitsugiri, Y., *Yogyo Kyokai Shi* **80**(11), 438–441 (1972) (Japan.).
340. Notin, M., Guillien, R., and Nabet, P., *Ann. Biol. Clin. (Paris)* **30**(2), 193–197 (1972) (Fr.).
341. Oehme, F., *DECHEMA (Deut. Ges. Chem. Apparatewesen) Monogr.* **67** (Pt. 2) 727–750 (1971) (Ger.).
342. Oehme, F., and Dolezalova, L., *Fresenius' Z. Anal. Chem.* **264**(2), 168–173 (1973) (Ger.).
343. Ogata, N., *Bunseki Kagaku* **21**(6), 780–787 (1972) (Japan.).
344. Okada, M., and Tanamachi, T., *Soda To Ense* **23**(5), 159–164 (1972) (Japan.).
345. Olderman, G. M., *Diss. Abstr. Int. B* **33**(11), 5230–5231 (1973).
346. Oliver, R. T., and Mannion, R. F., *Anal. Instrum.* **8**, VIII-3, 1–7 (1970).
347. Oliver, R. T., and Mannion, R. F., *Water Qual. Instrum.* **1**, 74–88 (1972).
348. Omang, S. H., *Kjemi* **32**(4), 12–14 (1972) (Norweg.).
349. Ottmers, W. W., *Diss. Abstr. Int. B* **32**(8), 4459 (1972).
350. Paillard, M., *J. Physiol.* **223**, 297 (1972).
351. Palmer, T. A., *Talanta* **19**(10), 1141–1145 (1972).
352. Panday, K. K., and Agrawal, V. K., *J. Indian Chem. Soc.* **48**(8), 775–777 (1971).
353. Papariello, G. J., Mikherji, A. K., and Shearer, C. M., *Anal. Chem.* **45**(4), 790–792 (1973).
354. Papay, M. K., Toth, K., Izvekov, V., and Pungor, E., *Anal. Chim. Acta* **64**(3), 409–415 (1973).
355. Papp, J., *Sv. Papperstidn.* **75**(16), 677–679 (1972).
356. Paskalev, N., *Izv. Otd. Khim. Nauki Bulg. Akad. Nauk.* **5**(2), 181–189 (1972).
357. Paskalev, N., *Izv. Otd. Khim. Nauki Bulg. Akad. Nauk.* **5**(2), 191–197 (1972).
358. Pearson, J. T., and Gray, C. M., *J. Hosp. Pharm.* **31**(1), 20–21, 23–25 (1973).
359. Pelloux, A., Fabry, P., and Deportes, C., *C.R. Acad. Sci. Ser. C* **276**(3), 241–244 (1973) (Fr.).
360. Penland, J. L., and Fischer, G., *Metalloberflaeche-Angew. Electrochem.* **26**(10), 391–394 (1972) (Ger.).
361. Peter, F., and Rosset, R., *Anal. Chim. Acta* **64**(3), 397–408 (1973) (Fr.).
362. Phang, S., and Steel, B. J., *Anal. Chem.* **44**(13), 2230–2232 (1972).
363. Pick, J., Toth, K., and Pungor, E., *Anal. Chim. Acta* **61**(2), 169–175 (1972).

364. Pick, J., Toth, K., Vasak, M., Pungor, E., and Simon, W., *Ion-Selective Electrodes Symposium* 1972 (E. Pungor, ed.), pp. 245–252, Akad. Kiado, Budapest (1973).
365. Pick, J., *Kem. Kozlem.* **39**(1), 19–33 (1973) (Hung.).
366. Pick, J., Toth, K., Pungor, E., Vasak, M., and Simon, W., *Anal. Chim. Acta* **64**(3), 477–480 (1973).
367. Pick, J., Toth, K., and Pungor, E., *Anal. Chim. Acta* **65**(1), 240–244 (1973).
368. Polesuk, J., *Am. Lab.* **1970** (May), 27–28, 30, 32, 35–36, 38–41.
369. Polozova, I. P., Peshekhonova, N. V., and Shul'ts, M. M., *Vestn. Leningrad Univ. Fiz. Khim.* **1972**(2), 93–100 (Russ.).
370. Polozova, I. P., Peshekhonova, N. V., and Shul'ts, M. M., *Vestn. Leningrad Univ. Fiz. Khim.* **1973**(1), 120–125 (Russ.).
371. Pommez, P., and Stachenko, S., U.S. Dept. Agr., Agr. Res. Serv. (Rep.) 1971, ARS 72–90, 82–102.
372. Popescu, I. C., Liteanu, C., and Ciovirnache, V., *Rev. Roum. Chim.* **18**(1), 145–153 (1973).
373. Potmann, W., and Dahmen, E. A. M. F., *Mikrochim. Acta* **1972**(3), 303–312.
374. Pretsch, E., Scholer, R., Kahr, G., and Wuhrmann, H. R., *Umschau* **73**(8), 244 (1973) (Ger.).
375. Provorov, V. N., Volkova, S. A., Khazanova, O. A., Emel'yanova, L. B., Zaitseva, V. S., Ukov, S. B., Root, T. F., Pirogova, N. I., and Kryaneva, L. M., *Ref. Zh. Khim.*, Abstr. No. 9S960 (1972).
376. Pungor, E., *Hem. Pregl.* **12**(2), 34–36 (1971) (Serb.).
377. Pungor, E., Feber, Zs., and Nagy, G., *Acta Chim. (Budapest)* **70**(3), 207–214 (1971).
378. Pungor, E., *DECHEMA (Deut. Ges. Chem. Apparatewesen) Monogr.* **67** (Pt. 2), 751–768 (1971).
379. Pungor, E., and Toth, K., *Pure Appl. Chem.* **31**(4), 521–535 (1972).
380. Pungor, E., and Toth, K., *Pure Appl. Chem.* **34**(1), 105–137 (1973).
381. Pungor, E., and Toth, K., *Kem. Kozlem.* **39**(1), 1–18 (1973) (Hung.).
382. Radde, I. C., Hoeffken, B., Parkinson, D. K., Sheppers, J., and Luckham, A., *Clin. Chem.* **17**(10), 1002–1006 (1971).
383. Raider, S. I., Gregor, L. V., and Flitsch, R., *J. Electrochem. Soc.* **120**(3), 425–431 (1973).
384. Raman, A., *Clin. Biochem.* **5**(4), 208–213 (1972).
385. Randell, A. W., and Linklater, P. M., *Aust. J. Dairy Technol.* **27**(2), 51–53 (1972).
386. Rauzen, F. V., and Solov'eva, Z. Ya., *Radiokhimiya* **15**(1), 101–106 (1973) (Russ.).
387. Reboiras, M. D., Riande, E., and Guzmann, G. M., *An. Quim.* **69**(1), 47–54 (1973) (Span.).
388. Rechnitz, G. A., and Eyal, E., *Anal. Chem.* **44**(2), 370–372 (1972).
388a. Rechnitz, G. A., Fricke, G. H., and Mohan, M. S., *Anal. Chem.* **44**(6), 1098–1099 (1972).
389. Reinsfelder, R. E., and Schultz, F. A., *Anal. Chim. Acta* **65**(2), 425–435 (1973).
390. Riseman, J. M., *Anal. Instrum.* **8**, V-3, 1–6 (1970).
391. Riseman, J. M., *Water Qual. Instrum.* **1**, 89–98 (1972).
392. Riseman, J. H., *Am. Lab.* **4**(12), 63–64, 66–67 (1972).
393. Rittner, R. C., and Ma, T. S., *Mikrochim. Acta* **1972**(3), 404–409.
394. Ross, J. W., and Frant, M. S., U.S. Patent 3,691,047 (12 September 1972).
395. Russell, A. M., and Brown, A. M., *Science* **175**, 1475 (1972).
395a. Ruzicka, J., *Mikrochim. Acta* **1973**(5), 823–828 (Ger.).
395b. Ruzicka, J., Lamm, C. G., and Tjell, J. Chr., *Anal. Chim. Acta* **62**(1), 15–28 (1972).
395c. Ruzicka, J., and Hansen, E. H., *Anal. Chim. Acta* **63**(1), 115–128 (1973).
396. Ryba, O., Knizakova, E., and Petranek, J., *Collect. Czech. Chem. Commun.* **38**(2), 497–502 (1973).

397. Ryba, O., and Petranek, J., *J. Electroanal. Chem. Interfacial Electrochem.* **44**(3), 425–430 (1973).
398. Rybak, B., Fr. Patent 1,604,408 (17 December 1971).
399. Sakai, K., *Keiso* **15**(5), 64–68 (1972) (Japan.).
400. Sapio, J. P., and Braun, R. D., *Chemistry* **46**(6), 14–17 (1973).
401. Saunders, A. M., U.S. Patent 3,709,811 (9 January 1973).
402. Savich, V. I., and Budagova, A. A., *Dokl. TSKHA* (*Timiryazev. Sel'skokhoz. Akad.*) **169**, 135–137 (1971) (Russ.).
403. Savvin, N. I., Dobrazhanskii, G. F., Gordievskii, A. V., Syrchenkov, A. Ya., Shterman, V. S., and Komar, T. V., *Ref. Zh. Khim.*, Abstr. No. 3B1327 (1971).
404. Savvin, N. I., Shterman, V. S., Gordievskii, A. V., and Syrchenkov, A. Ya., *Zavod. Lab.* **37**(9), 1025–1027 (1971) (Russ.).
405. Schachtschabel, P., *Z. Pflanzenernaehr. Bodenk.* **130**(1), 37–43 (1971) (Ger.).
406. Schick, A. L., *J. Ass. Offic. Anal. Chem.* **56**(4), 798–802 (1973).
407. Schmidt, E., and Pungor, E., in *Proceedings of the 2nd Conference on Applied Physical Chemistry*, 1971 (I. Buzas, ed.), Vol. 1, pp. 615–624, Akad. Kiado, Budapest (1971).
408. Schmidt, E., and Pungor, E., *Anal. Lett.* **4**(10), 641–652 (1971).
409. Scholer, R., and Simon, W., *Helv. Chim. Acta* **55**(5), 1801–1809 (1972) (Ger.).
410. Seal, B. K., Tewari, C., and Basu, A. S., *J. Indian Chem. Soc.* **49**(2), 195–197 (1972).
411. Sedysheva, L. P., Vrevskii, B. M., and Roskin, E. S., *Zh. Prikl. Khim.* (*Leningrad*) **45**(10), 2331–2332 (1972) (Russ.).
412. Selig, W., *Mikrochim. Acta* **1973**(3), 453–466.
413. Selig, W., *Mikrochim. Acta* **1973**(1), 87–100.
414. Selmer-Olsen, A. R., and Oien, A., *Analyst* (*London*) **98**(1167), 412–415 (1973).
415. Shikanova, N. S., Stefanova, O. K., and Materova, E. A., *Vestn. Leningrad Univ. Fis. Khim.* **1972**(2), 101–103 (Russ.).
416. Shirahama, K., *Kolloid-Z. Z. Poly.* **250**(6), 620–621 (1972).
417. Siddiqi, F. A., Lakshminarayanaiah, N., and Beg, M. N., *J. Polym. Sci. Part A*-1, **9**(10), 2869–2875 (1971).
418. Silva, F. M., and Olarte, L. I., *Suelos Ecuatoriales* **3**(1), 43–65 (1971) (Span.).
419. Simard, S., and Dorval, M., *Union Med. Can.* **102**(3), 551–555 (1973) (Fr.).
420. Simpson, R. J., *Metal Finish. J.* **18**(212), 265–267, 269 (1972).
421. Siska, E., and Pungor, E., *Fresenius' Z. Anal. Chem.* **257**(1), 12–18 (1971).
422. Siska, E., and Pungor, E., *Talanta* **19**(5), 715–716 (1972).
423. Siska, E., and Pungor, E., *Hung. Sci. Instrum.* **1972**(24), 11–14.
424. Slanina, J., Vermeer, P., Agterdenbos, J., and Griepink, B., *Mikrochim. Acta* **1973**(4), 607–614 (Ger.).
425. Smith, M. J., and Manahan, S. E., *Am. Chem. Soc. Div. Water. Air Waste Chem. Gen. Pap.* **11**(2), 8–17 (1971).
426. Smith, M. J., and Manahan, S. E., *Anal. Chem.* **45**(6), 836–839 (1973).
427. Stahr, H. M., and Clardy, D. O., *Anal. Lett.* **6**(3), 211–216 (1973).
428. Stefanova, O. K., and Alagova, Z. S., *Vestn. Leningrad Univ. Fiz. Khim.* **1972**(4), 112–114 (Russ.).
429. Steinhardt, A., U.S. Patent 3,743,591 (3 July 1973).
430. Stucky, G. L., Ger. Offen. Patent 2,117,869 (28 October 1971).
431. Subryan, V. L., Popovtzer, M. M., Parks, S. D., and Reeve, E. B., *Clin. Chem.* **18**(12), 1459–1462 (1972).
432. Sucha, L., Suchanek, M., and Urner, Z., in *Proceedings of the 2nd Conference on Applied Physical Chemistry*, 1971 (I. Buzas, ed.), Vol. 1, pp. 651–660, Akad. Kiado, Budapest (1971).
433. Suffet, I. N., Radziul, J. V., and Goff, D. R., *Water Qual. Instrum.* **1**, 11–29 (1972).

434. Susini, J., Rouault, M., and Kerkeb, A., *Cah. ORSTOM. Ser. Pedol.* **10**(3), 309–318 (1972) (Fr.).
435. Sutter, E., *Staub-Reinhalt. Luft* **33**(3), 114–117 (1973) (Ger.).
436. Suzuki, S., and Aizawa, M., *Denki Kagaku* **40**(2), 84–90 (1972) (Japan.).
437. Szabo, G., Eisenman, G., Krasne, S., Ciani, S., and Laprade, R., *Membranes—A Series of Advances* (G. Eisenman, ed.), Vol. II, Chap. 3, Marcel Dekker, New York (1973).
438. Szucs, Z., *Banyasz. Kut. Intez. Kozlem.* **15**(1), 45–53 (1971) (Hung.).
439. Takaki, U., Esch, T. E. H., and Smid, J., *J. Am. Chem. Soc.* **93**(25), 6760–6766 (1971).
440. Tamura, Y., *Niigata Igakkai Zasshi* **86**(1), 21–25 (1972) (Japan.).
441. Tamate, T., Yagi, E., and Ikeda, H., U.S. Patent 3,677,925 (18 July 1972).
442. Tarasyants, R. R., Potsepkina, R. N., Roze, V. P., and Bondarevskaya, E. A., *Zh. Anal. Khim.* **27**(4), 808–811 (Russ.).
443. Tateda, A., *Mem. Fac. Sci. Kyushu Univ. Ser. C* 1973, **8**(2), 213–218 (1973).
444. Tenygl, J., *MTP* (*Med. Tech. Publ. Co.*) *International Reviews of Science, Physical Chemistry, Ser. One* 1973 (T. S. West, ed.), Vol. 12, pp. 123–159, Butterworth, London (1973).
445. Thomas, R. C., *J. Physiology* **220**, 55 (1972).
446. Thompson, C. R., Farrah, G. H., Haff, L. V., Hillman, W. S., Hook, A. W., Schneider, E. J., Strauther, J. D., and Weinstein, L. H., *Health Lab. Sci.* **9**(4), 304–307 (1972).
447. Thompson, H., and Rechnitz, G. A., *Chem. Instrum.* **4**(4), 239–253 (1973).
448. Tobias, F., *Aluminum* (*Duesseldorf*) **48**(6), 431–432 (1972) (Ger.).
449. Tobias, F., *Galvanotechnik* **63**(7), 644–648 (1972) (Ger.).
450. Tomcsanyi, L., and Lanyi, G., *Anal. Chim. Acta* **62**(2), 377–384 (1972).
451. Tosteson, D. C., U.S. Patent 3,657,095 (18 April 1972).
452. Toth, K., and Pungor, E., *Hung. Sci. Instrum.* **25**(November), 15–21 (1972).
453. Toth, K., *Ion-Selective Electrodes Symposium* 1972 (E. Pungor, ed.), pp. 145–164, Akad. Kiado, Budapest (1973).
454. Totsuka, A., Toba, M., Namba, Y., and Kobuyama, Y., *Nippon Jozo Kyokai Zasshi* **67**(2), 146–150 (1972) (Japan.).
455. Turner, D. L., *J. Food Sci.* **37**(5), 791–792 (1972).
456. Tusl, J., *Anal. Chem.* **44**(9), 1693–1694 (1972).
457. Ujec, E., Vit, Z., Vyskocil, F., and Kralik, O., *Physiol. Bohemoslov.* **22**(3), 329–336 (1973).
458. Vajda, L., and Kovacs, J., *Hung. Sci. Instrum.* **20**, 31–35 (1971).
459. Vallo, F., *Banyasz. Kohasz. Lapok Kohasz.* **105**(3), 137–141 (1972) (Hung.).
460. Vandevenne, L., and Oudewater, J., *Trib. CEBEDEAU* (*Cent. Belge. Etude Doc. Eaux Air*) **26**(352), 127–136 (1973) (Fr.).
461. Van de Winkel, P., Mertens, J., De Baenst, G., and Massart, D. L., *Anal. Lett.* **5**(8), 567–577 (1972).
462. Varenko, E. S., Galushko, V. P., and Pirog, L. P., *Zh. Prikl. Khim.* (*Leningrad*) **45**(9), 2081–2083 (1972) (Russ.).
463. Vasadz, G. Sh., Chilaya, S. M., Dolidze, V. A., Kolelishvili, R. I., and Landan, I. N., *Krovoobrashchenie* **6**(1), 10–13 (1973) (Russ.).
464. Vesely, J., *Collect. Czech. Chem. Commun.* **36**(9), 3364–3369 (1971) (Ger.).
465. Vesely, J., Gregr, J., and Jindra, J., Czech. Patent 143,144 (15 October 1971).
466. Vesely, J., Jensen, O. J., and Nicolaisen, B., *Anal. Chim. Acta* **62**(1), 1–13 (1972).
467. Vesely, J., *J. Electroanal. Chem. Interfacial Electrochem.* **41**(1), 134–136 (1973).
468. Vickroy, D. G., and Gaunt, G. L., Jr., *Tob. Sci.* **174**(4), 50–52, 54 (1971).
469. Vilaverde, J. P., Picard, G., and Vedel, J., *C.R. Acad. Sci. Ser. C* **276**(8), 699–671 (1973) (Fr.).
470. Vorob'ev, L. N., and Khitrov, Yu. A., *Fiziol. Rast.* **18**(5), 1054–1059 (1971) (Russ.).

471. Vrbsky, J., and Fogl, J., *Chem. Prum.* **22**(5), 241–242 (1972) (Czech.).
472. Vyskocil, F., and Kriz, N., *Pfluegers Arch.* **337**(3), 265–276 (1972).
473. Warner, T. B., *Am. Chem. Soc. Div. Water Air Waste Chem. Gen. Pap.* **11**(1), 36–40 (1971).
474. Warner, T. B., *J. Mar. Technol. Soc.* **6**(2), 24–33 (1972).
475. Weber, S. J., *Am. Lab.* **1970** (July), 15–18, 20–23.
476. Weber, S. J., *Int. Lab.* **1971** (September–October), 28–33, 35–36.
477. Weiss, D., *Fresenius' Z. Anal. Chem.* **262**(1), 28–29 (1972).
478. Weiss, D., *Chem. Listy* **66**(8), 858–867 (1972) (Czech.).
479. Whitfield, M., *Limnol. Oceanogr.* **16**(5), 829–837 (1971).
480. Wikby, A., *J. Electroanal. Chem. Interfacial Electrochem.* **33**(1), 145–159 (1971).
481. Wikby, A., *J. Electroanal. Chem. Interfacial Electrochem.* **38**(2), 429–440 (1972).
482. Wikby, A., *J. Electroanal. Chem. Interfacial Electrochem.* **38**(2), 441–444 (1972).
483. Wilde, H. E., *Anal. Chem.* **45**(8), 1526–1528 (1973).
484. Wilson, M. F., *Suom. Kemistilehti A* **46**(2), 25–38 (1973).
485. Winship, D. H., and Caflisch, C. R., *Biochim. Biophys. Acta* **291**(1), 280–286 (1973).
486. Wise, W. M., U.S. Patent 3,671,413 (20 June 1972).
487. Wise, W. M., U.S. Patent 3,723,281 (27 March 1973).
488. Wood, W. W., *J. Res. U.S. Geol. Surv.* **1**(2), 237–241 (1973).
489. Wright, F. S., and McDougal, W. S., *Yale J. Biol. Med.* **45**(3–4), 373–383 (1972).
490. Yasuda, H., *Kagaku Sochi* **14**(12), 74–79 (1972) (Japan.).
491. Yoshida, M., Saita, T., Miyakawa, K., and Mizuno, Z., *Jpn. J. Appl. Phys.* **12**(6), 830–840 (1973).
492. Zarechenskii, M. A., and Rybak, Yu. V., *Ref. Zh. Khim.*, Abstr. 1G38 (1972).
493. Zentner, H., *Chem. Ind. (London)* **1973**(10), 480–481.
494. Zhilin, A. A., Parfenov, A. I., and Shult's, M. M., *Vestn. Leningrad Univ. Fiz. Khim.* **1972**(3), 95–101 (Russ.).

"D" REFERENCES (1974–1975)

1. Alexander, P. W., and Rechnitz, G. A., *Anal. Chem.* **46**, 250 (1974).
1a. Alexander, P. W., and Rechnitz, G. A., *Anal. Chem.* **46**, 860 (1974).
2. Alexander, P. W., and Rechnitz, G. A., *Anal. Chem.* **46**, 1253 (1974).
3. Ammann, D., Pretsch, E., and Simon, W., *Anal. Lett.* **7**, 23 (1974).
4. Anfalt, T., Graneli, A., and Jagner, D., *Anal. Chim. Acta* **76**, 253 (1975).
5. Asano, Y., Japan. Patent 75 11,289 (5 February 1975); Appl. 73 61,326 (31 May 1973).
6. Babcock, R. H., *J. Am. Water Works Assoc.* **67**, 26 (1975).
7. Bach, H., and Baucke, F. G. K., *Phys. Chem. Glasses* **15**, 123 (1974).
8. Bailey, P. L., and Riley, M., *Analyst (London)* **100**(1188), 145 (1975).
9. Balasubramanian, D., and Misra, B. C., *Biomembranes: Archit., Biog., Bioenerg., Differ., Proc. Int. Symp.*, 1973 (L. Packer, ed.), Vol. 185, p. 185, Academic Press, New York (1974).
10. Band, D. M., and Kratochvil, J., *J. Physiol.* **239**(1), 10 p (1974).
11. Barry, E. F., Butler, R. A., and Lavrakas, V., *Radiochem. Radioanal. Lett.* **21**, 105 (1975).
12. Baucke, F. G. K., *J. Non-Cryst. Solids* **14**, 13 (1974).
13. Baucke, F. G. K., *Mass Transport Phenomena in Ceramics* (A. R. Cooper and A. H. Heuer, eds.), Plenum Press, New York (1975).

14. Baumann, E. W., *Anal. Chem.* **47**. 959 (1975).
15. Belyustin, A. A., *Neorg. Ionoobmen. Mater.* **1**, 209 (1974) (Russ.).
16. Berge, H., and Hartmann, P., Ger. Patent (East Ger.), 103,152 (12 January 1974); Appl. 168,023 (29 December 1972).
17. Berndt, A. F., arnd Stearns, R. I., *Anal. Chim. Acta* **74**, 446 (1975).
18. Berthier, P., *Analusis* **2**, 722 (1974).
19. Berthold, R., and Gamer, G., Ger. Patent 2,353,451 (7 May 1975); Appl. P 23 53451.2 (25 October 1973).
20. Birch, B. J., and Clarke, D. E., *Anal. Chim. Acta* **67**, 387 (1973).
21. Birch, B. J., and Clarke, D. E., *Anal. Chim. Acta* **69**, 473 (1974).
22. Birch, B. J., Clarke, D. E., Lee, R. S., and Oakes, J., *Anal. Chim. Acta* **70**, 417 (1974).
23. Blaedel, W. J., and Dinwiddie, D. E., *Anal. Chem.* **46**, 873 (1974).
24. Blaedel, W. J., and Dinwiddie, D. E., *Anal. Chem.* **47**, 1070 (1975).
25. Boles, J. H., and Buck, R. P., *Anal. Chem.* **45**, 2057 (1973).
26. Bormsov, B. M., Mashevskii, G. N., and Roi, N. I., in *Kontrol Ionnogo Sostava Rudn. Pul'py Flotatsii* (*Dokl. Vses. Nauchn. Soveshch*) (N. A. Suvorovskaya and A. M. Okolovich, eds.), Vol. 84 (1972), Nauka, Moscow (1974).
27. Brand, M. J. D., *Proc. Soc. Anal. Chem.* **10**, 118 (1973).
28. Briggs, C. C., and Lilley, T. H., *J. Chem. Thermodyn.* **6**, 599 (1974).
29. Brown, J. F., Jr., U.S. Patent 3,900,382 (19 August 1975); Appl. 519,796 (1 November 1974).
30. Buck, R. P., *Anal. Chem.* **45**, 654 (1973).
31. Buck, R. P., Boles, J. H., Porter, R. D., and Margolis, J. A., *Anal. Chem.* **46**, 255 (1974).
32. Burnel, D., Carnielo, M., Guillien, R., Luberda, M., and Nabet, P., French Patent 2,207,602 (14 June 1974); Appl. 72 40,931 (17 November 1972).
33. Camoes, M. F., and Covington, A. K., *Anal. Chem.* **46**, 1547 (1974).
34. Canh, T. M., and Broun, G., *Anal. Chem.* **47**, 1359 (1975).
35. Celis, H., Estrada-O, S., and Montal, M., *J. Membr. Biol.* **18**, 187 (1974).
36. Cormos, D. C., Haiduc, I., and Stetiu, P., *Rev. Roum. Chim.* **20**, 259 (1975).
37. Cornish, D. C., and Wirtsch, Z., *Fertigung* **69**, 447 (1974).
38. Covington, A. K., *Electrochemistry* **3**, 1 (1973).
39. Covington, A. K., *Crit. Rev. Anal. Chem.* **3**, 355 (1974).
40. Crombie, D. J., Moody, G. J., and Thomas, J. D. R., *Talanta* **21**, 1094 (1974).
41. Crombie, D. J., Moody, G. J., and Thomas, J. D. R., *Anal. Chim. Acta* **80**, 1 (1975).
42. Cullen, L. F., Rusling, J. F., Schleifer, A., and Papariello, G. J., *Anal. Chem.* **46**, 1955 (1974).
43. Danesi, P. R., Meider-Gorican, H., Chiarizia, R., Capuano, V., and Scibona, G., in *Proceedings of the International Solvent Extracting Conference* (G. V. Jeffreys, ed.), Vol. 2, p. 1761, Society of Chemical Industry, London (1974).
44. Davis, D. G., *Biochem. Biophys. Res. Commun.* **63**, 786 (1975).
45. Dawson, A. P., and Selwyn, M. J., *Biochem. Soc. Trans.* **1**, 832 (1973).
46. Deol, B. S., Bermingham, M. A., Still, J. L., Haydon, D. A., and Gale, E. F., *Biochem. Biophys. Acta* **330**, 192 (1973).
47. Eastman, M. P., *J. Chem. Soc. Chem. Commun.* **1974**(19), 789.
48. Efstathiou, C. H., and Hadjiioannou, T. P., *Anal. Chem.* **47**, 864 (1975).
49. Eisenmann, G., ed., *Glass Electrode for Hydrogen and Other Cations*, Marcel Dekker, Inc., New York (1967).
50. Eisenman, G., in *Mod. Tech. Physiol. Science* (J. F. Gross, ed.), Vol. 245, Academic Press, London (1973).
51. Eriksson, T., Ger. Patent 2,357,039 (30 May 1974); Appl. 14,887/1972 (16 November 1972).

52. Estrada-O, S., Celis, H., Calderon, E., Gallo, G., and Montal, M., *J. Membr. Biol.* **18**, 201 (1974).
53. Fabiani, C., Danesi, P. R., Scibona, G., and Scuppa, B., *J. Phys. Chem.* **78**, 2370 (1974).
54. Fiedler, U., and Ruzicka, J., *Anal. Chim. Acta* **67**, 179 (1973).
55. Fiedler, U., Hansen, E. H., and Ruzicka, J., *Anal. Chim. Acta* **74**, 423 (1975).
56. Fletcher III, K. S., U.S. Patent 3,855,098 (17 December 1974); Appl. 064,952 (19 August 1970).
57. Fletcher III, K. S., U.S. Patent 3,856,633 (24 December 1974); Appl. 104,661 (7 January 1971).
58. Fogg, A. G., Kumar, J. L., and Burns, D. T., *Anal. Lett.* **7**, 629 (1974).
59. Fogg, A. G., Pathan, A. S., and Burns, D. T., *Anal. Chim. Acta* **73**, 220 (1974).
60. Frant, M. S., Riseman, J. H., and Krueger, J. A., U.S. Patent 3,859,191 (7 January 1975); Appl. 349,224 (9 April 1973).
61. Fuchs, C., Dorn, D., Fuchs, C. A., Henning, H. V., McIntosh, C., Scheler, F., and Stennert, M., *Clin. Chim. Acta* **60**, 157 (1975).
62. Gallo, R., and Mussini, T., Ital. Patent 903,290 (15 January 1972); Appl. (27 April 1970).
63. Gavach, C., and Savajols, A., *Electrochim. Acta* **19**, 575 (1974).
64. Geyer, R., and Preuss, I., *Z. Chem.* **14**, 29 (1974).
65. Geyer, R., and Preuss, I., *Z. Chem.* **14**, 368 (1974).
66. Gibson, K., and Guilbault, G. G., *Anal. Chim. Acta* **76**, 245 (1975).
67. Gordievskii, A. V., Zhukov, A. F., Shterman, V. S., Savvin, N. I., and Urusov, Y. I., *Zh. Anal. Khim.* **29**, 1414 (1974) (Russ.).
68. Gordievskii, A. V., Zhukov, A. F., Urusov, Y. I., and Shterman, V. S., *Zh. Anal. Khim.* **29**, 1298 (1974) (Russ.).
69. Gordievskii, A. V., Savvin, N. I., Shterman, V. S., Syrchenkov, A. Y., Zhukov, A. F., Levin, A. S., and Gotgelf, Y. E., USSR Patent 336,584 (5 July 1973); Appl. 1,452,785 (5 June 1970).
70. Gordievskii, A. V., Shterman, V. S., Urusov, Y. I., Savvin, N. I., Syrchenkov, A. Y., and Zhukov, A. F., USSR Patent 397,832 (17 September 1973); Appl. 1,751,251 (21 February 1972).
71. Gordievskii, A. V., Syrchenkov, A. Y., Sergievskii, V. V., Savvin, N. I., Chizhevskii, S. V., Shukov, A. F., Urusov, Y. I., and Kozhukhova, G. G., USSR Patent 411,364 (15 January 1974); Appl. 1,751,252 (21 February 1972).
72. Gray, D. N., and Breno, P. J., U.S. Patent 3,806,440 (23 April 1974); Appl. 346,957 (2 April 1973).
73. Grell, E., Funck, T., and Eggers, F., *Proceedings of the Symposium on Molecular Mechanisms of Antibiotic Action on Protein Biosynthesis and Membranes*, 1971 (E. Munoz, F. Garcia-Ferrandiz, and D. Vazquez, eds.), p. 646, Elsevier, Amsterdam (1972).
74. Grell, E., and Funck, T., *T. Supramolecular Struct.* **1**, 307 (1973).
75. Guilbault, G. G., Shu, F. R., and Von Storp, H., U.S. Patent 3,857,777 (31 December 1974); Appl. 387,569 (10 August 1973).
76. Guilbault, G. G., and Stokbro, W., *Anal. Chim. Acta* **76**, 237 (1975).
77. Guilbault, G. G., and Tarp, M., *Anal. Chim. Acta* **73**, 355 (1974).
78. Hakoila, E. J., Lukkari, U. O., Lukkari, H. K., and Takala, A., *Suom. Kemistilehti* **46**, 359 (1973).
79. Hattori, M., and Maeda, T., Ger. Patent 2,356,719 (6 June 1974); Appl. 72 113,350 (10 November 1972).
80. Hattori, M., Maeda, T., and Kuwahara, K., Japan. Patent 74 74,996 (19 July 1974); Appl. 72 115,804 (17 November 1972).
81. Hattori, M., Maeda, T., Nakamura, Y., and Kuwahara, K., Ger. Patent 2,431,288 (19 June 1975); Appl. 73 138,488 (10 December 1973).

82. Hattori, M., Maeda, T., and Kuwahara, K., Japan. Patent 75 19,496 (28 February 1975); Appl. 73 70,298 (20 June 1973).
83. Haydon, G. F., Williams, H. R., and Ahern, C. R., *Queensl. J. Agric. Anim. Sci.* **31**, 43 (1974).
84. Hayes, D. H., and Pressman, B. C., *J. Membr. Biol.* **16**, 195 (1974).
85. Herman, H. B., and Rechnitz, G. A., *Science* **184**, 1074 (1974).
86. Herman, H. B., and Rechnitz, G. A., *Anal. Lett* **8**, 147 (1975).
87. Herman, H. B., and Rechnitz, G. A., *Anal. Chim. Acta* **76**, 155 (1975).
88. Higuchi, T., U.S. Patent 3,843,490 (22 October 1974); Appl. 50,070 (30 June 1970).
89. Higuchi, T., U.S. Patent 3,843,505 (22 October 1974); Appl. 50,070 (26 June 1970).
90. Hirata, H., and Ayuzawa, M., *Chem. Lett.* 1451 (1974).
91. Hirata, H., Japan. Patent 74 27,478 (18 July 1974); Appl. 70 22,832 (17 Mar. 1970).
92. Hirata, H., and Toyama, K., Japan. Patent 74 38,477 (17 October 1974); Appl. 69 86,289 (27 October 1969).
93. Hirata, H., and Higashiyama, K., Japan. Patent 74 41,590 (9 November 1974); Appl. 69 89,450 (8 November 1969).
94. Hladky, S. B., *Biochim. Biophys. Acta* **375**, 327 (1975).
95. Hladky, S. B., *Biochim. Biophys. Acta* **375**, 350 (1975).
96. Hofton, M. E., and Davey, J., U.S.N.T.I.S., PB Rep. 1974, No. 236357/OGA, *Gov. Rep. Announc.* (*U.S.*) **74**(25), 85 (1974).
97. Hopirtean, E., and Popescu, I. C., *Rev. Chim.* (*Bucharest*) **25**, 679 (1974).
98. Hopirtean, E., and Liteanu, C., *Rev. Roum. Chim.* **19**, 145 (1974).
99. Hopirtean, E., and Stefaniga, E., *Rev. Roum. Chim.* **19**, 1265 (1974).
100. Hopirtean, E., and Predescu, M., *Stud. Univ. Babes-Bolyai Ser. Chem.* **19**, 53 (1974).
101. Hussein, W. R., and Guilbault, G. G., *Anal. Chim. Acta* **76**, 183 (1975).
102. Jasinski, R., Trachtenberg, I., and Andrychuk, D., *Anal. Chem.* **46**, 364 (1974).
103. Johansson, G., and Edstrom, K., *Talanta* **19**, 1623 (1972).
104. Jyo, A., Torikai, M., and Ishibashi, N., *Bull. Chem. Soc. Jpn.* **47**, 2862 (1974).
105. Kazaryan, N. A., and Syrykh, T. M., *Tr. Mosk. Khim.-Teknol. Inst.* **75**, 182 (1973) (Russ.).
106. Kina, K., Fukushima, H., and Ishibashi, N., *Kyushu Daigaku Kogaku Shuho* **47**, 787 (1974) (Japan.).
107. Kojima, T., Seo, Y., and Sato, J., *Bunseki Kagaku* **23**, 1389 (1974) (Japan.).
108. Kojima, T., Ichise, M., and Seo, Y., *Bunseki Kagaku* **24**, 7 (1975) (Japan.).
109. Kopylova, V. D., Karapetyan, L. P., Asambadze, G. D., and Saldadze, K. M., *Zh. Fiz. Khim.* **49**, 168 (1975) (Russ.).
110. Koryta, J., *Ion-Selective Electrodes*, Cambridge Monographs in Physical Chemistry, No. 2, Cambridge University Press, Cambridge, Massachusetts (1975).
111. Krasnoperov, V. M., and Moskvin, L. N., *Zh. Fiz. Khim.* **48**, 3052 (1974) (Russ.).
112. Krupskii, N. K., Aleksandrova, A. M., and Gubareva, D. N., *Agrokhimiya* **1975**(2), 133 (Russ.).
113. Lanza, P., and Buldini, P. L., *Anal. Chim. Acta* **75**, 149 (1975).
114. LeBlanc, O., U.S. Patent 3,896,020 (22 July 1975); Appl. 493,863 (2 August 1974).
115. Lechner, J. F., and Sekerka, I., *J. Electroanal. Chem. Interfac. Electrochem.* **57**, 317 (1974).
116. Lehn, J. M., Moradpour, A., and Behr, J. P., *J. Am. Chem. Soc.* **97**, 2532 (1975).
117. Lesourd, J. B., Peneloux, A., and Doucet, Y., *J. Chim. Phys. Phys.-Chim. Biol.* **72**, 153 (1975).
118. Levanovich, V. V., *Lab. Delo*, 301 (1975) (Russ.).
119. Liberman, E. A., in: *Biologicheskie Membrany*, Vol. 48, P. V. Sergeev, ed., Meditsina, Moscow (1973).
120. Light, T. S., and Fletcher III, K. S., U.S. Patent 3,865,708 (11 February 1975); Appl. 887,092 (22 December 1969).

121. Liteanu, C., Popescu, I. C., and Clovirnache, V., *Stud. Univ. Babes-Bolyai Ser. Chim.* **18**, 53 (1973).
122. Liteanu, C., Popescu, I. C., and Mocanu, A., *Rev. Roum. Chim.* **18**, 1467 (1973).
123. Liteanu, C., Hopirtean, E., and Mioscu, M., *Rev. Roum. Chim.* **18**, 1643 (1973).
124. Llenado, R. A., *Anal. Chem.* **47**, 2243 (1975).
125. Llenado, R. A., and Rechnitz, G. A., *Anal. Chem.* **46**, 1109 (1974).
126. Madeira, V. M. C., *Biochem. Biophys. Res. Commun.* **64**, 870 (1975).
127. Makhlina, A. M., and Maksimov, G. B., *Lab. Delo*, 295 (1975) (Russ.).
128. Malone, T. L., and Christian, G. D., *Anal. Lett.* **7**, 33 (1974).
128a. Manning, D. L., Stokely, J. R., and Magouyrk, D. W., *Anal. Chem.* **46**, 1116 (1974).
129. Mansfield, J. R., *Proc. Soc. Anal. Chem.* **9**, 10 (1972).
130. Mascini, M., and Liberti, A., *Anal. Chim. Acta* **64**, 63 (1973).
131. Mascini, M., and Pallozzi, F., *Anal. Chim. Acta* **73**, 375 (1974).
132. Mascini, M., and Napoli, A., *Anal. Chem.* **46**, 447 (1974).
133. Materova, E. A., Alagova, Z. S., and Zhesko, V. P., *Elektrokhimiya* **10**, 1568 (1974) (Russ.).
134. Materova, E. A., Alagova, Z. S., and Mamadieva, G. R., *Vestn. Leningrad Univ. Fiz. Chim.* 147 (1974) (Russ.).
135. Materova, E. A., Muchovikov, V. V., and Grigoreva, M. G., *Anal. Lett.* **8**, 167 (1975).
136. McCallum, C., and Meares, P., *Electrochim. Acta* **19**, 537 (1974).
137. Meredith, W. D., *Chem. Ind.* (*London*), 764 (1974).
138. Mioscu, M., Hopirtean, E., and Liteanu, C., *Rev. Roum. Chim.* **18**, 1637 (1973).
139. Montalvo, J. G., Jr., U.S. Patent 3,869,354 (4 March 1975); Appl. 348,673 (6 April 1973).
140. Moody, G. J., and Thomas, J. D. R., *Chem. Ind.* (*London*), 644 (1974).
141. Moody, G. J., and Thomas, J. D. R., *Sel. Ann. Rev. Analyt. Sci.* **3**, 59 (1973).
142. Mor, E., Scotto, V., Marcenaro, G., and Alabiso, G., *Anal. Chim. Acta* **75**, 159 (1975).
143. Morf, W. E., Ammann, D., Pretsch, E., and Simon, W., *Pure Appl. Chem.* **36**, 421 (1973).
144. Morf, W. E., Kahr, G., and Simon, W., *Anal. Lett.* **7**, 9 (1974).
145. Morf, W. E., Ammann, D., and Simon, W., *Chimia* **28**, 65 (1974).
146. Motohashi, R., Japan. Patent 74 35,718 (25 September 1974); Appl. 74 71,758 (18 August 1970).
147. Nakagawa, G., Wada, H., and Hayakawa, T., *Bull. Chem. Soc. Jpn.* **48**, 424 (1975).
148. Nakamura, T., *Bull. Chem. Soc. Jpn.* **48**, 1447 (1975).
149. Nanjo, M., and Guilbault, G. G., *Anal. Chim. Acta* **73**, 367 (1974).
150. Nanjo, M., Rohm, T. J., and Guilbault, G. G., *Anal. Chim. Acta* **77**, 22 (1975).
151. Neti, R. M., Sawa, K. B., and Bing, C. C., U.S. Patent 3,787,309 (22 January 1974); Appl. 68,587 (31 August 1970).
152. Newcomb, M., Helgeson, R. C., and Cram, D. J., *J. Am. Chem. Soc.* **96**, 7367 (1974).
153. Nikolskii, B. P., Materova, E. A., Grekovich, A. L., and Yurinskaya, V. E., *Zh. Anal. Khim.* **29**, 205 (1974) (Russ.).
154. Ovchinnikov, Y. A., *Usp. Sovrem. Biol.* **77**, 103 (1974) (Russ.).
155. Papastathopoulos, D. A., and Rechnitz, G. A., *Anal. Chem.* **47**, 1792 (1975).
156. Petranek, J., and Ryba, O., *Anal. Chim. Acta* **72**, 380 (1974).
157. Polozova, I. P., Sheshukova, G. E., Peshekhonova, N. V., and Shults, M. M., *Vestn. Leningrad Univ. Fiz. Khim.* **4**, 94 (1974) (Russ.).
158. Popescu, I. C., Liteanu, C., and Savici, L., *Rev. Roum. Chim.* **18**, 1451 (1973).
159. Popescu, I. C., Ciorvimache, V., Savici, L., and Liteanu, C., *Rev. Roum. Chim.* **18**, 1459 (1973).
160. Popescu, I. C., Hopirtean, E., Savici, L., and Vlad, R., *Rev. Roum. Chim.* **20**, 993 (1975).

161. Pressman, B. C., and Heeb, M. J., in *Proceedings of the Symposium on Molecular Mechanisms of Antibiotic Action on Protein Biosynthesis and Membranes*, 1971 (E. Munoz, F. Garcia-Ferrandiz, and D. Vazquez, eds.), p. 603, Elsevier, Amsterdam (1972).
162. Pressman, B. C., in *Proceedings of the 4th International Biophysics Congress, 1972* (G. M. Frank and L. P. Kayushin, eds.), Vol. 2, Part 1, p. 407, Akad Nauk, SSSR (1973).
163. Puskin, J., and Gunter, T. E., *Biochemistry* **14**, 187 (1975).
164. Rais, J., Kyrs, M., and Kadlecova, L., in *Proceedings of the International Solvent Extraction Conference, 1974* (G. V. Jeffreys, ed.), Vol. 2, p. 1705, Society of Chemical Industry, London (1974).
165. Rao, K. J. M., Pelavin, M. H., and Morgenstern, S. in *Advances in Automated Analysis, Technicon International Congress*, 1972, Vol. 1, p. 33, Futura Publishing Company (1972).
166. Riande, E., in *Physical Electrolytes* (J. Hladik, ed.), Vol. 1, p. 401, Academic Press, London (1972).
167. Riseman, J. H., Kreuger, J., and Frant, M. S., U.S. Patent 3,830,718 (20 August 1974); Appl. 343,868 (22 March 1973).
168. Riseman, J. H., and Frant, M. S., U.S. Patent 3,846,257 (5 November 1974); Appl. 218,881 (19 January 1972).
169. Ross, J. W., Riseman, J. H., and Krueger, J. A., *Pure Appl. Chem.* **36**, 473 (1973).
170. Rothschild, K. J., and Stanley, H. E., *Am. J. Clin. Pathol.* **63**, 695 (1975).
171. Rouchouse, A., Mesplede, J., and Porthault, M., *Anal. Chim. Acta* **74** 155 (1975).
172. Rousseau, J. C., *Analusis* **2**, 718 (1974).
173. Ruzicka, J., and Lamm., G. G., *Anal. Chim. Acta* **54**, 1 (1971).
174. Ruzicka, J., and Hansen, E. H., *Anal. Chim. Acta* **69**, 129 (1974).
175. Ruzicka, J., and Hansen, E. H. *Anal. Chim. Acta* **72**, 215 (1974).
176. Sandifer, J. R., and Buck, R. P., *J. Electroanal. Chem. Interfacial Electrochem.* **56**, 385 (1974).
177. Santos, M. M., Pinto, J. D. L., Ferreira, J. B., and Guedes de Carvalho, R. A., *J. Solution Chem.* **4**, 31 (1975).
178. Sarukhanova, E. P., Dolidze, V. A., Belyustin, A. A., and Shults, M. M., *Neorg. Ionoobmen. Mater.* **1**, 222 (1974) (Russ.).
179. Sata, T., and Izuo, R., Japan. Patent 74 64,585 (22 June 1974); Appl. 72 105,775 (25 October 1972).
180. Sawada, K., Nakano, M., Mori, H., and Tanaka, M., *Bull. Chem. Soc. Jpn.* **48**, 2282 (1975).
181. Schmidt, H. J., *Fortschr. Wasserchem. Ihrer Grenzgeb.* **15**, 59 (1973).
182. Schullery, S. E., *Chem. Phys. Lipids* **14**, 49 (1975).
183. Schultz, F. A., and Mathis, D. E., *Anal. Chem.* **46**, 2253 (1974).
184. Sharp, M., *Anal. Chim. Acta* **76**, 165 (1975).
185. Shigematsu, T., Ota, A., and Matsui, M., *Bull. Inst. Chem. Res. Kyoto Univ.* **51**, 268 (1973).
186. Shohami, E., and Ilani, A., *Biophys. J.* **13**, 1242 (1973).
187. Simpson, R. J., *Int. Dyer Text. Printer Bleacher Finish.* **153**, 131 (1975).
188. Skobets, E. M., Makovetskaya, L. I., and Makovetskii, Y. P., *Zh. Anal. Khim.* **29**, 2354 (1974) (Russ.).
189. Smith, A. C., and Hahn, C. E. W., *Br. J. Anaesth.* **47**, 553 (1975).
190. Smith, M. J., and Manahan, S. E., *Anal. Chem.* **45**, 836 (1973).
191. Sollner, K., *J. Dent. Res.* **53**, 267 (1974).
192. Sma, R. F., Report 1974 COM-74-11795/3GA; avail. NTIS, *Gov. Rep. Announce. (U.S.)* **75**, 60 (1975).
193. Stepanova, I., and Vadura, R., *Chem. Listy* **68**, 853 (1974) (Czech.).
194. Szczepaniak, W., and Ren, K., *Wiad. Chem.* **28**, 561 (1974) (Pol.).

195. Szczepaniak, W., and Ren, K., *Chem. Anal.* (*Warsaw*) **20**, 91 (1975) (Pol.).
196. Tacussel, J., Fr. Patent 2,203,518 (10 May 1974); Appl. 36,999 (13 October 1972).
197. Tateda, A., and Murakami, H., *Bull. Chem. Soc. Jpn.* **47**, 2885 (1974).
198. Tateda, A., Matsubara, A., and Suenaga, T., *Mem. Fac. Sci., Kyushu Univ. Ser. C* **9**, 9 (1974).
199. Thoma, A. P., Cimermann, Z., Fiedler, U., Bedekovic, D., Guggi, M., Jordan, P., May, K., Pretsch, E., Prelog, V., and Simon, W., *Chimia* **29**, 344 (1975).
200. Thomas, J. D. R., *Proc. Soc. Anal. Chem.* **11**, 340 (1974).
201. Thompson, H., and Rechnitz, G. A., *Anal. Chem.* **46**, 246 (1974).
202. Torma, L., *J. Assoc. Off. Anal. Chem.* **58**, 477 (1975).
203. Tosteson, D. C., in *Proceedings of the Symposium on Molecular Mechanisms of Antibiotic Action on Protein Biosynthesis and Membranes* 1971 (E. Munoz, F. Garcia-Ferrandiz, and D. Vazquez, eds.), p. 615, Elsevier, Amsterdam (1972).
204. Toyama, K., and Hirata, H., Japan. Patent 74 38,478 (17 Ocrober 1974); Appl. 70, 6768 (26 January 1970).
204a. Tseng, P. K., and Gutnecht, W. F., *Anal. Chem.* **47**, 2316 (1975).
205. Van der Neut-Kok, E. C. M., De Gier, J., Middelbeek, E. J., and Van Deenen, L. L. M., *Biochem. Biophys. Acta* **332**, 97 (1974).
205a. Van Osch, G. W. S., and Griepink, B., *Fresenius' Z. Anal. Chem.* **273**, 271 (1975).
206. Vanko, M., and Meola, J., *Adv. Autom. Anal. Technicon International Congress*, 1972, Vol. 1, p. 37 (1973).
207. Venkateswarlu, P., *Anal. Chem.* **46**, 878 (1974).
208. Vesely, J., *Collect. Czech. Chem. Commun.* **39**, 710 (1974).
209. Wawro, R., and Rechnitz, G. A., *Anal. Chem.* **46**, 806 (1974).
210. Weelink, A. M. H., Van Osch, G. W. S., and Van Houwelingen, J., U.S. Patent 3,824,170 (16 July 1974); Appl. 323,697 (15 January 1973).
211. Weiss, D., *Chem. Listy* **69**, 202 (1975) (Czech.).
212. Wikby, A., and Karlberg, B., *Electrochim. Acta* **19**, 323 (1974).
213. Wikby, A., *Electrochim. Acta* **19**, 329 (1974).
214. Wikby, A., *Phys. Chem. Glasses* **15**, 37 (1974).
215. Wilson, M. F., Haikala, E., and Kivalo, P., *Anal. Chim. Acta* **74**, 395 (1975).
216. Wilson, M. F., Haikala, E., and Kivalo, P., *Anal. Chim. Acta* **74**, 411 (1975).
217. Wong, K. H., Yagi, K., and Smid, J., *J. Membr. Biol.* **18**, 379 (1974).
218. Yoshida, N., and Ishibashi, N., *Chem. Lett.*, 493 (1974).

"E" REFERENCES (1976–1977)

1. Akimoto, N., and Hozumi, K., *Bunseki Kagaku* **25**, 554 (1976) (Japan.).
2. Amblard, G., and Gavach, C., *Biochim. Biophys. Acta* **448**, 284 (1976).
3. Ammann, D., Bissig, R., Gueggi, M., Pretsch, E., and Simon, W. *Helv. Chim. Acta* **58**, 1535 (1975).
4. Ammann, D., Gueggi, M., Pretsch, E., and Simon, W., *Anal. Lett.* **8**, 709 (1975).
5. Anghel, D., and Ciocan, N., *Colloid Polym. Sci.* **254**, 114 (1976).
6. Anghel, D. F., and Ciocan, N., *Anal. Lett.* **10**, 423 (1977).
7. Antonov, V. F., Ivanov, A. S., Korpanova, E. A., and Petrov, V. V., *Itogi Nauki Tekh. Biofiz.* **5**, 166 (1975) (Russ.).
8. Antson, J., and Suntola, T., Ger. Patent 2,626,277 (30 December 1976); Appl. 75/1773 (13 June 1975).
9. Armstong, W. M., Wojtkowski, W., and Bixenman, W. R., *Biochim. Biophys. Acta* **465**, 165 (1977).

10. Bach, H., *J. Non-Cryst. Solids* **19**, 65 (1975).
11. Bailey, P. L., *Analysis With Ion-Selective Electrodes*, Heyden, London (1976), 240 pp.
12. Baiulescu, G. E., and Cosofret, V. V., *Applications of Ion-Selective Membrane Electrodes in Organic Analysis*, Halsted, New York (1977), 250 pp.
13. Baiulescu, G. E., and Ciocan, N., *Talanta* **24**, 37 (1977).
14. Baiulescu, G. E., Cosofret, V. V., and Cristescu, C., *Rev. Chim.* (*Bucharest*) **26**, 429 (1975) (Rom.).
15. Baiulescu, G. E., and Cosofret, V. V., *Rev. Chim.* (*Bucharest*) **26**, 1051 (1975) (Rom.).
16. Baiulescu, G. E., and Cosofret, V. V., *Talanta* **23**, 677 (1976).
17. Baiulescu, G. E., and Cosofret, V. V., *Rev. Chim.* (*Bucharest*) **27**, 158 (1976) (Rom.).
18. Band, D. M., and Treasure, T., *J. Physiol.* **266**, 12P (1977).
19. Band, D. M., Kratochvil, J., and Treasure, T., *J. Physiol.* **265**, 5P (1977).
20. Baucke, F. G. K., *J. Non-Cryst. Solids* **19**, 75 (1975).
21. Baucke, F. G. K., *J. Electroanal. Chem.* **67**, 277 (1976).
22. Baucke, F. G. K., *J. Electroanal. Chem.* **67**, 291 (1976).
23. Baucke, F. G. K., in *Ion and Enzyme Electrodes in Biology and Medicine* (M. Kessler, L. C. Clark, D. W. Lubbers, and I. A. Silver, eds.), p. 177, Urban and Schwarzberg or University Park Press, University Park, Maryland (1976).
24. Baucke, F. G. K., in *4th International Conference on the Physics of Non-Crystalline Solids, 1976* (G. H. Frischat, ed.), p. 503, Trans. Tech. (1977).
25. Baumann, E. W., *Anal. Chem.* **48**, 548 (1976).
26. Beguin, C. G., and Coulombeau, C., *Anal. Chim. Acta* **90**, 237 (1977) (Fr.).
27. Berge, H., Gruenke, U., Hartmann, P., Scheutze, R., and Stojan, D., Ger. Patent (East. Ger.) 112314 (5 April 1975); Appl. 178,536 (16 May 1974).
28. Bloch, R., and Lobel, E., *Membrane Separation Processes* (P. Meares, ed.), p. 477, Elsevier, Amsterdam (1976).
29. Bobrov, V. S., Murshova, V. V., and Shults, M. M., *Zh. Prikl. Khim.* (*Leningrad*) **50**, 1257 (1977) (Russ.).
30. Bray, P., Clark, G. C. F., Moody, G. J., and Thomas, J. D. R., *Clin. Chim. Acta* **77**, 69 (1977).
31. Bray, P. T., Clark, G. C. F., Moody, G. J., and Thomas, J. D. R., *Clin. Chim. Acta* **80**, 333 (1977).
32. Brown, A. M. and Kunze, D. L., *Adv. Exp. Med. Biol.* **50** (*Ion-Selective Microelectrodes*), 57–73 (1974).
33. Buck, R. P., *Crit. Rev. Anal. Chem.* **5**, 323 (1975).
34. Buck, R. P., *J. Electroanal. Chem.* **83**, 379 (1977).
35. Buck, R. P., Stover, F. S., and Mathis, D. E., *J. Electroanal. Chem.* **82**, 345 (1977).
36. Buechi, R., Pretsch, E., and Simon, W., *Helv. Chim. Acta* **59**, 2327 (1976) (Ger.).
37. Buechi, R., Pretsch, E., and Simon, W., *Tetrahedron Lett.* **1976**(20), 1709.
38. Burgermeister, W., and Winkler-Oswatitsch, R., "Topics in Current Chemistry 69," in *Inorganic Biochemistry*, Vol. 2, p. 91, Springer Verlag, New York (1977).
39. Burman, J. O., and Johansson, G., *Anal. Chim. Acta* **80**, 215 (1975).
40. Burrell, R. E., *Proceedings of the Special Conference on Ind. Pollution Control Meas. Instrum.*, 1976, p. 263.
41. Bychkov, A. S., Petrukhin, O. M., Zarinskii, V. A., and Zolotov, Y. A., *Zh. Anal. Khim.* **30**, 2213 (1975) (Russ.).
42. Bychkov, A. S., Petrukhin, O. M., Zarinskii, V. A., Zolotov, Y. A., Bakhtinova, L. V., and Shanina, G. G., *Zh. Anal. Khim.* **31**, 2114 (1976) (Russ.).
43. Cakrt, M., Bercik, J., and Hladky, Z., *Fresenius' Z. Anal. Chem.* **281**, 295 (1976).
44. Carmack, G. D., and Freiser, H., *Anal. Chem.* **49**, 1577 (1977).
45. Cattrall, R. W., and Pui, C. P., *Anal. Chem.* **48**, 552 (1976).
46. Cattrall, R. W., and Pui, C. P., *Anal. Chim. Acta* **83**, 355 (1976).

47. Cattrall, R. W., and Pui, C. P., *Anal. Chim. Acta* **87**, 419 (1976).
48. Cattrall, R. W., and Pui, C. P., *Anal. Chim. Acta* **88**, 185 (1977).
49. Chopin-Dumas, J., Mauger, R., and Zanchetta, J. V., *J. Chim. Phys. Phys.-Chim. Biol.* **73**, 946 (1976) (Fr.).
50. Clysters, H., Adams, F., and Verbeek, F., *Anal. Chim. Acta* **83**, 27 (1976).
51. Ciocan, N., and Anghel, D., *Tenside Deterg.* **13**, 188 (1976); *Chem. Abstr.* **85**, 145164w (1976).
52. Coetzee, C. J., and Basson, A. J., *Anal. Chim. Acta* **83**, 361 (1976).
53. Cordonnier, M., Lawny, F., Chapot, D., and Thomas, D., *FEBS Lett.* **59**, 263 (1975).
54. Cosofret, V. V., *Rev. Chim. (Bucharest)* **27**, 240 (1976) (Rom.).
55. Cosofret, V. V., Blasnic, M., and Panaitescu, T., *Rev. Chim. (Bucharest)* **28**, 68 (1977) (Rom.).
56. Covington, A. K., and Karesh, N., *Anal. Chim. Acta* **85**, 175 (1976).
57. Craggs, A., Moody, G. J., and Thomas, J. D. R., *Proc. Anal. Div. Chem. Soc.* **12**, 64 (1975).
58. Craggs, A., Keil, L., Moody, G. J., and Thomas, J. D. R., *Talanta* **22**, 907 (1975).
59. Craggs, A., Moody, G. J., Thomas, J. D. R., and Willcox, A., *Talanta* **23**, 799 (1976).
60. Crowe, W., Mayevsky, A., and Mela, L., *Am. J. Physiol.* **233**, C56 (1977).
61. Cserfalvi, T., and Guilbault, G. G., *Anal. Chim. Acta* **84**, 259 (1976).
62. Czaban, J. D., and Rechnitz, G. A., *Anal. Chem.* **48**, 277 (1976).
63. Czerwenka, G., and Scheubeck, E., *Fresenius' Z. Anal. Chem.* **276**, 37 (1975).
64. D'Angiuro, L., *Ind. Carta* **14**, 205 (1976) (Ital.); *Chem. Abstr.* **85**, 197537q (1976).
65. Delouma, J. P., Fombon, J. J., Lancelot, F., Paris, J., Roubin, M., Tacussel, J., and Verdier, J. C., Fr. Patent 2,268,264 (14 November 1975); Appl. 74/13272 (17 April 1974).
66. Dobson, J. V., Ger. Patent 2,513,613 (2 October 1975); Appl. Brit. 14,074/74 (29 March 1974).
67. Dobson, J. V., and Dickinson, T., Ger. Patent 2,538,739 (18 March 1976); Brit. Appl. 38,748/74 (4 September 1974).
68. Doktor, E., Havas, J., Kecskes, L., Patko, M., and Pungor, E., Ger. Patent 2,530,646 (6 May 1976); Hung. Appl. 623 (31 October 1974).
69. D'Orazio, P., and Rechnitz, G. A., *Anal. Chem.* **49**, 41 (1977).
70. D'Orazio, P., and Rechnitz, G. A., *Anal. Chem.* **49**, 2083 (1977).
71. Duff, E. J., and Stuart, J. L., *Analyst (London)* **100**, 739 (1975).
72. Duff, E. J., and Stuart, J. L., *Talanta* **22**, 823 (1975).
73. Duff, E. J., and Stuart, J. L., *Talanta* **22**, 901 (1975).
74. Durlist, H., Comtat, M., Mahenc, J., and Baudras, A., *J. Electroanal. Chem.* **66**, 73 (1975) (Fr.).
75. Durst, R. A., *Adv. Exp. Med. Biol.* **50** (*Ion-Selective Microelectrodes*), 13 (1974).
76. Durst, R. A., in *International Corrosion Conference Series 1971*, NACE-3 Localized Corrosion, p. 151 (1974).
77. Durst, R. A., *Clin. Chim. Acta* **80**, 225 (1977).
78. Efstathiou, C. E., and Hadjiioannou, T. P., *Anal. Chem.* **49**, 414 (1977).
79. Efstathiou, C. E., and Hadjiioannou, T. P., *Anal. Chim. Acta.* **89**, 55 (1977).
80. Elbakai, A. M., Kakabadse, G. J., Kayat, M. N., and Tyas, D., *Proc. Anal. Div. Chem. Soc.* **12**, 83 (1975).
81. Enfors, S. O., Molin, N. L., Mosbach, K. H., and Nilsson, H. J., Ger. Patent 2,612,433 (14 October 1976); Swed. Appl. 75/3652 (27 March 1975).
82. Evseeva, N. K., Kremenskaya, I. N., Golubev, V. N., and Timofeeva, S. K., *Zh. Anal. Khim.* **31**, 822 (1976) (Russ.).
83. Fiedler, U., *Anal. Chim. Acta* **89**, 101 (1977).
84. Fiedler, U., *Anal. Chim. Acta* **89**, 111 (1977).

85. Fischer, D. J., Kunz, H. J., and Norby, T. E., U.S. Patent 3,923,625 (2 December 1975); Appl. 158,448 (30 June 1971).
86. Fleet, B., Bound, G. P., and Sandbach, D. R., *Bioelectrochem. Bioenerg.* **3**, 158 (1976).
87. Flinn, D. R., and Stern, K. H., *J. Electrochem. Soc.* **123**, 978 (1976).
88. Fogg, A. G., Al-Sibaai, A. A., and Burgess, C., *Anal. Lett.* **8**, 129 (1975).
89. Fogg, A. G., and Al-Sibaai, A. A., *Anal. Lett.* **9**, 39 (1976).
90. Fogg, A. G., and Al-Sibaai, A. A., *Anal. Lett.* **9**, 33 (1976).
91. Fogg, A. G., Al-Sibaii, A. A., and Yoo, K. S., *Anal. Lett.* **10**, 173 (1977).
92. Fray, D. J., Brit. Patent 1,470,558 (14 April 1977); Appl. 74/29,490 (3 July 1974).
93. Freiser, H., *Res./Dev.* **27**, 28–30, 32–33 (1976).
94. Frost, L. W., U.S. Patent 3,956,136 (11 May 1976); Appl. 430,452 (3 January 1974).
95. Fuchs, C., Dorn, D., McIntosh, C., Scheler, F., and Kraft, B., *Clin. Chim. Acta* **67**, 99 (1976).
96. Fuchs, C., *Ion-Selective Electrodes in Medicine*, Thieme, Stuttgart (1976), 151 pp.
97. Fuchs, C., and McIntosh, C., *Clin. Chem.* (*Winston-Salem, N.C.*) **23**, 610 (1977).
98. Fujimoto, M., and Kubota, T., *Jpn. J. Physiol.* **26**, 631 (1976).
99. Fukamachi, K., and Ishibashi, N., *Bunseki Kagaku* **26**, 69 (1977) (Japan.).
100. Gavach, C., Seta, P., and D'Epenoux, B., *J. Electroanal. Chem.* **83**, 225 (1977).
101. Geyer, R., and Stein, P., *Wiss. Z. Tech. Hochsch. "Carl Schorlemmer," Leuna-Mersburg* **17**, 364 (1975).
102. Gilligan, T. J., Cussler, E. L., and Evans, D. F., *Biochim. Biophys. Acta* **497**, 627 (1977).
103. Golubev, V. N., Evseeva, N. K., Kremenskaya, I. N., and Timofeeva, S. K., *Elektrokhimiya* **12**, 263 (1976) (Russ.).
104. Gordievskii, A. V., Zhukov, A. F., Snakin, V. V., Urusov, Y. I., and Shterman, V. S., U.S.S.R. Patent 486,266 (30 September 1975); Appl. 1,896,716 (22 March 1973); *Chem. Abstr.* **84**, 38365v (1976).
105. Gordievskii, A. V., Vishnyakov, A. V., Zhukov, A. F., Shterman, V. S., and Urusov, Y. I., U.S.S.R. Patent 49,721 (30 November 1975); Appl. 2,004,519 (13 March 1974); *Chem. Abstr.* **85**, 13468p (1976).
106. Gordievskii, A. V., Urusov, Y. I., Syrchenkov, A. Y., Sergievskii, V. V., Zhukov, A. F., and Pokidyshev, A. M., U.S.S.R. Patent 495,597 (15 December 1975); Appl. 2,038,410 (28 June 1974).
107. Gorton, L., and Fiedler, U., *Anal. Chim. Acta* **90**, 233 (1977).
108. Gotow, T., Ohba, M., and Tomita, T., *IEEE Trans. Biomed. Eng.* **24**, 366 (1977).
109. Gray, D. N., and Guilbault, G. G., U.S. Patent 3,929,609 (30 December 1975); Appl. 328,360 (31 January 1973).
110. Gray, D. N., and Young, C. C., Brit. Patent 1,425,912 (25 February 1976); Appl. 26,859/74 (18 June 1974).
111. Gueggi, M., Fiedler, U., Pretsch, E., and Simon, W., *Anal. Lett.* **8**, 857 (1975).
112. Gueggi, M., Oehme, M., Pretsch, E., and Simon, W., *Helv. Chim. Acta* **59**, 2417 (1978).
113. Gueggi, M., Pretsch, E., and Simon, W., *Anal. Chim. Acta* **91**, 107 (1977).
114. Guilbault, G. G., and Cserfaivi, T., *Anal. Lett.* **9**, 277 (1976).
115. Guillot, C., Vanuex, D., and Grimaud, C., *Pathol. Biol.* **24**, 431 (1976).
116. Haiduc, I., Belcu, S., and Liteanu, C., *Rev. Roum. Chim.* **22**, 427 (1977).
117. Haiduc, I., Cormos, D. C., and Mandrea, M., *Stud. Univ. Babes-Bolyai Ser. Chem.* **21**, 56 (1976) (Rom.).
118. Hansen, A. C., Engle, K., Kildeberg, P., and Wamberg, S., *Clin. Chim. Acta* **79**, 507 (1977).
119. Hassan, S. S. M., *Mikrochim. Acta* **1**, 405 (1977).
120. Hazamoto, N., Kamo, N., and Kobatake, Y., *J. Assoc. Offic. Anal. Chem.* **59**, 1097 (1976).
121. Herbert, N. C., *Adv. Exp. Med. Biol.* **50**, (*Ion-Selective Microelectrodes*) 23–38 (1974).

122. Hiro, K., Moody, G. J., and Thomas, J. D. R., *Talanta* **22**, 918 (1975).
123. Hildebrandt, W. A., and Pool, K. H., *Talanta* **23**, 469 (1976).
124. Hirsch, R. F., and Olderman, G. M., *Anal. Chem.* **48**, 771 (1976).
125. Hirst, A. D., Gay, P., Richardson, P., and Howorth, P. J. N., *Natl. Bur. Stand. (U.S.) Spec. Publ.* **450**, 311 (1977).
126. Hjemdahl-Monsen, C. E., Papastathopoulos, D. S., and Rechnitz, G. A., *Anal. Chim. Acta* **88**, 253 (1977).
127. Hopirtean, E., Chereches, M., and Liteanu, C., *Chem. Anal (Warsaw)* **20**, 965 (1975).
128. Hopirtean, E., Jurat, M., and Masa, A. I., *Rev. Chim. (Bucharest)* **26**, 872 (1975) (Rom).
129. Hopirtean, E., Liteanu, C., and Kormos, F., *Rev. Chim. (Bucharest)* **28**, 378 (1977) (Rom.).
130. Hopirtean, E., Liteanu, C., and Vlad, R., *Talanta* **22**, 912 (1975).
131. Hopirtean, E., Preda, M., and Liteanu, C., *Chem. Anal. (Warsaw)* **21**, 861 (1976).
132. Hopirtean, E., Preda, M., and Liteanu, C., *Fresenius' Z. Anal. Chem.* **286**, 65 (1977).
133. Hopirtean, E., and Stefaniga, E., *Rev. Roum. Chim.* **20**, 863 (1975).
134. Hopirtean, E., and Stefaniga, E., *Rev. Roum. Chim.* **21**, 305 (1976).
135. Hopirtean, E., Stefaniga, E., and Liteanu, C., *Chem. Anal. (Warsaw)* **21**, 867 (1976).
136. Hopirtean, E., Stefaniga, E., Liteanu, C., and Gusan, I., *Rev. Chim. (Bucharest)* **27**, 346 (1976) (Rom.).
137. Horner, J. E., and Selby, C. R., U.S. Patent 3,959,107 (25 May 1976); Appl. 477,076 (6 June 1974).
138. Hsiung, C. P., Kuan, S. S., and Guilbault, G. G., *Anal. Chim. Acta* **90**, 45 (1977).
139. Hulanicki, A., Trojanowicz, M., and Cichy, M., *Talanta* **23**, 47 (1976).
140. Jaber, A. M. Y., Moody, G. J., and Thomas, J. D. R., *Analyst (London)* **101**, 179 (1976).
141. Jaber, A. M. Y., Moody, G. J., and Thomas, J. D. R., *Proc. Anal. Div. Chem. Soc.* **11**, 328 (1976).
142. Jaber, A. M. Y., Moody, G. J., and Thomas, J. D. R., *J. Inorg. Nucl. Chem.* **39**, 1689 (1977).
143. Jacobson, J. S., and Heller, L. I., *J. Assoc. Offic. Anal. Chem.* **58**, 1129 (1975).
144. Janata, J. A., and Janata, J., U.S. Patent 3,966,580 (29 June 1976); Appl. 506,464 (16 September 1974).
145. Jerrold-Jones, P., and Krull, I. H., U.S. Patent 4,012,308 (15 March 1977); Appl. 629,833 (7 November 1975).
146. Jyo, A., Fukamachi, K., Koga, W., and Ishibashi, N., *Bull. Chem. Soc. Jpn.* **50**, 670 (1977).
147. Jyo, A., Imato, T., Fukamachi, K., and Ishibashi, N., *Chem. Lett.* **1977**(7), 815.
148. Karlberg, B., *Talanta* **22**, 1023 (1975).
149. Kataoka, M., and Kambara, T., *J. Electroanal. Chem.* **73**, 279 (1976).
149a. Kataoka, M., Shin, M., and Kambara, T., *Talanta* **24**, 261 (1977).
150. Kawashima, T., and Rechnitz, G. A., *Anal. Chim. Acta* **83**, 9 (1976).
151. Kazaryan, N. A., Bykova, L. N., and Chernova, N. S., *Zh. Anal. Khim.* **31**, 334 (1976) (Russ.).
152. Kedem, O., Perry, M., and Bloch, R., *Charged React. Polym.* **4** (*Charged Gels Membr. Part 2*), 125 (1976).
153. Keil, L., Moody, G. J., and Thomas, J. D. R., *Analyst (London)* **102**, 274 (1977).
154. Kereichuk, A. S., and Mokhnatova, N. V., *Zh. Neorg. Khim.* **21**, 1195 (1976) (Russ.).
155. Kessler, M., Clark, L. C., Lübbers, D. W., Silver, I. A., and Simon, W., eds., *Ion and Enzyme Electrodes in Biology and Medicine*, Urban and Schwarzenberg or University Park Press, University Park, Maryland (1976).

156. Kessler, M., Hajek, K., and Simon, W., in *Ion and Enzyme Electrodes in Biology and Medicine* (M. Kessler, L. C. Clark, D. W. Lübbers, I. A. Silver, and W. Simon, eds.), p. 136, Urban and Schwarzenberg or University Park Press, University Park, Maryland (1976).
157. Kiang, C., Kuan, S. S., and Guilbault, G. G., *Anal. Chim. Acta* **80**, 209 (1975).
158. Kimura, K., and Yukio, H., Ger. Patent 2,634,562 (17 February 1977); Japan Appl. 75/9 (August 1975); *Chem. Abstr.* **86**, 135603y (1977).
159. Kirsch, N. N. L., and Simon, W., *Helv. Chim. Acta* **59**, 357 (1976) (Ger.).
160. Kirchner, S. J., and Fermando, Q., *Anal. Chem.* **49**, 1636 (1977).
161. Kivalo, P., Virtanen, R., Wickstrom, K., Wilson, M., Pungor, E., Toth, K., and Sundholm, G., *Anal. Chim. Acta* **87**, 387 (1976).
162. Kivalo, P., Virtanen, R., Wickstrom, K., Wilson, M., Pungor, E., Horvai, G., and Toth, K., *Anal. Chim. Acta* **87**, 401 (1976).
163. Kobos, R. K., and Rechnitz, G. A., *Arch. Biochem. Biophys.* **175**, 11 (1976).
164. Kobos, R. K., and Rechnitz, G. A., *Biochem. Biophys. Res. Commun.* **71**, 762 (1976).
165. Koryta, J., *Anal. Chim. Acta* **91**, 1 (1977).
166. Kosov, A. E., Novikov, P. D., Krylov, O. T., and Nesterova, M. P., *Okeanologiya (Moscow)* **16**, 815 (1976) (Russ.).
167. Kotani, H., Kunifusa, T., and Sasaki, K., Ger. Patent 2,544,360 (8 April 1976); Japan. Appl. 74/114970 (4 October 1974).
168. Kraig, R. P., and Nicholson, C., *Science* **194**, 725 (1976).
169. Kraus, W. R., Kelch, D. J., and Jensen, G. A., U.S. Patent 4,040,928 (9 August 1977); Appl. 235,055 (16 March 1972).
170. Lakshminarayanaiah, N., *Electrochemistry* **5**, 132 (1975).
171. Lakshminarayanaiah, N., *Membrane Electrodes*, Academic Press, New York (1976).
172. LeBlanc, O. H., and Grubb, W. T., *Anal. Chem.* **48**, 1658 (1976).
173. Lewis, S. B., and Buck, R. P., *Anal. Lett.* **9**, 439 (1976).
174. Linch, A. L., *Health Lab. Sci.* **12**, 182 (1975).
175. Liteanu, C., Hopirtean, E., and Stefaniga, E., *Rev. Roum. Chim.* **22**, 653 (1977).
176. Luca, C., Pleniceanu, M., and Muresan, N., *Rev. Chim. (Bucharest)* **27**, 1088 (1976) (Rom.).
177. Lyalin, O. O., and Turaeva, M. S., *Zh. Anal. Khim.* **31**, 1879 (1976) (Russ.).
178. Lyalin, O. O., and Turaeva, M. S., *Elektrokhimiya* **13**, 256 (1977) (Russ.).
179. Mackay, R. A., Hermansky, C., and Agarwal, R., *Colloid Interface Sci. (Proc. Int. Conf.)* **2**, 289 (1976).
180. Malev, V. V., *Elektrokhimiya* **12**, 710 (1976) (Russ.).
181. Malik, W. U., Srivastava, S. K., Razdan, P., and Kumar, S., *J. Electroanal. Chem.* **72**, 111 (1976).
182. Malissa, H., Grasserbauer, M., Pungor, E., Toth, K., Papay, M. K., and Polos, L., *Anal. Chim. Acta* **80**, 223 (1975).
183. Marsoner, H. J., and Harnoncourt, K., *Aerztl. Lab.* **23**, 327 (1977) (Ger.).
184. Materova, E. A., Grekovich, A. L., and Garbuzova, N. V., *Ionnyi Obmen Ionometriya* **1**, 137 (1976) (Russ.).
185. Materova, E. A., and Ovchinnikova, S. A., *Zh. Anal. Khim.* **32**, 331 (1977) (Russ.).
186. Matsubara, A., and Nomura, K., *Mem. Fac. Sci. Kyushu Univ. Ser. C* **10**, 37 (1976).
187. Mauger, R., Chopin-Dumas, J., and Pariaud, J. C., *J. Electroanal. Chem.* **83**, 251 (1977).
188. Mertens, J., Van den Winkel, P., and Massart, D. L., *Anal. Chem.* **48**, 272 (1976).
189. Meyerhoff, M., and Rechnitz, G. A., *Anal. Chim. Acta* **85**, 277 (1976).
190. Meyerhoff, M., and Rechnitz, G. A., *Science* **195**, 494 (1977).
191. Midgley, D., *Anal. Chim. Acta* **87**, 7 (1976).
192. Midgley, D., *Anal. Chim. Acta* **87**, 19 (1976).

193. Moeller, W., and Stamm, O., U.S. Patent 3,973,555 (10 August 1976); Appl. Swiss 73/14,659 (16 October 1973).
194. Moody, G. J., and Thomas, J. D. R., in *Introduction to Bio-Inorganic Chemistry* (D. R. Williams, ed.), p. 220, Thomas, Springfield, Illinois (1976).
195. Moody, G. J., and Thomas, J. D. R., *Analyst* (*London*) **100**, 609 (1975).
196. Moody, G. J., and Thomas, J. D. R., *J. Sci. Food Agric.* **27**, 43 (1976).
197. Morf, W. E., Oehme, M., and Simon, W., in *Ionic Actions on Vascular Smooth Muscle with Special Regard to Brain Vessels* (*Symp.*), *2nd* 1976 (E. Betz, ed.), Springer, New York (1977).
198. Morf, W. E., Wuhmann, P., and Simon, W., *Anal. Chem.* **48**, 1031 (1976).
199. Morie, G. P., *Beitr. Tabakforsch.* **9**, 19 (1977).
200. Muratsugu, M., Kamo, N., Kunihara, K., and Kobatake, Y., *Biochim. Biophys. Acta* **464**, 613 (1977).
201. Mussini, T., *Chim. Ind.* (*Milan*) **58**, 179 (1976) (Ital.).
202. Napoli, A., and Mascini, M., *Anal. Chim. Acta* **89**, 209 (1977).
203. Naray, M., *Munkavedelem* **21**, 31 (1975) (Hung.); *Chem. Abstr.* **84**, 147108j (1976).
204. Newman, D. P., U.S. Patent 3,979,274 (7 September 1976); Appl. 616,326 (24 September 1975).
205. Niedrach, L. W., and Stoddard, W. H., U.S. Patent 3,923,627 (2 December 1975); Appl. 519,797 (1 November 1974).
206. Nielsen, H. J. I., and Hansen, E. H., Ger. Patent 2,652,703 (26 May 1977); Appl. 75/5217 (20 November 1975).
207. Oehme, M., Kessler, M., and Simon, W., *Chimia* **30**, 204 (1976).
208. Oehme, M., and Simon, W., *Anal. Chim. Acta* **86**, 21 (1976).
209. Oglesby, G. B., Duer, W. C., and Millero, F. J., *Anal. Chem.* **49**, 877 (1977).
210. Olson, V. K., Carr, J. D., Hargens, R. D., and Force, R. K., *Anal. Chem.* **48**, 1228 (1976).
211. Ovchinnikova, S. A., Materova, E. A., and Karavan, V. S., *Vest. Leningrad Univ. Fiz. Khim.* **1977**(2), 117 (Russ.).
212. Pace, S. J., and Brand, M. J. D., *Natl. Bur. Stand.* (*U.S.*) **450**, 227 (1977).
213. Papastathopoulos, D. S., and Rechnitz, G. A., *Anal. Chim. Acta* **79**, 17 (1975).
214. Papastathopoulos, D. S., and Rechnitz, G. A., *Anal. Chem.* **48**, 862 (1976).
215. Pataki, L., Harka, K., Havas, J., and Keomley, G., *Radiochem. Radioanal. Lett.* **24**, 305 (1976).
216. Pataki, L., Harka, K., Havas, J., and Keomley, G., *Radiochem. Radioanal. Lett.* **25**, 205 (1976).
217. Pataki, L., Harka, K., Havas, J., and Keomley, G., *Radiochem. Radioanal. Lett.* **26**, 223 (1976).
218. Pataki, L., Harka, K., Havas, J., and Keomley, G., *Radiochem. Radioanal. Lett.* **27**, 287 (1976).
219. Pataki, L., Harka, K., Havas, J., and Keomley, G., *Radiochem. Radioanal. Lett.* **27**, 385 (1976).
220. Perry, M., Lobel, E., and Bloch, R., *J. Memb. Sci.* **1**, 223 (1976).
221. Petranek, J., and Ryba, O., Czech Patent 165,835 (15 November 1976); Appl. 72/5529 (8 August 1972).
222. Pick, J., *Hung. Sci. Instrum.* **34**, 41 (1975).
223. Popa, G., Birsen, G., and Luca, C., *Rev. Chim.* (*Bucharest*) **26**, 512 (1975) (Rom.).
224. Prat, J., *Tec. Lab.* **44**, 346 (1975) (Span.).
225. Pungor, E., Toth, K., and Nagy, G., *Hung. Sci. Instrum.* **35**, 1 (1975).
226. Pungor, E., ed., *Ion-Selective Electrodes, Proceedings of the Matrafured Conference, October* 18–21, 1976, Akademiai Kiado, Budapest (1977).
227. Pusch, W., *Chem.-Ing.-Tech.* **47**, 914 (1975) (Ger.).
228. Rajput, A. R., Kataoka, M., and Kambara, T., *J. Electroanal. Chem.* **66**, 67 (1975).

229. Rechnitz, G. A., Nogle, G. J., Bellinger, M. R., and Lees, H., *Clin. Chim. Acta* **76**, 295 (1977).
230. Riechel, T. L., and Rechnitz, G. A., *Biochem. Biophys. Res. Commun.* **74**, 1377 (1977).
231. Rix, C. J., Bond, A. M., and Smith, J. D., *Anal. Chem.* **48**, 1236 (1976).
232. Rock, P. A., Eyrich, T. L., and Styer, S., *J. Electrochem. Soc.* **124**, 530 (1977).
233. Rottenberg, H., *J. Bioenerg.* **7**, 61 (1975).
234. Ryan, D. E., and Cheung, M. T., *Anal. Chim. Acta* **82**, 409 (1976).
235. Ryba, O., and Petranek, J., *J. Electroanal. Chem.* **67**, 321 (1976).
236. Scheide, E. P., and Durst, R. A., *Anal. Lett.* **10**, 55 (1977).
237. Schindler, J. G., and Riemann, W., Ger. Patent 2,503,176 (29 July 1976); Appl. (27 January 1975).
238. Schwartz, H. D., *Clin. Chim. Acta* **64**, 227 (1975).
239. Schwartz, H. D., *Clin. Chem.* (*Winston-Salem, N.C.*) **22**, 461 (1976).
240. Schwartz, H. D., *Clin. Chem.* (*Winston-Salem, N.C.*) **23**, 610 (1977).
241. Secor, G. E., McDonald, G. M., and McReady, R. M., *J. Assoc. Office Anal. Chem.* **59**, 761 (1976).
242. Sekerka, I., and Lechner, J. F., *J. Electroanal. Chem.* **69**, 339 (1976).
243. Sekerka, I., and Lechner, J. F., *Anal. Lett.* **9**, 1099 (1976).
244. Sekerka, I., Lechner, J. F., and Harrison, L., *J. Assoc. Offic. Anal. Chem.* **60**, 625 (1977).
245. Selig, W., *Microchem. J.* **22**, 1 (1977).
246. Semler, M., and Adametzova, H., *Chem. Prum.* **25**, 377 (1975) (Czech.); *Chem. Abstr.* **84**, 11802y (1976).
247. Sharp, M., *Anal. Chim. Acta* **85**, 17 (1976).
248. Shechter, H., and Gruener, N., *J. Am. Water Works Assoc.* **68**, 543 (1976).
249. Sherken, S., *J. Assoc. Offic. Anal. Chem.* **59**, 971 (1976).
250. Shikanova, M. S., Pukshanskii, M. D., Rodichev, A. G., Sirota, A. G., Krunchak, V. G., and Karkozova, G. F., U.S.S.R. Patent 515,060 (25 May 1976); Appl. 2,083,476 (16 December 1974).
251. Shults, M. M., and Stefanova, O. K., *Vestn. Leningrad. Univ. Fiz. Khim.* **1976**(1), 88 (Russ.).
252. Simon, W., Morf, W. E., Pretsch, E., and Wuhrmann, P., in *Proceedings of the International Symposium on Calcium Transport in Contraction and Secretion*, 1975 (E. Carafoli, F. Clementi, and W. Drabikowski, eds.), North-Holland, Amsterdam (1976).
253. Simon, W., Pretsch, E., Ammann, D., Morf, W. E., Gueggi, M., Bissig, R., and Kessler, M., *Pure Appl. Chem.* **44**, 613 (1975).
254. Singer, L., and Ophaug, R. H., *Anal. Chem.* **49**, 38 (1977).
255. Sternberg, J., Updike, S. J., and Lehane, D. P., *Microtech. Clin. Lab.* **1976**, 129.
256. Sugawara, M., Nakajima, T., and Kambara, T., *J. Electroanal. Chem.* **67**, 315 (1976).
257. Szczpaniak, W., Malicka, J., and Ren, K., *Chem. Anal.* (*Warsaw*) **20**, 1141 (1975).
258. Szczepaniak, W., Ren, K., and Mickiewicz, A., *Anal. Chim. Acta* **82**, 37 (1976).
259. Szonntagh, E. L., U.S. Patent 4,031,606 (28 June 1977); Appl. 552,284 (24 February 1975).
260. Takaishvili, O. G., Motsonelidze, E. P., Karachentseva, Y. M., and Lavitaya, P. I., *Zh. Anal. Khim.* **30**, 1629 (1975) (Russ.).
261. Tanikawa, S., Adachi, T., Shiraishi, N., Kakagawa, G., and Kodoma, K., *Bunseki Kagaku* **25**, 646 (1976) (Japan.).
262. Tanikawa, S., Kirihara, H., Shiraishi, N., Nakagawa, G., and Kodama, K., *Anal. Lett.* **8**, 879 (1975).
263. Tenygl, J., *Int. Rev. Sci. Phys. Chem., Ser. Two* **13**, 1 (1976) (A. D. Buckingham, ed., Butterworth, London).
264. Thoma, A. P., Viviani-Nauer, A., Arvanitis, S., Morf, W. E., and Simon, W., *Anal. Chem.* **49**, 1567 (1977).

265. Thomas, J. D. R., *Proc. Anal. Div. Chem. Soc.* **14**, 7 (1977).
266. Thomas, J. D. R., and Moody, G. J., *Proc. Anal. Div. Chem. Soc.* **12**, 48 (1975).
267. Thompson, D. E., and Danchik, R. S., *Anal. Lett.* **8**, 699 (1975).
268. Tong, S. L., and Rechnitz, G. A., *Anal. Lett.* **9**, 1 (1976).
269. Trachtenberg, I., in *Proceedings of the Symposium on Traces Heavy Metal Water Removal Processes Monitoring*, 1973 (J. E. Sabadell, ed.), p. 323, NTIS, Springfield, Virginia.
270. Treasure, T., and Band, D. M., *J. Med. Eng. Technol.* **1**, 271 (1977).
271. Tseng, P. K. C., and Gutknecht, W. F., *Anal. Lett.* **9**, 795 (1976).
272. Tseng, P. K. C., and Gutnecht, W. F., *Anal. Chem.* **48**, 1996 (1976).
273. Turaeva, M. S., and Lyalin, O. O., U.S.S.R. Patent 544,899 (30 January 1977); Appl. 1,950,247 (3 August 1973).
274. Urusov, Y. I., Sergievskii, V. V., Syrchenkov, A. Y., Zhukov, A. F., and Gordievskii, A. V., *Zh. Anal. Khim.* **30**, 1757 (1975) (Russ.).
275. Vallon, J. J., *Lyon Pharm.* **26**, 487 (1975) (Fr.).
276. Van De Leest, R. E., *Analyst (London)* **101**, 433 (1976).
277. Van De Leest, R. E., Beekmans, N M., and Heijne, L., Ger. Patent 2,600,846 (19 August 1976); Neth. Appl. 75/823 (24 January 1975).
278. Van De Leest, R. E., and Heijne, L., Ger. Patent 2,621,731 (9 December 1976); Neth. Appl. 75/6410 (30 May 1975).
279. Van den Heede, M. A., Heyndrickx, A. M., Van Peteghem, C. H., and Van Zele, W. A., *J. Assoc. Offic. Anal. Chem.* **58**, 1135 (1975).
280. Van den Winkel, P., Mertens, J., Boel, T., and Vereecken, J., *J. Electrochem. Soc.* **124**, 1338 (1977).
281. Van der Meer, J. M., Den Boef, G., and Van der Linden, W. E., *Anal. Chim. Acta* **79**, 27 (1975).
282. Van der Meer, J. M., Den Boef, G., and Van der Linden, W. E., *Anal. Chim. Acta* **85**, 317 (1976).
283. Verkhovskaya, M. L., Imasheva, E. S., Kurella, G. A., and Yaglova, L. G., *Biol. Nauki (Moscow)* **19**, 142 (1976) (Russ.).
284. Vesely, J., and Jindra, J., Czech Patent 163,358 (15 June 1976); Appl. 68/5344 (22 July 1968).
285. Warner, T. B., *Phys. Chem. Sci. Res. Rep.* **1975** (1) (Nat. Seawater, Rep. Dahlem Workshop 1975), 191.
286. Watson, B., and Breno, P. J., U.S. Publ. Pat. Appl. B, 511,002 (6 March 1976); Appl. 511,002 (1 October 1974).
287. Wawro, R., and Rechnitz, G. A., *J. Membr. Sci.* **1**, 143 (1976).
288. Whitfield, M., *Proc. Anal. Div. Chem. Soc.* **12**, 56 (1975).
289. Wyvill, R. D., *Prod. Finish. (London)* **30**, 21 (1977).
290. Young, C. C., U.S. Patent 4,033,777 (5 July 1977); Appl. 511,720 (3 October 1974).
291. Zetter, M. S., Ger. Patent 2,646,314 (5 May 1977); U.S. Appl. 625,857 (28 October 1975).
292. Zhukov, A. F., Gordievskii, A. V., Shterman, V. S., Syrchenkov, A. Y., and Urusov, Y. I., U.S.S.R. Patent 515,984 (30 May 1976); Appl. 1,939,127 (3 July 1973).
293. Zhukov, A. F., Vishnyakov, A. V., Kharif, Y. L., Urusov, Y. I., Volynets, F. K., Rhyzhikov, E. I., and Gordievskii, A. V., *Zh. Anal. Khim.* **30**, 1761 (1975) (Russ.).

"F" REFERENCES (1978)

1. Adachi, T., Shiraishi, N., Nakagawa, G., Kodama, K., *Bunseki Kagaku* **26**(10), 658 (1977).

2. Afromowitz, M. A., and Yee, S. S., *J. Bioeng.* **1**(2), 55 (1977).
3. Aizawa, S., and Suzuki, S., *Hyomen* **16**(8), 513 (1978).
4. Akimoto, N., Ozaki, M., and Hozumi, K., *Bunseki Kagaku* **26**(6), 417 (1977).
5. Anon, *Res. Discl.* **161**, 29 (1977).
6. Antson, J., and Suntola, T., *Tutkimus Tek.* **1975**(9), 26.
7. Aomi, T., *Denki Kagaku Oyobi Kogyo Butsuri Kagaku* **46**(5), 259 (1978).
8. Aomi, T., *Denki Kagaku Oyobi Kogyo Butsuri Kagaku* **46**(6), 343 (1978).
9. Baiulescu, G. E., and Cosofret, V. V., in *Ion-Selective Electrodes Conference*, 1977 (E. Pungor and I. Buzas, eds.), pp. 207–214, Elsevier, Amsterdam (1978).
10. Bates, R., and Robinson, R. A., in *Ion-Selective Electrodes Conference*, 1977 (E. Pungor and I Buzas, eds.), pp. 3–19, Elsevier, Amsterdam (1978).
11. Battaglia, C. J., Chang, J. C., Daniel, D. S., Hamblen, D. P., Glover, C. P., and Kim, S. H., Ger. Offen. 2,722,627 (1 December 1977); U.S. Appl. 687,725 (19 May 1976).
12. Baumann, E. W., *Anal. Chim. Acta* **99**(2), 247 (1978).
13. Beider, T. B., and Solomentseva, K. R., *Methody Anal. Kontrolya Proizvod. Khim. Prom-sti.* **7**, 44 (1977).
14. Belyustin, A. A., Shults, M. M., Peshnekhonova, N. V., Ivanova, L. I., Shugalova, A. L., Moglieva, V. V., and Ivanovskaya, I. S., *Dokl. Akad. Nauk SSSR* **240**(6), 1376 (1978).
15. Bhattacharjee, A. P., and Guha, S. K., *Cent. Glass Ceram. Res. Inst. Bull.* **24**(1), 4 (1977).
16. Bissig, R., Oeach, U., Pretsch, E., Morf, W. E., and Simon, W., *Helv. Chim. Acta* **61**(5), 1531 (1978).
17. Bixler, J. W., Nee, R., and Perone, S. P., *Anal. Chem.* **99**(2), 225 (1978).
18. Blaine, L. L., and Toman, F. R., *Trans. Ky. Acad. Sci.* **39**(3-4), 160 (1978).
19. Boeke, J., Ger. Offen. 2,726,271 (22 December 1977); U.S. Appl. 694,655 (10 June 1976).
20. Bohnke, C., Malugani, J. P., and Robert, G., *C.R. Hebd. Séances Acad. Sci. Ser. C* **286**(1), 13 (1978).
21. Boksay, Z., in *Ion-Selective Electrodes Conference*, 1977 (E. Pungor and I. Buzas, eds.), pp. 245–267, Elsevier, Amsterdam (1978).
22. Boksay, Z., Csakvari, B., Havas, J., and Patko, M., Hung. Halasztott 2,210 (28 March 1977).
23. Boksay, Z., Csakvari, B., Havas, J., and Patko, M., in *Ion-Selective Electrodes Conference*, 1977 (E. Pungor and I. Buzas, eds.), pp. 269–280, Elsevier, Amsterdam (1978).
24. Boniface, H. J., and Jenkins, R. H., *Analyst (London)* **102**(1219), 739 (1977).
25. Bray, P. T., Clark, G. C. F., Moody, G. J., and Thomas, J. D. R., *Clin. Chim. Acta* **80**(2), 333 (1977).
26. Breno, P. J., Ger. Offen. 2,706,942 (29 September 1977); U.S. Appl. 667,788 (17 March 1976).
27. Breno, P. J., and Watson, B., Ger. Offen. 2,612,428 (6 October 1977).
28. Bresnahan, W. T., Grant, C. L., and Weber, J. H., *Anal. Chem.* **50**(12), 1674 (1978).
29. Bruckenstein, S., and Sherwood, W. G., U.S. 4,057,478 (8 November 1977).
30. Brunfelt, A. O., *Bull. Soc. R. Sci. Liege* **46**(3-4), 133 (1977).
31. Buck, R. P., *Proc. Anal. Div. Chem. Soc.* **14**(11), 332 (1977).
32. Buck, R. P., in *Ion-Selective Electrodes Conference*, 1977 (E. Pungor and I. Buzas, eds.), pp. 21–55, Elsevier, Amsterdam (1978).
33. Buck, R. P., *Anal. Chem.* **50**(5), 17R (1978).
34. Buck, R. P., *J. Electroanal. Chem. Interfacial Electrochem.* **83**(2), 379 (1977).
35. Bykova, L. N., Kazaryan, N. A., Pungor, E., and Chernova, N. S., in *Ion-Selective Electrodes Conference*, 1977 (E. Pungor and I. Buzas, eds.), pp. 281–287, Elsevier, Amsterdam (1978).
36. Cammann, K., *Anal. Chem.* **50**(7), 936 (1978).
37. Campanella, L., Ferri, T., and Gozzi, D., *Rev. Roum. Chim.* **23**(2), 281 (1978).

38. Campanella, L., Ferri, T., Gozzi, D., and Scorcelletti, G., in *Ion-Selective Electrodes Conference*, 1977 (E. Pungor and I. Buzas, eds.), pp. 307–316, Elsevier, Amsterdam (1978).
39. Campanella, L., DeAngelis, G., Gozzi, D., and Ferri, T., *Analyst* (*London*) **102**(1219), 723 (1977).
40. Cattrall, R. W., Martin, A. R., and Tribuzio, S., *J. Inorg. Nucl. Chem.* **40**(4), 687 (1978).
41. Cavagnaro, D., Report 1977 NTIS/PS-77/0876.
42. Cavagnaro, D., Report 1977 NTIS/PS-77/0875.
43. Cheng, K. L., and Chao, E. E., U.S. Patent 4,071,427 (31 January 1978).
44. Childs, P. E., *Sch. Sci. Rev.* **59**(208), 503 (1978).
45. Christensen, J. J., Lamb, J. D., Izatt, S. R., Starr, S. E., Weed, G. C., Astin, M. S., Stitt, B. D., and Izatt, R. M., *J. Am. Chem. Soc.* **100**(10), 3219 (1978).
46. Clysters, H., and Adams, F., *Anal. Chim. Acta* **92**(2), 251 (1977).
47. Coetzee, C. J., and Basson, A. J., *Anal. Chim. Acta* **92**(2), 399 (1977).
48. Cocofret, V. V., Zugravescu, P. G., and Baiulescu, G. E., *Talanta* **24**(7), 461 (1977).
49. Cosofret, V. V., Cristescu, C., and Zugravescu, P. G., in *Ion-Selective Electrodes Conference*, 1977 (E. Pungor and I. Buzas, eds.), pp. 325–333, Elsevier, Amsterdam (1978).
50. Cosofret, V. V., and Zugravescu, P. G., *Rev. Chim.* (*Bucharest*) **28**(8), 785 (1977).
51. Craggs, A., Moody, G. J., Thomas, J. D. R., and Birch, B. J., in *Ion-Selective Electrodes Conference*, 1977 (E. Pungor and I. Buzas, eds.), pp. 335–338, Elsevier, Amsterdam (1978).
52. Cutler, S. G., Meares, P., and Hall, D. G., *J. Chem. Soc. Faraday Trans. 1* **74**(7), 1758 (1978).
53. Cutler, S. G., Meares, P., and Hall, D. G., *J. Electroanal. Chem. Interfacial Electrochem.* **85**(1), 145 (1977).
54. Diamandis, E. P., and Hadjiioannou, T. P., *Mikrochem. Acta* **2**(3-4), 255 (1977).
55. Diamandis, E. P., Koupparis, M. A., and Hadjiioannou, T. P., *Microchem. J.* **22**(4), 498 (1977).
56. Dobson, J. V., Brit. Patent 1,481,510 (3 August 1977).
57. Dojlido, J., and Taboryska, B., *Wiad. Meteorol. Gospod. Wodnej.* **3**(1), 65 (1976).
58. D'Orzaio, P., Meyerhoff, M. E., and Rechnitz, G. A., *Anal. Chem.* **50**(11), 1531 (1978).
59. Dorsett, L. P., and Mulcahy, D. E., *Anal. Lett.* **A11**(1), 53 (1978).
60. Durst, R. A., *Clin. Chim. Acta* **80**(1), 225 (1977).
61. Durst, R. A., *Anal. Lett.* **10**(12), 961 (1977).
62. Durst, R. A., *Natl. Bur. Stand.* (*U.S.*) *Spec. Publ.* **464**, 229–232 (1977).
63. Ebock, V., and Neiser, C., *Z. Chem.* **18**(9), 343 (1978).
64. Eicken, D., *Vom. Wasser* **49**, 139–171, 1977 (Pub. 1978).
65. Fjeldly, T. A., Johannessen, J. S., and Nagy, K., *Proc. Int. Vac. Congr., 7th* 1977, **3**, 2327–2330.
66. Fogg, A. G., and Yoo, K. S., in *Ion-Selective Electrodes Conference*, 1977 (E. Pungor and I. Buzas, eds.), pp. 369–372, Elsevier, Amsterdam (1978).
67. Foraison, D., and Souquet, L., *Electrochim. Acta* **23**(5), 457 (1978).
68. H. Freiser, ed., *Ion-Selective Electrodes in Analytical Chemistry*, Vol. I, Plenum Press, New York (1978).
69. Fukumachi, K., and Ishibashi, N., *Bunseki Kagaku* **27**(3), 152 (1978).
70. Gaarenstroom, P. D., English, J. C., Perone, S. P., and Bixler, J. W., *Anal. Chem.* **50**(6), 811 (1978).
71. Gava, S. A., Poleuktov, N. S., and Koroleva, G. N., *Zh. Anal. Khim.* **33**(3), 506 (1978).
72. Goina, T., and Hobai, S., *Rev. Med.* (*Tirgu-Mures Rom.*) **23**(1), 70 (1977).

73. Golubev, V. N., Serebrennikova, N. V., Mikhailov, V. A., and Osipov, V. V., *Tezisy Dokl.-Vses. Konf. Ekstr.* **3**, 13 (1977).
74. Gorobenko, F. P., and Sodirenko, E. V., *Metody Poluch. Anal. Mater. Elektron Tekh.*, 106 (1976).
75. Gordievskii, A. V., and Zeinalova, E. A., *Ref. Zh. Khim.* **1978**, Abstr. No. 11240.
76. Gozzi, D., and Scorcelletti, G., *J. Electroanal. Chem. Interfac. Electrochem.* **93**(2), 109 (1978).
77. Grekovich, A. L., Materova, E. A., Pletneva, V. V., Khutsishvili, A. N., and Bart, T. Y., U.S.S.R. Patent 562,765 (25 June 1977).
78. Gruenke, U., and Hartmann, P., *Hermsdorfer Tech. Mitt.* **18**(51), 1619 (1978).
79. Gyenge, R., Toth, K., Pungor, E., and Koros, E., *Anal. Chim. Acta* **94**(1), 111 (1977).
80. Haberich, F. J., Ger. Offen. 2,627,038 (22 December 1977).
81. Hadjiioannou, T. P., and Diamandis, E. P., *Anal. Chim. Acta* **94**(2), 443 (1977).
82. Hadjiioannou, T. P., Koupparis, M. A., and Diamandis, E. P., *Proc. Anal. Div. Chem. Soc.* **15**(3), 78 (1978).
83. Hansen, E. H., Krug, F. J., Ghose, A. K., and Ruzicka, J., *Analyst (London)* **102**(1219), 714 (1977).
84. Hansen, E. H., Ghose, A. K., and Ruzicka, J., *Analyst (London)* **102**(1219), 705 (1977).
85. Hansen, E. H., Ruzicka, J., and Ghose, A. K., *Anal. Chim. Acta* **100**, 151 (1978).
86. Hastings, D. F., *Anal. Biochem.* **83**(2), 416 (1977).
87. Hato, M., *Sen'i Kobunshi Zairo Kenkyusho Hokoku* **113**, 15 (1976).
88. Havas, J., and Kecskes, L., *Magy. Kem. Foly.* **83**(12), 529 (1977).
89. Havas, J., and Patko, M., Hung. Teljes 14,005 (28 September 1977).
90. Havas, J., Kecskes, L., and Somodi, R., *Orv. Tech.* **15**(2), 41 (1977).
91. Havas, J., Kecskes, L., and Somodi, R., *Hung. Sci. Instrum.* **41**, 47 (1977).
92. Haynes, S. J., *Talanta* **25**(2), 85 (1978).
93. Heijne, G. J. M., and Van der Linden, W. E., *Anal. Chim. Acta* **93**, 99 (1977).
94. Heijne, G. J. M., and Van der Linden, W. E., *Anal. Chim. Acta* **96**(1), 13 (1978).
95. Heijne, G. J. M., Van der Linden, W. E., and Den Boef, G., *Anal. Chim. Acta* **98**(2), 221 (1978).
96. Heijne, G. J. M., Van de Linden, W. E., and Den Boef, G., *Anal. Chim. Acta* **100**, 193 (1978).
97. Herrmann, F., and Bilal, B. A., *Fresenius' Z. Anal. Chem.* **291**(2), 113 (1978).
98. Hopirtean, E., Veress, E., and Muresan, V., *Rev. Roum. Chim.* **22**(8), 1243 (1977).
99. Hopirtean, E., *Rev. Roum. Chim.* **22**(9-10), 1385 (1977).
100. Hopirtean, E., and Stefaniga, E., *Rev. Roum. Chim.* **23**(1), 137 (1978).
101. Hopirtean, E., and Stefaniga, E., *Chem. Anal. (Warsaw)* **22**(5), 845 (1977).
102. Hopirtean, E., and Veress, E., *Rev. Roum. Chim.* **23**(2), 273 (1978).
103. Hopirtean, E., and Kormos, F., *Stud. Univ. Babes-Bolyai (Ser.) Chem.* **22**(2), 35 (1977).
104. Hopkins, D. M., *J. Res. U.S. Geol. Surv.* **5**(5), 589 (1977).
105. Hrabeczy-Pall, A., Toth, K., and Pungor, E., in *Ion-Selective Electrodes, 2nd Symposium*, 1976 (E. Pungor and I. Buzas, eds.), pp. 127–137, Akademiai Kiado, Budapest (1977).
106. Hrabeczy-Pall, A., Vallo, F., Toth, K., and Pungor, E., *Hung. Sci. Instrum.* **41**, 55 (1977).
107. Hulanicki, A., Augustowska, Z., and Trojanowicz, M., *Chem. Anal. (Warsaw)* **22**(5), 955 (1977).
108. Hulanicki, A., Trojanowicz, M., and Krawczynski, T., *Water Res.* **11**(8), 627 (1977).
109. Hulanicki, A., and Trojanowicz, in *Ion-Selective Electrodes, 2nd Symposium*, 1976 (E. Pungor and I. Buzas, eds.), pp. 139–150, Akademiai Kiado, Budapest (1977).
110. Inlow, R. O., Report 1977, NBL-287, New Brunswick Lab., ERDA.
111. Ishikawa, Y., Yamazaki, K., Awano, K., Chiba, T., Kojima, H., Funaki, H., and Koriyama, T., *Miyagi-ken Eisei Kenkyusho Nempo*, 140 (1975).

112. Ito, J., and Ueno, K., *Kagaku (Kyoto)* **32**(5), 402 (1977).
113. Jaber, A. M. Y., Moody, G. J., and Thomas, J. D. R., in *Ion-Selective Electrodes Conference*, 1977 (E. Pungor and I. Buzas, eds.), pp. 411–417, Elsevier, Amsterdam (1978).
114. Jaber, A. M. Y., Moody, G. J., and Thomas, J. D. R., *Analyst (London)* **102**(1221), 943 (1977).
115. Jaber, A. M. Y., Moody, G. J., Thomas, J. D. R., and Willcox, A., *Talanta* **21**(10), 655 (1977).
116. Jain, A. K., Srivastava, S. K., Singh, R. P., and Agrawal, S., *J. Appl. Chem. Biotechnol.* **27**(12), 680 (1977).
117. Johansson, G., and Ogren, L., in *Ion-Selective Electrodes, 2nd Symposium*, 1976 (E. Pungor and I. Buzas, eds.), pp. 93–100, Akademiai Kiado, Budapest (1977).
118. Jones, M. H., and Woodcock, J. T., *Trans. Inst. Min. Metall. Sec. C* **87**(June), 99 (1978).
119. Jones, M. H., and Woodcock, J. T., *Proc. Australas. Inst. Min. Metall.* **266**, 11 (1978).
120. Joshi, K. M., and Ganu, G. M., *Indian J. Technol.* **16**(2), 53 (1978).
121. Karasawa, T., and Tsujimoto, T., *Ibaraki-ken Kogai Gijutsu Senta Nempo* **8**, 188 (1975).
122. Karube, I., and Suzuki, S., *Kagaku Kogaku* **42**(9), 496 (1978).
123. Kambara, T., and Kataoka, M., *Kagaku (Kyoto)* **33**(5), 400 (1978).
124. Kataoka, M., Tsukamoto, M., and Kambara, T., *Denki Kagaku Oyubi Kogyo Butsuri Kagaku* **45**(2), 100 (1977).
125. Kauranen, P., *Anal. Lett.* **10**(6), 451 (1977).
126. Keil, L., Moody, G. J., and Thomas, J. D. R., *Anal. Chim. Acta* **96**(1), 171 (1978).
127. Khristova, R., Stefanova, R., and Mac Van Men, *God. Sofii. Univ. Khim. Fak.* **68**, 109 (1977).
128. Kim, G. O., Li, A. K., Kim, S. N., and Yun, S. D., *Punsok Hwahak* **15**(3), 41 (1977).
129. Kirpichnikova, Z. F., Nikolaishvili, T. N., Khutsishvili, A. N., and Timakov, S. G., U.S.S.R. Patent 562,766 (25 June 1977).
130. Kisaki, K., *Kaijo Hoan Daigakko Kenkyo Hokoku Dai-2-Bu* **19**, 21 (1973).
131. Kivalo, P., and Virtanen, R., in *Ion-Selective Electrodes, 2nd Symposium*, 1976 (E. Pungor and I. Buzas, eds.), pp. 151–158, Akademiai Kiado, Budapest (1977).
132. Klosinska-Rycerska, B., *Przem. Ferment. Rolny* **21**(8), 23 (1977).
133. Kobos, R. K., and Rechnitz, G. A., *Anal. Lett.* **10**(10), 751 (1977).
134. Kitakyushu Tech. Coll., *Kitakyushu Kogyo Koto Semmon Gakko Kenkyo Hokoku* **11**, 159 (1978).
135. Koupparis, M. A., and Hadjiioannou, T. P., *Mikrochim. Acta* **2**(3-4), 267 (1978).
136. Koupparis, M. A., and Hadjiioannou, T. P., *Microchem. J.* **23**(2), 178 (1978).
137. Leontev, V. M., Guk, L. P., and Khokhlova, G. K., *Biosfera Chel., Mater. Vses. Simp., 1st* 1973 (V. A. Kovda, ed.), pp. 286–287, Nauka, Moscow (1975).
138. Leontevskaya, P. K., Pendin, A. A., Trofimov, M. A., and Shults, I. M., Vzaimodeistviya v Rastvorakh Okislitelno-vosstanovit. Sistem., No. 12B1569 (1977).
139. Liberti, L., and Pinto, A., *Anal. Chem.* **49**(14), 2377 (1977).
140. Lindner, E., Wuhrmann, P., Simon, W., and Pungor, E., in *Ion-Selective Electrodes, 2nd Symposium*, 1976 (E. Pungor and I. Buzas, eds.), pp. 159–167, Akademiai Kiado, Budapest (1977).
141. Lindner, E., Toth, K., Pungor, E., Morf, W. E., and Simon, W., *Anal. Chem.* **50**(12), 1627 (1978).
142. Liteanu, C., Popescu, I. C., and Domesa, S. B., Rom. 59,059 (30 September 1975).
143. Liteanu, C., and Hopirtean, E., *Fresenius' Z. Anal. Chem.* **288**(1-2), 59 (1977).
144. Loscombe, C. R., and Dalziel, J. A. W., *Proc. Anal. Div. Chem. Soc.* **14**(9), 260 (1977).
145. Lubrano, G. J., and Guilbault, G. G., *Anal. Chim. Acta* **97**(2), 229 (1978).
146. Manakova, L. I., Bausova, N. V., Moiseev, V. E., Bamburov, V. G., and Sivoplyas, A. P., *Zh. Anal. Khim.* **33**(8), 1517 (1978).

147. Manley, W., *Proceedings of the Annual Conference Am. Water Works Association, Ontario Section, Pollution Control Association of Ontario*, 1975, pp. 200–209, Environ. Prod. Ser., Environment Canada, Ottawa, Ontario, Canada (1976).
148. Materova, E. A., Grekovich, A. L., Didina, S. E., Dolodze, V. A., Knutsishvili, A. N., Sarakhanova, E. P., and Siradze, T. A., U.S.S.R. Patent 575,558 (5 October 1977).
149. Materova, E. A., and Garbuzova, N. V., *Elektrokhimiya* **13**(12), 1846 (1977).
150. Materova, E. A., Ovchinnikova, S. A., and Smekalova, S. A., *Elektrokhimiya* **14**(1), 71 (1978).
151. Materova, E. A., Alagova, Z. S., Kustanovich, I. V., and Shumilova, G. I., *Vestn. Leningr. Univ. Fiz. Khim.* **1978**(2), 103.
152. Meier, P. C., Oehme, M., and Simon, W., *Bibl. Anat.* **15** (*Recent Adv. Basic Microcirc. Res.*), 120 (1977).
153. Mertens, J., Van den Winkel, P., and Vereecken, J., *J. Electroanal. Chem. Interfacial Electrochem.* **85**(2), 277 (1977).
154. Mikhailov, V. A., Osipov, V. V., and Serebrennikova, N. V., *Zh. Anal. Khim.* **33**(6), 1154 (1978).
155. Mirkin, V. A., Goncharuk, V. G., and Bakanina, V. V., *Khimiya i Khim. Tekhnol.* **1976**(20), 64.
156. Misniakiewicz, W., and Raszka, K., in *Ion-Selective Electrodes Conference*, 1977 (E. Pungor and I. Buzas, eds.), pp. 467–475, Elsevier, Amsterdam (1978).
157. Mohan, M. S., Bates, R. G., Hiller, J. M., and Brand, M. J., *Clin. Chem.* (*Winston-Salem, N.C.*) **24**(4), 580 (1978).
158. Moody, G. J., and Thomas, J. D. R., *Lab. Pract. Rev. Curr. Tech. Instrum.*, 1–20 (1977).
159. Moody, G. J., and Thomas, J. D. R., *Prog. Med. Chem.* **14**, 51 (1977).
160. Moody, G. J., and Thomas, J. D. R., *Lab. Pract.* **27**(4), 285 (1978).
161. Moody, G. J., Nassory, N. S., and Thomas, J. D. R., *Analyst* (*London*) **103**(1222), 68 (1978).
162. Morf, W. E., and Simon, W., *Hung. Sci. Instrum.* **41**, 1 (1977).
163. Morf, W. E., and Simon, W., in *Ion-Selective Electrodes, 2nd Symposium*, 1976 (E. Pungor and I. Buzas, eds.), pp. 25–40, Akademiai Kiado, Budapest (1977).
164. Morf, W. E., and Simon, W., in *Ion-Selective Electrodes Conference*, 1977 (E. Pungor and I. Buzas, eds.), pp. 149–159, Elsevier, Amsterdam (1978).
165. Moskvin, L. N., and Krasnoperov, V. M., *Zh. Anal. Khim.* **32**(8), 1559 (1977).
166. Muramatsu, K., Japan. Kokai 78, 30,387 (22 March 1978).
167. Nagy, G., Feher, Z., Toth, K., and Pungor, E., *Hung. Sci. Instrum.* **41**, 27 (1977).
168. Nagy, G., Feher, Z., Toth, K., and Pungor, E., *Anal. Chim. Acta* **91**(2), 97 (1977).
169. Nagy, G., Feher, Z., Toth, K., and Pungor, E., *Anal. Chim. Acta* **91**(2), 87 (1977).
170. Nagy, K., Fjeldly, T. A., and Johannessen, J. S., in *Ion-Selective Electrodes Conference*, 1977 (E. Pungor and I. Buzas, eds.), pp. 491–502, Elsevier, Amsterdam (1978).
171. Nakahara, K., Japan. Kokai 78, 27,490 (15 March 1978).
172. Nanjo, M., *Tohoku Daigaku Senko Seiren Kenkyusho* **32**(2), 127 (1976).
173. Neshkova, M., and Sheytanov, H., in *Ion-Selective Electrodes Conference*, 1977 (E. Pungor and I. Buzas, eds.), pp. 503–510, Elsevier, Amsterdam (1978).
174. Niki, E., *Asahi Garasu Kogyo Gijntsu Shoreikai Kenkyu Hokoku* **29**, 225 (1976).
175. Niki, E., and Shirai, H., *Elektrokhimiya* **14**(5), 714 (1978).
176. Nikolskii, B. P., Materova, E. A., and Grekovich, A. L., in *Ion-Selective Electrodes, 2nd Symposium*, 1976 (E. Pungor and I. Buzas, eds.), pp. 171–176, Akademiai Kiado, Budapest (1977).
177. Nikolskii, B. P., Materova, E. A., Timofeev, S. V., and Arkhangelskii, L. K., *Elektrokhimiya* **14**(1), 68 (1978).
178. Norov, S. K., Palchevskii, V. V., and Gureev, E. S., *Zh. Anal. Khim.* **32**(12), 2394 (1977).
179. Novozamsky, I., and Van Riemsdijk, W. H., Neth. Appl. 75, 15,205 (4 July 1977).

180. Oehme, F., *Taschenb. Abwasserbehandl. Metallverarb. Ind.* **2**, 509 (1977).
181. Papeschi, G., Bordi, S., and Carla, M., *J. Electrochem. Soc.* **125**(11), 1807 (1978).
182. Pataki, L., Harka, K., Havas, J., and Keomley, G., *Radiochem. Radioanal. Lett.* **30**(3), 219 (1977).
183. Pataki, L., in *Ion-Selective Electrodes, 2nd Symposium*, 1976 (E. Pungor and I. Buzas, eds.), pp. 177–184, Akademiai Kiado, Budapest (1977).
184. Philbert, F. J., Smith, M. N., and El Kei, O., in *Advançes in Automated Analysis, Technicon International Congress, 7th*, 1976, Vol. 2, p. 43, Futura Publishing Company (1977).
185. Pick, J., *Hung. Sci. Instrum.* **37**, 37 (1976).
186. Pick, J., *Hung. Sci. Instrum.* **41**, 71 (1977).
187. Pick, J., *Hung. Sci. Instrum.* **42**, 37 (1978).
188. Pick, J., *Hung. Sci. Instrum.* **43**, 31 (1978).
189. Pokorna, H., *Chem. Prum.* **28**(5), 238 (1978).
190. Pretsch, E., Buchi, R., Ammann, D., and Simon, W., in *Essays in Analytical Chemistry* (Wanninen, E., ed.), p. 321, Pergamon Press, Oxford (1977).
191. Privalova, M. M., Tulina, M. D., Sheyanova, E. M., Selivanova, B. D., and Polyakova, L. F., *Zh. Anal. Khim.* **32**(10), 1969 (1977).
192. Pucacco, L. R., and Carter, N. W., *Anal. Biochem.* **89**(1), 151 (1978).
193. Pui, C. P., Rechnitz, G. A., and Miller, R. F., *Anal. Chem.* **50**(2), 330 (1978).
194. Pungor E. (ed.), *Ion Selective Electrodes Conference*, 1977, Elsevier, Amsterdam (1968).
195. Pungor, E., and Toth, K., *Rev. Anal. Chem., 2nd Euroanal. Conf.*, 1975 (W. Fresenius, ed.), Masson, Paris (1977) (Fr.).
196. Pungor, E., Toth, K., Nagy, G., and Feher, Z., in *Ion-Selective Electrodes, 2nd Symposium*, 1976 (E. Pungor and I. Buzas, eds.), pp. 67–91, Akademiai Kiado, Budapest (1977).
197. Pungor, E., in *Ion-Selective Electrodes Conference*, 1977 (E. Pungor and I. Buzas, eds.), pp. 161–173, Elsevier, Amsterdam (1978).
198. Pungor, E., Toth, K., and Nagy, G., in *Essays in Analytical Chemistry* (Wanninen, E., ed.), pp. 331–334, Pergamon, Oxford (1977).
199. Pungor, E., Toth, K., and Nagy, G., *Mikrochim. Acta* **1**(5-6), 531 (1978).
200. Puxbaum, H., and Simeonov, V., *Hung. Sci. Instrum.* **41**, 17 (1977).
201. Ramamoorthy, S., and Kushner, D. J., *Natl. Bur. Stand.* (*U.S.*) *Spec. Pub.* **464**, 467 (1977).
202. Rechnitz, G. A., Kobos, R. K., Riechel, S. J., and Gebauer, C. R., *Anal. Chim. Acta* **94**(2), 357 (1977).
203. Rechnitz, G. A., Riechel, T. L., Kobos, R. K., and Meyerhoff, M. E., *Science* **199**(4327), 440 (1978).
204. Rock, P. A., in *Special Topics in Electrochemistry* 1977 (P. Rock, ed.); Elsevier, Amsterdam (1977).
205. Ruzicka, J., and Hansen, E. H., Ger. Offen. 2,740,570 (16 March 1978).
206. Saharay, S. K., and Basu, A. S., *J. Indian Chem. Soc.* **54**(12), 1120 (1977).
207. Savenko, V. S., *Okeanologiya* (*Moscow*) **17**(6), 1123 (1977).
208. Schaefer, O. F., Ger. Offen. 2,700,567 (13 July 1978).
209. Schindler, J. G., Dennhardt, R., and Simon, W., *Chimia* **31**(10), 404 (1977).
210. Schindler, J. G., Stork, G., Strueh, H. J., and Schael, W., *Fresenius' Z. Anal. Chem.* **290**(1), 45 (1978).
211. Schindler, J. G., Schindler, R. G., and Aziz, O., *J. Clin. Chem. Clin. Biochem.* **16**(8), 447 (1978).
212. Sekerka, I., and Lechner, J. F., *Anal. Chim. Acta* **93**, 129 (1977).
213. Selinger, K., and Staroscik, R., *Pharmazie* **33**(4), 208 (1978).
214. Semler, M., in *Ion-Selective Electrodes Conference*, 1977 (E. Pungor and I. Buzas, eds.), pp. 529–537, Elsevier, Amsterdam (1977).

215. Serjeant, E. P., and Warner, A. G., *Anal. Chem.* **50**(12), 1724 (1978).
216. Shavnya, Y. V., and Chikin, Y. M., *Elektrokhimiya* **14**(2), 336 (1978).
217. Shavnya, Y. V., Bychkov, A. S., Petrukhin, O. M., Zarinskii, V. A., Bakhtinova, L. V., and Zolotov, Y. A., *Zh. Anal. Khim.* **33**(8), 1531 (1978).
218. Sheina, N. M., Izvekov, V. P., Papay, M. K., Toth, K., and Pungor, E., *Anal. Chim. Acta* **92**(2), 261 (1977).
219. Shinbo, T., Kamo, N., Kurihara, K., and Kobatake, Y., *Arch. Biochem. Biophys.* **187**(2), 414 (1978).
220. Shono, T., and Kimura, K., *Kagaku* (*Kyoto*) **32**(4), 314 (1977).
221. Siemroth, J., Hennig, I., and Hartmann, P. Ger. (East) Patent 127,372 (21 September 1977).
222. Siemroth, J., Gruenke, U., Hartmann, P., Schuetze, R., and Stojan, D., Ger. (East) Patent 125,857 (25 May 1977).
223. Siemroth, J., Hennig, I., and Claus, R., in *Ion-Selective Electrodes, 2nd Symposium*, 1976 (E. Pungor and I. Buzas, eds.), pp. 185–197, Akademiai Kiado, Budapest (1977).
224. Simon, W., Ammann, D., Osswald, H. F., Meier, P. C., and Dohner, R. E., in *Advances in Automated Analysis, Technicon International Congress, 7th*, 1976, Vol. 1, pp. 56–62 (1977).
225. Simon, W., Morf, W. E., and Thoma, A. P., in *Ion-Selective Electrodes, 2nd Symposium*, 1976 (E. Pungor and I. Buzas, eds.), pp. 13–24, Elsevier, Amsterdam (1977).
226. Simon, W., *Adv. Chem. Phys.* **39** (*Mol. Movements Chem. React. Cond. Membrane, Enzymes Other Macromolecules* 1976), 287 (1978).
227. Smith, G. D., Beswick, G., and Rosie, D. A., *Fluoride* **12**(3), 142 (1978).
228. Solsky, R. L., and Rechnitz, G. A., *Anal. Chim. Acta* **99**(2), 241 (1978).
229. Srivastava, S. K., Jain, A. K., Agrawal, S., and Singh, R. P., *Talanta* **25**(3), 157 (1978).
230. Srivastava, S. K., Jain, A. K., Agrawal, S., and Singh, R. P., *J. Electroanal. Chem. Interfacial Electrochem.* **90**(2), 291 (1978).
231. Starobinets, G. L., Rakhmanko, E. M., and DelToro, D. R., *Vestsi. Akad. Navuk BSSR Ser. Khim. Navuk* **1978**(4), 75.
232. Staroscik, R., and Malecki, F., *Acta Pol. Pharm.* **34**(6), 643 (1977).
233. Stefanova, O. K., and Rusina, I. V., *Elektrokhimiya* **14**(6), 882 (1978).
234. Stover, F. S., and Buck, R. P., *J. Phys. Chem.* **81**(22), 2105 (1977).
235. Suzuki, S., Aisawa, M., Ishiguro, I., Shinohara, R., and Nagamura, Y., Japan. Kokai 78, 47,517 (28 April 1978).
236. Sykut, K., and Nowakowska, E., *Biul. Lubel. Tow. Nauk. Mat.-Fiz.-Chem.* **19**(1), 89 (1977).
237. Szepesvary, E., Pungor, E., and Szepesvary, P., in *Ion-Selective Electrodes, 2nd Symposium*, 1976 (E. Pungor and I. Buzas, eds.), pp. 217–224, Akademiai Kiado, Budapest (1977).
238. Tacussel, J., and Fombon, J. J., in *Ion-Selective Electrodes Conference*, 1977 (E. Pungor and I. Buzas, eds.), pp. 567–575, Elsevier, Amsterdam (1978).
239. Taddia, M., *Microchem. J.* **22**(3), 369 (1977).
240. Takaisvili, O. G., Motsonelidze, E. P., Fider, Z. N., Karachentseva, Y. M., and Davitaya, P. I., *Zh. Anal. Khim.* **32**(4), 727 (1977).
241. Tanaka, T., Hiiro, K., and Kawahara, A., Japan. Kokai 78,39,193 (10 April 1978).
242. Tassara, E., Ciurlo, R., Deferrari, G. B., and Andreoni, F., *Rass. Chim.* **29**(4), 167 (1977).
243. Tassara, E., Ciurlo, R., and Deferrari, G. B., *Chim. Ind.* (*Milan*) **59**(4), 301 (1977).
244. Teh, G. H., Fung, K. W., and Hosking, K. F. G., *Geol. Soc. Malays. Bull.* **8**, 151 (1977).
245. Thomas, J. D. R., in *Ion-Selective Electrodes Conference*, 1977 (E. Pungor and I. Buzas, eds.), pp. 175–198, Elsevier, Amsterdam (1978).
246. Thurnau, R. C., in *Proceedings of the AWWA Water Quality Technology Conference*, 1976, 3A-3a, 15 pp., American Waterworks Association (1977).

247. Timofeev, S. V., Materova, E. A., Nikolskii, B. P., and Arkhangelskii, L. K., *Vestn. Leningr. Univ. Fiz. Khim.* **1978**(2), 135.
248. Toteva, N. M., and Havas, J., Hung. Halasztott 2,209 (28 March 1977).
249. Trojanowicz, M., and Hulanicki, A., *Chem. Anal.* (*Warsaw*) **22**(4), 615 (1977).
250. Troll, G., Farzaneh, A., and Cammann, K., *Chem. Geol.* **20**(4), 295 (1977).
251. Uchida, M., Akiba, M., Wada, S., and Kashima, T., *Kyoritsu Yakka Daigaku Kenkyu Nempo* **22**, 9 (1977).
252. Van de Leest, R. E., *Analyst* (*London*) **102** (1216), 509 (1977).
253. Van de Leest, R. E., Renaat, E., and Heijne, L., Ger. Offen. 2,716,646 (10 November 1977).
254. Van de Leest, R. E., and Geven, A., *J. Electroanal. Chem. Interfacial Electrochem.* **90**(1), 97 (1978).
255. Van der Linden, W. E., and Heijne, G. J. M., in *Ion-Selective Electrodes Conference*, 1977 (E. Pungor and I. Buzas, eds.), pp. 445–452, Elsevier, Amsterdam (1978).
256. Virtanen, R., in *Ion-Selective Electrodes Conference*, 1977 (E. Pungor and I. Buzas, eds.), pp. 589–595, Elsevier, Amsterdam (1978).
257. Vishnyakov, A. V., Zhukov, A. F., Lyubchak, T. A., Urusov, Y. I., and Gordievskii, A. V., *Zh. Anal. Khim.* **32**(4), 840 (1977).
258. Vladimirskaya, T. N., and Gorskaya, A. P., *Metody Anal. Kontolya Proizvod. Khim. Prom-sti.* **11**, 53 (1977).
259. Vlasov, Y. G., and Kocheregin, S. B., in *Ion-Selective Electrodes Conference*, 1977 (E. Pungor and I. Buzas, eds.), pp. 597–601, Elsevier, Amsterdam (1978).
260. Vlasov, Y. G., Kocheregin, S. B., and Ermolenko, Y. E., *Zh. Anal. Khim.* **32**(9), 1843 (1977).
261. Weiss, D., *Sklar Keram.* **28**(7), 214 (1978).
262. Vesely, J., Weiss, D., and Stulik, K., *Analysis With Ion-Selective Electrodes*, Wiley, New York (1977).
263. Wright, J. A., and Bailey, P. L., in *Ion-Selective Electrodes Conference*, 1977 (E. Pungor and I. Buzas, eds.), pp. 603–609, Elsevier, Amsterdam (1978).
264. Yang, H.-Y., and Pu, K.-K., *Huan Ching K'o Hsueh* **1978**(2), 42.
265. Yoshida, N., and Ishibashi, N., *Bull. Chem. Soc. Jpn.* **50**(12), 3189 (1977).
266. Zadeczky, S., Kuttel, D., Havas, J., and Kecskes, L., *Acta Pharm. Hung.* **48**(3), 131 (1978).
267. Zarinskii, V. A., Petrukhin, O. M., Bychkov, A. S., Shpigun, L. K., and Zolotov, Y. A., in *Ion-Selective Electrodes, 2nd Symposium*, 1976 (E. Pungor and I. Buzas, eds.), pp. 245–253, Akademiai Kiado, Budapest (1977).
268. Zarinskii, V. A., Shpigun, L. K., Trepalina, V. M., and Volobueva, I. V., *Zavod. Lab.* **43**(8), 941 (1977).
269. Zimmermann, R. L., Jr., and Bertrand, H. G., *Anal. Lett.* **A11**(7), 569 (1978).
270. Zykina, G. K., Bystritskaya, T. L., Materova, E. A., Grekovich, A. L., and Volkova, V. V., *Pochv. Biogeotsenol. Issled. Priazove.* **1**, 102 (1975).

Index